Dual Sets of Envelopes
and
Characteristic Regions
of Quasi-Polynomials

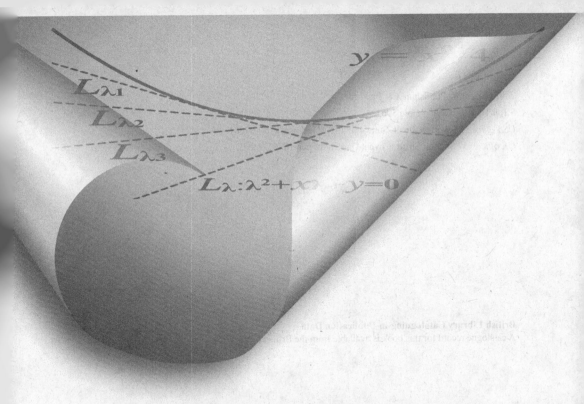

Dual Sets of Envelopes
and
Characteristic Regions
of Quasi-Polynomials

Sui Sun Cheng
National Tsing Hua University, R. O. China

Yi-Zhong Lin
Fujian Normal University, P. R. China

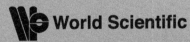 **World Scientific**

NEW JERSEY · LONDON · SINGAPORE · BEIJING · SHANGHAI · HONG KONG · TAIPEI · CHENNAI

Published by

World Scientific Publishing Co. Pte. Ltd.

5 Toh Tuck Link, Singapore 596224

USA office: 27 Warren Street, Suite 401-402, Hackensack, NJ 07601

UK office: 57 Shelton Street, Covent Garden, London WC2H 9HE

British Library Cataloguing-in-Publication Data
A catalogue record for this book is available from the British Library.

ISBN-13 978-981-4277-27-3
ISBN-10 981-4277-27-4

Printed in Singapore

Preface

Existence and/or nonexistence of roots of functions have been the subjects of numerous investigations. In some cases, the function under investigation may have one or more parameters, to which the corresponding existence problem then depends on. A simple example is the following. We are given a quadratic polynomial of the form $p(\lambda|\alpha, \beta) = \lambda^2 + \alpha\lambda + \beta$, where α, β are real numbers, and we are required to find the exact conditions on the parameters α and β such that the quadratic polynomial p does not have any real roots. Such a problem is easy since a necessary and sufficient condition for all roots of p to be nonreal numbers is that its discriminant is negative, that is, $\alpha^2 - 4\beta < 0$.

More generally, given a nonlinear function $F(\lambda|\alpha_1, \alpha_2, ..., \alpha_n)$ with n parameters $\alpha_1, ..., \alpha_n$, it is desired to determine the exact region containing these parameters such that for each $(\alpha_1, ..., \alpha_n)$ in this region, none, some or all of the roots of F lie in a specified subregion of the domain of F. Unfortunately a general answer to this problem is unknown.

In some situations, however, the corresponding problems can be transformed into the existence and nonexistence of tangents of curves associated with the function F. Such an approach has been exploited by the authors and collaborators recently for characteristic functions associated with difference and differential equations. A formal theory, however, has only been developed by the first author in the last two years. In this book, we present this Cheng-Lin envelope method in a systematic manner, introduce sufficiently many technical tools related to this method, and show how they are used in handling characteristic functions involving a reasonable number of parameters.

- The book begins with an elementary example involving the quadratic polynomials. Basic definitions, symbols and results are then introduced which will be used throughout the book.
- In Chapter 2, envelopes of families of straight lines and dual points of order m of envelopes are introduced.
- In Chapter 3, quasi-tangent lines of curves are introduced. Then sweeping functions are used to obtain information on the number of tangent lines of

curves and the distribution of dual points.

- In Chapter 4, quasi-polynomials associated with ordinary difference equations and ordinary functional differential equations are introduced.
- In Chapter 5, necessary and sufficient conditions are established for real quintic polynomials to have nonpositive roots.
- In Chapter 6, quasi-polynomials related to difference equations are considered, and necessary as well as sufficient conditions for the nonexistence of positive roots are found.
- In Chapter 7, quasi-polynomials related to functional differential equations are considered, and necessary as well as sufficient conditions for the existence of real roots are found.
- In the appendix, we collect some useful distribution maps of dual sets of order 0 of curves made up of several pieces.

This book is self-contained in the sense that only basic Calculus and elementary properties of convex functions are needed. Therefore this book will be useful to college students who want to see immediate applications of first semester analysis they have just learned. On the other hand, the Cheng-Lin envelope method announced here is new, therefore this book will also be useful to graduate students who wish to pursue research in the area of equations involving several parameters. Since we have derived numerous necessary and sufficient conditions for the existence and/or nonexistence of roots of characteristic functions, our book can also be used as a reference for scientists who are concerned with qualitative properties of functional equations.

Our thanks go to Shao Yuan Huang who helped in the preparation of the graphs and in suggesting some of the results in this book, and also to the National Science Council of R. O. China for the financial supports of the first author during the last 30 years.

We tried our best to eliminate any errors. If there are any that have escaped our attention, your comments and corrections will be much appreciated.

Sui Sun Cheng and Yi-Zhong Lin

Contents

Chapter 1

Prologue

1.1 An Example

A basic problem in mathematics is to make sure roots of a given function exist or do not exist. There are now numerous means for handling such problems. Yet the idea of transforming the root seeking problem into one that counts the number of tangent lines of a related function does not seem to have drawn much attention. In this book, we intend to explain in great detail how this idea can be realized for functions that involve a reasonable number of parameters.

To motivate what follows, let us consider the familiar real quadratic polynomial

$$f(\lambda|a, b) = \lambda^2 + a\lambda + b,$$

involving two real parameters a and b.

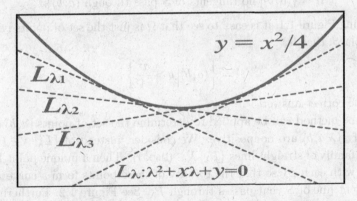

$$y = x^2/4$$

L_{λ_1}

L_{λ_2}

L_{λ_3}

$$L_\lambda : \lambda^2 + x\lambda + y = 0$$

Fig. 1.1

Suppose we want to determine the necessary and sufficient condition satisfied by the coefficients a and b in order that all the roots of f are nonreal. This is an easy question since the roots of f are nonreal if, and only if, the discriminant $a^2 - 4b < 0$. We may however, approach this problem in a different manner. Let us collect all the desired pairs (a, b) into the plane set Ω.

Note first that,

(c, d) is in the complement of Ω

$\Rightarrow \lambda^2 + c\lambda + d = 0$ has a real root $\lambda_* \in \mathbf{R}$

$\Rightarrow (c, d)$ lies on the straight line $L_{\lambda_*} : \lambda_* x + y = -\lambda_*^2$.

This motivates us to draw all possible straight lines defined by

$$L_\lambda : \lambda x + y = -\lambda^2, \ \lambda \in \mathbf{R}. \tag{1.1}$$

The resulting figure (see Figure 1.1) suggests the existence of a curve which 'touches' each of the possible straight lines at exactly one point. Such a curve is called the envelope of the family of straight lines $\{L_\lambda : \lambda \in \mathbf{R}\}$. Furthermore, this curve, as can be checked by computer experiments (and by analysis to be given later), coincides with the parabola

$$S : y = \frac{x^2}{4}, \ x \in \mathbf{R}.$$

Thus

(c, d) is in the complement of $\Omega \Rightarrow (c, d)$ lies on a tangent line of S.

Furthermore, by reversing the above arguments, it is easily seen that the converse is also true! In other words,

complement of $\Omega = \{(c, d)|$ there is a tangent of S passing through $(c, d)\}$,

or,

$\Omega = \{(a, b)|$ no tangents of S pass through $(a, b)\}$.

By inspecting Figure 1.1, it is easy to see that Ω is just the set of points lying above the parabola S, that is,

$$\Omega = \left\{ (a, b) \,\Big|\, b > \frac{a^2}{4} \right\},$$

which is the correct answer!

The same method can be applied to determine the set of points (a, b) such that all roots of $f(\lambda|a, b)$ are nonpositive. We consider, instead of $\{L_\lambda| \lambda \in (-\infty, \infty)\}$ above, the family of straight lines $\{L_\lambda| \lambda \in (0, \infty)\}$. Then a unique point P_λ can be associated with each L_λ so that the totality of these points form a curve S and L_λ is the tangent line of S that passes through P_λ. See Figure 1.2. Furthermore, S is the graph of the function $y = x^2/4$ over the interval $(-\infty, 0)$ (instead of $(-\infty, \infty)$ as in the previous case). Again, it is clear from the Figure 1.2 that (a, b) is in the required region if, and only if, $a < 0$ and $b > a^2/4$, or, $a \geq 0$ and $b \geq 0$.

The same conclusion can easily be checked by manipulating the well known formula for the roots of $f(\lambda|a, b)$. Indeed, for each point (a, b) in the lower half plane, $f(0|a, b) = b < 0$. Since $\lim_{\lambda \to \infty, \lambda \in \mathbf{R}} f(\lambda|a, b) = +\infty$, we see that $f(\lambda|a, b)$ must have a positive root. Therefore, the lower half plane cannot be part of the

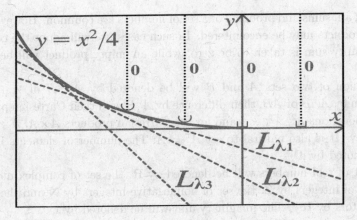

Fig. 1.2

required region. So, we consider the closed upper half plane. For each (a, b) which lies strictly above the parabola defined by $y = x^2/4$, since $a^2 - 4b < 0$, $f(\lambda|a, b)$ has two conjugate roots (which are definitely nonpositive). Suppose now (a, b) lies below the parabola and above the x-axis. Since $a^2 - 4b \geq 0$, $f(\lambda|a, b)$ has two real nonzero roots

$$\lambda_\pm = \frac{-x \pm \sqrt{a^2 - 4b}}{2}.$$

Furthermore, if $\lambda_+, \lambda_- \leq 0$, then

$$-x = \lambda_+ + \lambda_- \leq 0 \text{ and } y = \lambda_+ \lambda_- \geq 0,$$

and conversely, if $x \geq 0, y \geq 0$, then

$$\lambda_- \leq \lambda_+ = \frac{-a + \sqrt{a^2 - 4b}}{2} \leq \frac{-a + a}{2} = 0.$$

We have thus verified that the region that we are seeking is defined by

$$x < 0 \text{ and } y > x^2/4, \text{ or, } x \geq 0 \text{ and } y \geq 0.$$

In this book, we intend to make clear the various concepts involved in the above examples, to develop the corresponding mathematical tools, and to illustrate our idea by studying in great details two types of functions (to be collectively called quasi-polynomials) that are much studied in the theory of ordinary difference equations and ordinary differential equations.

1.2 Basic Definitions

Basic concepts from Calculus will be assumed in this book. For the sake of completeness, we will, however, briefly go through some of these concepts and their related information. We will also introduce here some common notations and conventions which will be used in this book.

First of all, sums and products of a set of numbers are common. However, empty sums or products may be encountered. In such cases, we will adopt the convention that an empty sum is taken to be zero, while an empty product will be taken as one.

The union of two sets A and B will be denoted by $A \cup B$ or $A + B$, their intersection by $A \cap B$ or AB, their difference by $A \backslash B$, and their Cartesian product by AB. The notations $A^2, A^3, ...$, stand for the Cartesian products $A \times A, A \times A \times A, ...$, respectively. It is also natural to set $A^1 = A$. The number of elements in a set Ω will be denoted by $|\Omega|$.

The set of real numbers will be denoted by $\mathbf{R}$, the set of complex numbers by $\mathbf{C}$, the set of integers by $\mathbf{Z}$, the set of nonnegative integers by $\mathbf{N}$ and the set of all real m vectors by $\mathbf{R}^m$. The imaginary unit will be denoted by $\mathbf{i}$.

Therefore, the set of integers which are greater than or equal to $n \in \mathbf{R}$ is $\mathbf{Z}[n, \infty)$, the set of nonpositive complex numbers is $\mathbf{C} \backslash (0, \infty)$, the set of nonnegative complex numbers is $\mathbf{C} \backslash (-\infty, 0)$, the set of nonzero complex numbers is $\mathbf{C} \backslash \{0\}$, etc.

Real functions defined on intervals of $\mathbf{R}$ will usually be denoted by f, g, h, F, G, H, etc. However, the identity function will be denoted by Υ, that is $\Upsilon(x) = x$ for all $x \in \mathbf{R}$; while the constant function will be denoted by Θ_γ, that is, $\Theta_\gamma(x) = \gamma$ for $x \in \mathbf{R}$. In particular, the null function is denoted by Θ_0. Given an interval I in $\mathbf{R}$, the chi-function $\chi_I : I \to \mathbf{R}$ is defined by

$$\chi_I(x) = 1, \ x \in I. \tag{1.2}$$

The restriction of a real function f defined over an interval J (which is not disjoint from I) will be written as $f\chi_I$, so that $f\chi_I$ is now defined on $I \cap J$ and

$$(f\chi_I)(x) = f(x), \ x \in I \cap J.$$

Let f be a real function defined for points near $\alpha \in \mathbf{R}$. Recall that the derivative of f at α is $f'(\alpha) = \lim_{x \to \alpha}(f(x) - f(\alpha))/(x - \alpha)$, the right derivative of f at α is $f'_+(\alpha) = \lim_{x \to \alpha^+}(f(x) - f(\alpha))/(x - \alpha)$ and the left derivative of f at α is $f'_-(\alpha) = \lim_{x \to \alpha^-}(f(x) - f(\alpha))/(x - \alpha)$. For real functions of two or more independent variables, we may define partial derivatives in similar manners. In particular, let $g(x, y)$ be a real function of two variables, the partial derivative of g with respect to x at (α, β) is denoted by $g'_x(\alpha, \beta)$.

Let I be an interval which may be open or closed on either sides. As usual, $f : I \to \mathbf{R}$ is said to be continuous at an interior point ξ of I if $\lim_{x \to \xi} f(x) = f(\xi)$. If I is closed on the left (right) with the boundary point c (respectively d), then f is said to be continuous at c if $\lim_{x \to c^+} f(x) = f(c)$ (respectively $\lim_{x \to d^-} f(x) = f(d)$). f is said to continuous on I if it is continuous at every point of I. The derivative of f at an interior point ξ of I is $f'(\xi)$. If I is closed on the left with the boundary point c, then the *derivative* of f at c is taken to be the right derivative of f at c, i.e.,

$$f'(c) = \lim_{x \to c^+} \frac{f(x) - f(c)}{x - c} = f'_+(c);$$

while if I is closed on the right with the boundary point d, then the *derivative* of f at d is

$$f'(d) = \lim_{x \to d^-} \frac{f(x) - f(d)}{x - d} = f'_-(c).$$

Therefore, if f is a real function defined on an interval I (e.g. $[c, d)$), then f is said to be differentiable if $f'(x)$ exists for every $x \in I$, and in such a case, the derived function f' is the function defined on I with $f'(x)$ as its value at $x \in I$. f is said to be *smooth* on I if its derived function f' is continuous on I. A smooth function g defined on an interval I is denoted by $g \in C^1(I)$.

In the sequel, we need sufficient conditions for a function to be smooth on an interval I.

Example 1.1. Suppose $g : I \to R$, where $I = [c, d)$, is continuous. If $f : I \to R$ satisfies $f'(x) = g(x)$ for $x \in (c, d)$, then $f \in C^1(I)$. Indeed, it suffices to show that f' is continuous at c. This can be seen by using the mean value theorem:

$$\lim_{x \to c^+} \frac{f(x) - f(c)}{x - c} = \lim_{x \to c^+} q(\zeta_x)$$

where ζ_x is a number in $(0, x)$. Since $\lim_{x \to c^+} q(\zeta_x) = q(c)$, we see that $f'_+(c)$ exists, is equal to $q(c)$, and

$$\lim_{x \to c^+} f'(x) = \lim_{x \to c^+} q(x) = q(c) = f'_+(c).$$

In other words, f' is continuous at c.

Given a real function g of a real variable, the following limits (which may, or may not exist)

$$\lim_{x \to c^+} g(x), \ \lim_{x \to c^-} g(x), \ \lim_{x \to c^+} g'(x), \ \lim_{x \to c^-} g'(x),$$

$$\lim_{x \to -\infty} g(x), \ \lim_{x \to +\infty} g(x), \ \lim_{x \to -\infty} g'(x), \ \lim_{x \to +\infty} g'(x)$$

will be needed quite extensively for expressing various facts. For this reason, we will employ the following notations

$$g(c^+) = \lim_{x \to c^+} g(x), \ g(c^-) = \lim_{x \to c^-} g(x),$$

$$g'(c^+) = \lim_{x \to c^+} g'(x), \ g'(c^-) = \lim_{x \to c^-} g'(x),$$

$$g(-\infty) = \lim_{x \to -\infty} g(x), \ g(+\infty) = \lim_{x \to +\infty} g(x),$$

$$g'(-\infty) = \lim_{x \to -\infty} g'(x), \ g'(+\infty) = \lim_{x \to +\infty} g'(x).$$

The *graph* of a real function f defined on a set J of real numbers is the set

$$\left\{ (x, y) \in \mathbf{R}^2 | y = f(x), \ x \in J \right\}.$$

For the sake of convenience, *we will use the same notation to indicate a (real) function of a real variable and its graph.* Therefore, in the sequel, we will meet statements such as 'the set S is also the graph of a function $y = S(x)$ defined on the interval I ...'.

Now let g be a function defined on an interval I. If the derivative $g'(\lambda)$ exists, then the *tangent line* of the graph g through the point $(\lambda, g(\lambda))$ is taken to mean the graph of the function $L_{g|\lambda}$ defined by

$$L_{g|\lambda}(x) = g'(\lambda)(x - \lambda) + g(\lambda), \ x \in \mathbf{R}. \tag{1.3}$$

The so called 'vertical tangents' in some of the elementary analysis text books will not be regarded as tangent lines of our functions.

We say that a point (α, β) in the plane is strictly above (above, strictly below, below) the graph of a function g if α belongs to the domain of g and $g(\alpha) < \beta$ (respectively $g(\alpha) \le \beta$, $g(\alpha) > \beta$ and $g(\alpha) \ge \beta$). The corresponding notations[1] are $(\alpha, \beta) \in \vee(g)$, $(\alpha, \beta) \in \overline{\vee}(g)$, $(\alpha, \beta) \in \wedge(g)$ and $(\alpha, \beta) \in \underline{\wedge}(g)$.

Example 1.2. Let $g(x) = x^2/4$ for $x \in \mathbf{R}$. Then $(a, b) \in \vee(g)$ if, and only if, $b > a^2/4$, while $(a, b) \in \overline{\vee}(g)$ if, and only if, $b \ge a^2/4$.

We also need to handle the 'ordering' relations between points and several graphs in the plane. Suppose we now have two real functions g_1 and g_2 defined on real subsets I_1 and I_2 respectively. We say that $(\alpha, \beta) \in \vee(g_1) \oplus \vee(g_2)$ if $\alpha \in I_1 \cap I_2$ and $\beta > g_1(\alpha)$ and $\beta > g_2(\alpha)$, or, $\alpha \in I_1 \backslash I_2$ and $\beta > g_1(\alpha)$, or, $\alpha \in I_2 \backslash I_1$ and $\beta > g_2(\alpha)$. The notations $(\alpha, \beta) \in \overline{\vee}(g_1) \oplus \vee(g_2)$, $(\alpha, \beta) \in \overline{\vee}(g_1) \oplus \wedge(g_2)$, etc. are similarly defined.

If we now have n real functions $g_1, ..., g_n$ defined on intervals $I_1, ..., I_n$ respectively, we write $(\alpha, \beta) \in \vee(g_1) \oplus \vee(g_2) \oplus \cdots \oplus \vee(g_n)$ if $\alpha \in I_1 \cup I_2 \cup \cdots \cup I_n$, and if

$$\alpha \in I_{i_1} \cap I_{i_2} \cap \cdots \cap I_{i_m} \Rightarrow \beta > g_{i_1}(\alpha), \beta > g_{i_2}(\alpha), ..., \beta > g_{i_m}(\alpha), \ i_1, ..., i_m \in \{1, ..., n\}.$$

The notations $(\alpha, \beta) \in \overline{\vee}(g_1) \oplus \overline{\vee}(g_2) \oplus \cdots \oplus \overline{\vee}(g_n)$, etc. are similarly defined.

Example 1.3. Let $f(x) = x^2/4$ for $x < 0$ and $g(x) = x^2/4$ for $x \in (-\infty, 1]$. Then

$$(a, b) \in \vee(f) \oplus \overline{\vee}(\Theta_0) \text{ if, and only if, } a < 0 \text{ and } b > a^2/4, \text{ or, } a \ge 0 \text{ and } b \ge 0,$$

and

$$(a, b) \in \vee(g) \oplus \overline{\vee}(\Upsilon/4) \text{ if, and only if, } a < 1 \text{ and } b > a^2/4, \text{ or, } a \ge 1 \text{ and } b \ge a/4.$$

Example 1.4. Let G_1, G_2 and G_3 be respectively the functions (see Figure 1.3)

$$G_1(x) = 2\exp(1 - x) - 3, \ x \in (-\infty, 1),$$

$$G_2(x) = 2\exp(1 + x) - 3, \ x \in (-1, \infty),$$

and

$$G_3(x) = -x^2, \ x \in [-1, 1].$$

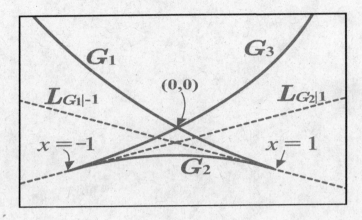

Fig. 1.3

Then

$$\{(x,y) \in \mathbf{R}^2 : y > \max\{G_1(x), G_2(x)\}\} = \vee(G_1) \oplus \vee(G_2) = \vee(G_1) \oplus \vee(G_2) \oplus \vee(G_3).$$

Note that $\vee(G_1) \oplus \vee(G_2)$ is a better description of the set on the left hand side than $\vee(G_1) \oplus \vee(G_2) \oplus \vee(g_3)$ since less work is needed to figure out its precise content. An even better one is as follows. We first note by solving $G_1(x) = G_2(x)$ that the x-coordinate of the point of intersection of the two graphs G_1 and G_2 is 0. Then $\vee(G_1) \oplus \vee(G_2)$ is equal to $\vee(G_1\chi_{(-\infty,0)}) \oplus \vee(G_2\chi_{[0,\infty)})$. Again, the latter description is preferred since it is 'precise' in the sense that the domains of $G_1\chi_{(-\infty,0)}$ and $G_2\chi_{[0,\infty)}$ have empty intersection. In spite of the precision obtained, however, cumbersome calculations (such as finding the point of intersection of the graphs G_1 and G_2) may be needed which are irrelevant to the illustration of the principles involved. Therefore in the sequel, we may use less precise formulas such as $\vee(G_1) \oplus \vee(G_2)$.

---

[1]$\nabla(g)$ is the epigraph of g. Such a term, however, is not needed in the sequel.

Chapter 2

Envelopes and Dual Sets

2.1 Plane Curves

The concept of a 'plane curve' naturally arises from observing a piece of string lying in the plane or from observing the motion of a particle moving in the plane. There are different ways to 'simulate' plane curves. For instance, a stretched string between the real numbers a and b, where $a < b$, can be simulated by the mathematical interval $[a, b]$ in the set $\mathbf{R}$ of real numbers. A more general plane curve can be regarded as a subset of the plane. For example, the unit circle can be described as

$$\left\{ (x, y) \in \mathbf{R}^2 | x^2 + y^2 = 1 \right\}.$$

Yet another way to simulate a plane curve is to regard it as the 'portrait' of a pair of 'parametric' functions. For instance, the unit circle can be regarded as a set of points (x, y) in the plane such that

$$x = \cos t, y = \sin t, \ t \in [0, 2\pi].$$

Undoubtedly, the reason for choosing a pair of functions of the form $x = \psi(t)$ and $y = \phi(t)$ for t in a closed interval $[a, b]$ and then describing our curve as

$$S = \left\{ (\psi(t), \phi(t)) \in \mathbf{R}^2 | t \in [a, b] \right\}$$

is that this set resembles our curve. Note that if we trace the points in S, and plot them in the plane. We may label each of the points with the independent variable t. In particular, the point $(\psi(a), \phi(a))$ is labeled by a, and $(\psi(b), \phi(b))$ by b. By this act, we have ordered the image points according to increasing values of t.

On the other hand, if we are given two continuous functions $x = u(t)$ and $y = v(t)$ defined on the interval $[a, b)$, the set of points

$$\{(u(t), v(t)) | t \in [a, b)\} \tag{2.1}$$

does not seem to resemble a physical string. However, it resembles the trajectory traced out by a moving particle between the time a and any time $c < b$. Since c is arbitrary, we may still (loosely) call (2.1) the curve[1] represented by $u(t)$ and

---

[1] also called the phase portrait of the parametric functions

$v(t)$ over $[a, b)$. Again, a direction is associated naturally with this curve so as to indicate the fact that our particle moves from the present time $t = a$ to an arbitrary future time $c < b$.

For the same reason, let I be a general real interval and let $x = \psi(t)$ and $y = \phi(t)$ be continuous functions defined on I. Then the set $\{(\psi(t), \phi(t)) | t \in I\}$ is called the curve generated by the pair on I. It is also natural to say that the point $(\psi(t_1), \phi(t_2))$ precedes the point $(\psi(t_2), \phi(t_2))$ if $t_1 < t_2$.

For example, the functions

$$\psi(t) = \lambda \cos t, \, \phi(t) = \lambda \sin t, \, \lambda > 0,$$

defined for $t \in [0, 2\pi)$ represents a 'counterclockwise' curve

$$\Gamma_\lambda = \{(\psi(t), \phi(t)) | t \in [0, 2\pi)\}. \tag{2.2}$$

Note that for each $\lambda > 0$, Γ_λ defined by (2.2) is a different curve. The totality of such curves is called a family. Note further that Γ_λ is equal to

$$\left\{(x, y) \in \mathbf{R}^2 | \, x^2 + y^2 = \lambda^2 \right\}. \tag{2.3}$$

The 'defining relation' in (2.3) is

$$x^2 + y^2 = \lambda^2,$$

which can also be put in the general form

$$F(x, y, \lambda) = 0, \tag{2.4}$$

where F is a function of three variables.

Note that the expression (2.4) is not 'standard' since the curves corresponding to λ are not 'explicitly' given by pairs of functions. In fact, in the general situation, 'implicit function theorems' in Advanced Calculus are needed to show that implicitly defined curves can also be described by functions. Nevertheless, there may be some easy situations. As an example, let $f, g, h : I \to \mathbf{R}$. Then for each fixed $\lambda \in I$, the equation

$$f(\lambda)x + g(\lambda)y = h(\lambda), \, (f(\lambda), g(\lambda)) \neq 0,$$

is the defining equation for a straight line L_λ in the x, y-plane.

One common type of curves consists of *graphs* of real functions of a real variable. More precisely, let g be a real function of a real variable defined on an interval I, then its graph

$$\{(x, g(x)) | x \in I\}$$

is a curve which can be generated by

$$x = t, \, y = g(t), \, t \in I.$$

The question then arises as to when a curve G described by a pair of continuous parametric functions $x = \psi(t)$ and $y = \phi(t)$ is also the graph of a real function g. One case is quite simple. Suppose ψ and ϕ are defined on the interval I and ψ is strictly increasing on I. Then the inverse ψ^{-1} exists and is defined on $\psi(I)$. Thus the function g defined by $g(x) = \phi(\psi^{-1}(x))$ for $x \in \psi(I)$ will do the job. Note that if ψ and ϕ are also smooth (i.e. continuously differentiable) on the interval I, say $I = (\alpha, \beta)$, then the fact that $\psi'(t) > 0$ on I is sufficient for ψ to be strictly increasing. We may, however, allow $\psi'(t) = 0$ for some point $\gamma \in (\alpha, \beta)$ for reaching the same conclusion. Indeed, if $\alpha < t < \gamma$, then $\psi(\gamma) - \psi(t) = \psi'(\xi)(\gamma - t)$ for some $\xi \in (t, \gamma)$, and hence $\psi(\gamma) > \psi(t)$. By similar reasoning, we see that ψ is strictly increasing on I. Hence G is also the graph of a function g which, in view of the chain rule,

$$\frac{dg}{dx} = \frac{\phi'(t)}{\psi'(t)}$$

is also smooth on the interval $(\psi(\alpha), \psi(\beta))$ except perhaps when $\psi'(t) = 0$. There may be situations such that g is smooth on the whole interval $\psi(I)$. For instance, if

$$\frac{\phi'(t)}{\psi'(t)} = q(t), \ t \in (\alpha, \beta)\backslash\{\gamma\}$$

but $q(t)$ is continuous at $t = \gamma$ (cf. Example 1.1), then

$$\lim_{x \to \psi(\gamma)^+} \frac{dg}{dx} = \lim_{t \to \gamma^+} q(t) = q(\gamma) = \lim_{t \to \gamma^-} q(t) = \lim_{x \to \psi(\gamma)^-} \frac{dg}{dx},$$

which shows that g is smooth on $\psi(I)$.

Theorem 2.1. *Let G be the curve described by a pair of smooth functions $x(t)$ and $y(t)$ on an interval I such that $x'(t) > 0$ for $t \in I$ except at perhaps one point γ. If q is a continuous function defined on I such that $y'(t)/x'(t) = q(t)$ for $t \in I\backslash\{\gamma\}$, then G is also the graph of a smooth function $y = G(x)$ defined on $x(I)$.*

For example, let

$$x(t) = 3t^2 + 2\alpha t, \ y(t) = 2t^3 + \alpha t^2, \ \lambda > 0,$$

where $\alpha < 0$. Note that

$$(x(0^+), y(0^+)) = (0, 0), \ (x(+\infty), y(+\infty)) = (+\infty, +\infty).$$

Since

$$x'(t) = 6t + 2\alpha, \ y'(t) = 6t^2 + 2\alpha t,$$

we have

$$\frac{y'(t)}{x'(t)} = t, \ t \in R\backslash\{-\alpha/3\}.$$

By Theorem 2.1, the curve G described by the parametric functions $x(t)$ and $y(t)$ over $(0, \infty)$ is the graph of a smooth function $y = G(x)$ defined over $(0, \infty)$ which is also strictly decreasing.

2.2   Envelopes

Given a family of curves G_λ which can be expressed in the implicit form (2.4), where λ belongs to a subset Λ of $\mathbf{R}$, we may sometimes be able to associate exactly one point P_λ in each G_λ such that the totality of these points form a curve S. Such an associated curve S is called an **envelope** of the family $\{G_\lambda | \lambda \in \Lambda\}$ if the curves G_λ and S share a common tangent line[2] at the common point P_λ. An example can be seen in Figure 1.1 in which the curve S defined by $y = x^2/4$ is the envelope of the family of straight lines L_λ.

There are several concepts which we need to clarify before envelopes can be determined. Let us first proceed in an informal manner. Let Λ be an interval in $\mathbf{R}$. Assume the curve S represented by $x = \psi(\lambda)$ and $y = \phi(\lambda)$ for $\lambda \in \Lambda$ is an envelope of the family of curves G_λ expressed in the form (2.4). For each point $(\psi(\lambda), \phi(\lambda))$ that lies in the curve S, the requirement that it is also some point in G_λ leads us to

$$F(\psi(\lambda), \phi(\lambda), \lambda) = 0,$$

and hence

$$F'_x(\psi(\lambda), \phi(\lambda), \lambda)\psi'(\lambda) + F'_y(\psi(\lambda), \phi(\lambda), \lambda)\phi'(\lambda) + F'_\lambda(\psi(\lambda), \phi(\lambda), \lambda) = 0. \quad (2.5)$$

If $(\psi'(\lambda), \phi'(\lambda)) = 0$ or $(F'_x(\psi(\lambda), \phi(\lambda), \lambda), F'_y(\psi(\lambda), \phi(\lambda), \lambda)) = 0$, then $F'_\lambda(\psi(\lambda), \phi(\lambda), \lambda) = 0$ by (2.5). Otherwise, we may suppose $(\psi'(\lambda), \phi'(\lambda)) \neq 0$ and $(F'_x(\psi(\lambda), \phi(\lambda), \lambda), F'_y(\psi(\lambda), \phi(\lambda), \lambda)) \neq 0$. Since the former is the tangent vector of S and the latter is the gradient of F (as a function of the variables x and y) at the point $(x, y) = (\psi(\lambda), \phi(\lambda))$, if we require further that the tangent lines coincide, then

$$(\psi'(\lambda), \phi'(\lambda)) = K(F'_y(\psi(\lambda), \phi(\lambda), \lambda), -F'_x(\psi(\lambda), \phi(\lambda), \lambda)) \quad (2.6)$$

for some nonzero constant K. By (2.5) and (2.6), we see that $F'_\lambda(\psi(\lambda), \phi(\lambda), \lambda) = 0$ holds again.

Note that in the above derivations, we have made assumptions that the various derivatives exist and that the tangent lines are well defined. The following is now clear.

Theorem 2.2. *Let Λ be an interval in $\mathbf{R}$. Suppose we have a family $\{G_\lambda | \lambda \in \Lambda\}$ of curves G_λ that can be expressed in the form (2.4), where F is differentiable. If the envelope of this family exists and can be expressed by the differentiable vector function $(\psi(\lambda), \phi(\lambda))$ over Λ. Then*

$$\begin{cases} F(\psi(\lambda), \phi(\lambda), \lambda) = 0, \\ F'_\lambda(\psi(\lambda), \phi(\lambda), \lambda) = 0, \end{cases} \quad (2.7)$$

for $\lambda \in \Lambda$.

---

[2]Strictly speaking, we need to define the concept of a tangent line of a plane curve described by implicit functions or by parametric functions. See the remarks in the Note section of this Chapter.

Note that the above result only provides a necessary condition. If there is a curve which satisfies the equations in (2.7), it may not yield the envelope of the family of the curves. As an example, let a family of curves be defined by

$$F(x, y, \lambda) = (x - \lambda)^2 - y^3 = 0, \lambda \in \mathbf{R}.$$

Solving (2.7), we see that $\psi(\lambda) = \lambda$ and $\phi(\lambda) = 0$ for $\lambda \in \mathbf{R}$. The corresponding curve is just the x-axis which is not the envelope since it is not tangent to any curve in the family.

However, in some special cases, the solution of (2.7) may yield the envelope. Indeed, let us consider the family of straight lines L_λ in the x, y-plane defined by

$$L_\lambda : f(\lambda)x + g(\lambda)y = h(\lambda), \ (f(\lambda), g(\lambda)) \neq 0, \tag{2.8}$$

where f, g, h are real differentiable functions defined on an interval I. Then the condition (2.7) becomes the system

$$\begin{cases} f(\lambda)\psi(\lambda) + g(\lambda)\phi(\lambda) = h(\lambda), \\ f'(\lambda)\psi(\lambda) + g'(\lambda)\phi(\lambda) = h'(\lambda), \end{cases} \tag{2.9}$$

which is solvable for $\psi(\lambda)$ and $\phi(\lambda)$ when the determinant $f(\lambda)g'(\lambda) - f'(\lambda)g(\lambda) \neq 0$.

Theorem 2.3. *Let f, g, h be real differentiable functions defined on the interval I such that $f(\lambda)g'(\lambda) - f'(\lambda)g(\lambda) \neq 0$ for $\lambda \in I$. Let Φ be the family of straight lines of the form (2.8). Let the curve S be defined by the functions $x = \psi(\lambda), y = \phi(\lambda)$:*

$$\psi(\lambda) = \frac{g'(\lambda)h(\lambda) - g(\lambda)h'(\lambda)}{f(\lambda)g'(\lambda) - f'(\lambda)g(\lambda)}, \quad \phi(\lambda) = \frac{f(\lambda)h'(\lambda) - f'(\lambda)h(\lambda)}{f(\lambda)g'(\lambda) - f'(\lambda)g(\lambda)}, \quad \lambda \in I. \tag{2.10}$$

If ψ and ϕ are differentiable functions over I such that $(\psi'(\lambda), \phi'(\lambda)) \neq 0$ for $\lambda \in I$, then S is the envelope of the family Φ.

Proof. Let $\lambda \in I$ and $P_\lambda = (\psi(\lambda), \phi(\lambda))$. Then P_λ is a point of L_λ since $(\psi(\lambda), \phi(\lambda))$ is a solution of (2.9). Next, we need to show that the curves L_λ and S share a common tangent at the common point P_λ. Since L_λ is a straight line, we only need to show that the straight line L_λ is also the tangent of the graph S at the point P_λ. To see this, note that $(\psi(\lambda), \phi(\lambda))$ is a solution of (2.9), hence we may see from the first equation in (2.9) that

$$f'(\lambda)\psi(\lambda) + f(\lambda)\psi'(\lambda) + g'(\lambda)\phi(\lambda) + g(\lambda)\phi'(\lambda) = h'(\lambda)$$

and hence in view of the second equation in (2.9),

$$f(\lambda)\psi'(\lambda) + g(\lambda)\phi'(\lambda) = 0. \tag{2.11}$$

This shows the 'velocity' vector $(\psi'(\lambda), \phi'(\lambda))$ of S is orthogonal to the nonzero vector $(f(\lambda), g(\lambda))$. Since the vector $(f(\lambda), g(\lambda))$ is also orthogonal to the straight line L_λ, we see that our assertion must hold. Note that P_λ is unique since any other point will yield another solution of (2.9) which is impossible. The proof is complete.

Example 2.1. For each $\lambda \in \mathbf{R}$, let the straight line L_λ in the x, y-plane be defined by

$$L_\lambda : \lambda x + y = -\lambda^2.$$

To find the envelope of the family $\{L_\lambda | \ \lambda \in \mathbf{R}\}$, note that the determinant of the following system of equations

$$f(\lambda|x, y) = \lambda^2 + \lambda x + y = 0,$$
$$f'_\lambda(\lambda|x, y) = 2\lambda + x = 0,$$

is equal to -1. Hence by Theorem 2.3, the parametric functions describing the envelope is given by

$$x(\lambda) = -2\lambda, \ y(\lambda) = \lambda^2, \ \lambda \in \mathbf{R}.$$

Thus the envelope S is described by the graph of the function

$$y = \frac{x^2}{4}, \ x \in \mathbf{R}.$$

We remark that the condition $(\psi'(\lambda), \phi'(\lambda)) \neq 0$ in the above result is not absolutely necessary for S to be the envelope of Φ. The reason can be seen from the following example. Let $g(x) = x^2$ for $x \in \mathbf{R}$. Then its graph has a 'tangent line' at the point $(0,0)$. Yet if we describe this graph by the parametric functions $x(\lambda) = \lambda^2$ and $y(\lambda) = \lambda^4$ for $\lambda \in \mathbf{R}$, then $x'(0) = 0 = y'(0)$ and hence a nonzero velocity vector at $(0,0)$ is not available[3]. For this reason, we provide a variant of Theorem 2.3 as follows.

Theorem 2.4. *Let f, g, h be real differentiable functions defined on the real interval I such that $f(\lambda)g'(\lambda) - f'(\lambda)g(\lambda) \neq 0$ and $g(\lambda) \neq 0$ for $\lambda \in I$ where I is closed on the right with boundary point d. Let Φ be the family of straight lines of the form*

$$L_\lambda : f(\lambda)x + g(\lambda)y = h(\lambda), \ \lambda \in I. \qquad (2.12)$$

Let S be the curve defined by the parametric functions in (2.10). If ψ and ϕ are smooth functions over I such that $\psi'(\lambda) \neq 0$ for $\lambda \in I \backslash \{d\}$ (but $\psi'(d)$ may equal 0) and $\lim_{\lambda \to d^-} \phi'(\lambda)/\psi'(\lambda)$ exists, then S is the envelope of the family Φ.

In view of the proof of Theorem 2.3, it suffices to consider what may happen for points near and to the left of d. Assume without loss of generality that $\psi'(\lambda) > 0$ for $\lambda \in (c, d)$ where $c \in I$ is sufficiently near d. Then ψ is strictly increasing on $(c, d]$ and hence S is the graph of a differentiable function $y = S(x)$ for x at $\psi(d)$ and for x near and to the left of $\psi(d)$. In view of Cauchy's mean value theorem

$$S'(\psi(d)) = \lim_{\lambda \to d^-} \frac{\phi(d) - \phi(\lambda)}{\psi(d) - \psi(\lambda)} = \lim_{\lambda \to d^-} \frac{\phi'(\mu_\lambda)}{\psi'(\mu_\lambda)}$$

---

[3]Although a nonzero velocity vector is not available, a 'tangent line' may still exists for a smooth parametric curve. See the remarks in the Note section of this Chapter.

where $\mu_\lambda \to d$ as $\lambda \to d^-$. By (2.11), we may also see that

$$\lim_{\lambda \to d^-} \frac{\phi'(\mu_\lambda)}{\psi'(\mu_\lambda)} = \lim_{\lambda \to d^-} \frac{-f(\mu_\lambda)}{g(\mu_\lambda)} = \frac{-f(d)}{g(d)}.$$

These show that the line L_d is the tangent line of S at the point $(\psi(d), \phi(d))$. The proof is complete.

We remark that the condition that $g(\lambda) \neq 0$ for $\lambda \in I$ implies every straight line of Φ is not vertical, so that we may treat the envelope as a graph whose tangent lines can be defined by derivatives.

We remark further that the above result may be extended so as to allow $\psi'(\lambda) = 0$ at finitely many points in I, but is presented alone since it will be used many times in the sequel. The proof of the following more general result is similar since we are dealing with local properties of the envelope.

Theorem 2.5. *Let* f, g, h *be real differentiable functions defined on the real interval* I *such that* $f(\lambda)g'(\lambda) - f'(\lambda)g(\lambda) \neq 0$ *and* $g(\lambda) \neq 0$ *for* $\lambda \in I$. *Let* Φ *be the family of straight lines of the form* (2.12). *Let* S *be the curve defined by the parametric functions in* (2.10). *If* ψ *and* ϕ *are smooth functions over* I *such that* $\psi'(\lambda) \neq 0$ *for* $I \backslash \{d_1, .., d_m\}$ *where each* $d_i \in I$ *and* $\lim_{\lambda \to d_i^-} \phi'(\lambda)/\psi'(\lambda)$ *as well as* $\lim_{\lambda \to d_i^+} \phi'(\lambda)/\psi'(\lambda)$ *exist and are equal, then* S *is the envelope of the family* Φ.

Example 2.2. Let

$$f(\lambda|p, q) = \lambda^2 - \lambda + \lambda^{-1}(\lambda - 1)p + q, \ \lambda > 0.$$

For each $\lambda \in (0, \infty)$, let L_λ be the straight line

$$L_\lambda : \lambda^2 - \lambda + \lambda^{-1}(\lambda - 1)x + y = 0.$$

Since

$$f_\lambda'(\lambda|x, y) = 2\lambda - 1 + \lambda^{-2}x,$$

we see that the determinant of the system $f(\lambda|x, y) = 0 = f_\lambda'(\lambda|x, y)$ is 1. Solving this system, we obtain

$$x(\lambda) = \lambda^2 - 2\lambda^3, \ y(\lambda) = 2\lambda\left(\lambda^2 - 2\lambda + 1\right), \ \lambda > 0.$$

Note that

$$x'(\lambda) = 2\lambda(1 - 3\lambda), \ y'(\lambda) = 2(\lambda - 1)(3\lambda - 1).$$

Thus $x'(\lambda)$ has the unique positive root $1/3$. Furthermore, $x(\lambda)$ is strictly increasing on $(0, 1/3)$ and strictly decreasing on $(1/3, \infty)$. Also,

$$\frac{dy}{dx} = \frac{1}{\lambda} - 1, \ \lambda \in (0, 1/3) \cup (1/3, \infty),$$

$$\frac{d^2y}{dx^2} = \frac{1}{2\lambda^3(3\lambda - 1)}, \ \lambda \in (0, 1/3) \cup (1/3, \infty).$$

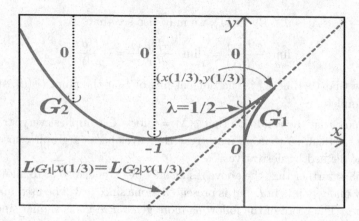

Fig. 2.1

These properties together with other easily obtained information allow us to depict the curve G described by the parametric functions $x(\lambda)$ and $y(\lambda)$ as shown in Figure 2.1.

It is composed of two pieces G_1 and G_2. The first piece G_1 corresponds to the case where $\lambda \in (0, 1/3)$ and the second G_2 to the case where $\lambda \in [1/3, \infty)$. Furthermore, G_1 is the graph of a strictly increasing, strictly concave and smooth function $y = G_1(x)$ defined on $(0, x(1/3))$ such that $G_1(0^+) = 0$ and $G_1'(0^+) = +\infty$; and G_2 is the graph of a function $y = G_2(x)$ which is strictly convex and smooth over $(-\infty, x(1/3)]$ such that $G_2(-\infty) = +\infty$ and $G_2'(-\infty) = -\infty$. By Theorem 2.3, G_1 is the envelope of the family $\{L_\lambda : \lambda \in (0, 1/3)\}$. Since

$$\lim_{\lambda \to 1/3} \frac{y'(\lambda)}{x'(\lambda)} = \lim_{\lambda \to 1/3} \frac{dy}{dx}(\lambda) = 2,$$

by Theorem 2.4, G_2 is the envelope of the family $\{L_\lambda : \lambda \in [1/3, \infty)\}$. Theorem 2.5 also shows that G is the envelope of the family $\{L_\lambda : \lambda \in (0, \infty)\}$.

2.3   Dual Sets of Plane Curves

Given a plane curve S described by parametric functions, a point in the plane is said to be a **dual point** of order m of S, where m is a nonnegative integer, if there exist exactly m mutually distinct tangents of S that also pass through it. The set of all dual points of order m of S in the plane is called the **dual** set of order m of S. We remark that $m = 0$ is allowed. In this case, there are no tangents of S that pass through the point in consideration.

Example 2.3. An easy example is the following. Let S be the graph of a real linear function g defined by

$$g(x) = ax + b, \ x \in \mathbf{R}.$$

Then, the only tangent line of S is itself. Hence, any point in S is a dual point of order 1 and any other point is a dual point of order 0 (see Figure 2.2). Note that if $\tilde{S}$ is the graph of the restriction $g\chi_I$ of g, where I is a real interval with positive length, then the only tangent line of $\tilde{S}$ is S. Hence, any point in S is a dual point of order 1 of $\tilde{S}$ and any other point is a dual point of order 0 of $\tilde{S}$.

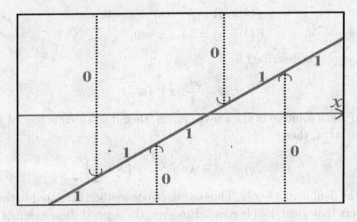

Fig. 2.2

Let Φ be the family of straight lines of the form (2.8), and let S be the envelope of the family Φ. If there is a point (a, b) in the plane such that there is a straight line L_{λ^*} in Φ which also passes through it, then

$$f(\lambda^*)a + g(\lambda^*)b = h(\lambda^*),$$

that is, the equation

$$f(\lambda)a + g(\lambda)b = h(\lambda) \tag{2.13}$$

has a solution $\lambda = \lambda^*$ in Λ. Conversely, if (2.13) has a solution $\lambda = \lambda^*$ in Λ, then the straight line L_{λ^*} in Φ passes through it. The set Φ, however, is just the set of tangents of S. Thus we have the following result.

Theorem 2.6. *Let Λ be an interval in $\mathbf{R}$, and f, g, h be real differentiable functions defined on Λ such that $f(\lambda)g'(\lambda) - f'(\lambda)g(\lambda) \neq 0$ for $\lambda \in \Lambda$. Let Φ be the family of straight lines of the form (2.8), and let the curve S be the envelope of the family Φ. Then the point (a, b) in the plane is a dual point of order m of S, if, and only if, the function $f(\lambda)a + g(\lambda)b - h(\lambda)$ has exactly m mutually distinct roots in Λ.*

The above result is fundamental in our later discussions since root seeking is now equivalent to counting the number of tangents.

2.4   Notes

Envelope is a standard concept in mathematics. An elementary exposition of related concepts can be found in [2]. See also [13]. The concept of a dual set, however, is new.

In Example 2.1, we solve the linear system

$$f(\lambda|x, y) = \lambda^2 + \lambda x + y = 0,$$
$$f'(\lambda|x, y) = 2\lambda + x = 0,$$

to find the envelope described by

$$y = \frac{x^2}{4}, \ x \in \mathbf{R}.$$

Note that if λ is a solution of the above system, then it is a double root of f. Indeed, if we let $y = x^2/4$, then

$$f(\lambda|x, y) = \lambda^2 + \lambda x + \frac{x^2}{4} = \left(\lambda + \frac{x}{2}\right)^2,$$

which has the double root $-x/2$. Thus our envelpoe method is related to the location of parameters that yield double roots. However, the general theory cannot be based on the simple idea of locating such parameters.

The plane curves described by implicit functions that we will encounter in the sequel are straight lines. Therefore we do not need to define the concept of a tangent line for such a curve. As for curves defined by parametric functions. We may employ the concept of contact (due to Lagrange) to define their tangent lines. For instance, let C be a plane curve and L a plane straight line with a point P in common. Take a point A in C near P and let AD be the distance to L. If

$$\lim_{A \to P, A \in C} \frac{AD}{AP} = 0,$$

then L is the tangent line of C at P. We have avoided such a definition since we only need tangent lines of graphs. As can be verified easily, the curve, defined by the parametric functions $x(\lambda) = \lambda^2$ and $y(\lambda) = \lambda^4$ for $\lambda \in \mathbf{R}$, has the tangent line $y = 0$ at the point $(0, 0)$.

Chapter 3

Dual Sets of Convex-Concave Functions

3.1 Quasi-Tangent Lines

Envelopes as curves can take on complicated forms. In some cases, however, they are graphs of convex functions or concave functions. In such cases, we may deduce some of the properties of the corresponding envelopes.

Recall that a real function g defined on interval I of $\mathbf{R}$ is said to be convex if

$$g(\omega\alpha + (1-\omega)\beta) \leq \omega g(\alpha) + (1-\omega)g(\beta), \ \omega \in (0,1); \ \alpha, \beta \in I,$$

and strictly convex if the above inequality is strict for $\alpha \neq \beta$. g is said to be concave (strictly concave) if $-g$ is convex (respectively strictly convex). We will assume some elementary facts about convex functions studied in elementary analysis (see e.g. the first Chapter of [32]). In particular, it is known that when g is convex on I, then it is continuous on the interior of I. The following is also true.

Lemma 3.1. *Suppose g is a real differentiable function on the interval I. Then g is strictly convex on I if, and only if, the derived function g' is strictly increasing on I. If in addition to the differentiability of g, $g''(x)$ exists and is positive for each x in the interior of I, then g is strictly convex on I.*

A proof can be found in the first Chapter of [32]. For the sake of completeness, we sketch it as follows. Let g be a real differentiable function on the interval I. Then

$$g(x) - g(d) = \int_d^x g'(t)dt, \ x \in I,$$

where $d \in I$. Let $\omega \in (0,1)$ and $\alpha, \beta \in I$ such that $\alpha < \beta$. If g' is strictly increasing, then

$$\omega g(\alpha) + (1-\omega)g(\beta) - g(\omega\alpha + (1-\omega)\beta)$$

$$= (1-\omega) \int_{\omega\alpha+(1-\omega)\beta}^{\beta} g'(t)dt - \omega \int_{\alpha}^{\omega\alpha+(1-\omega)\beta} g'(t)dt$$

$$> (1-\omega) \int_{\omega\alpha+(1-\omega)\beta}^{\beta} g'(\omega\alpha + (1-\omega)\beta)dt - \omega \int_{\alpha}^{\omega\alpha+(1-\omega)\beta} g'(\omega\alpha + (1-\omega)\beta)dt$$

$$= 0.$$

Conversely, let us take three points α, δ, β in I such that $\alpha < \delta < \beta$. Let $P = (\alpha, g(\alpha))$, $Q = (\delta, g(\delta))$ and $R = (\beta, g(\beta))$. If g is strictly convex on I, then

$$\text{slope } PQ < \text{slope } PR < \text{slope } QR.$$

Thus

$$f'(\alpha) \leq \frac{g(\delta) - g(\alpha)}{\delta - \alpha} < \frac{g(\beta) - g(\delta)}{\beta - \delta} \leq f'(\beta).$$

Finally, if $g''(x) > 0$ on the interior of I, then g' is strictly increasing on I (and hence g is strictly convex on I) since

$$g'(\beta) - g'(\alpha) = \int_\alpha^\beta g''(t)dt > 0$$

for any $\alpha, \beta \in I$ such that $\alpha < \beta$. The proof is complete.

Example 3.1. Let I be a real interval with $b \in I$. Let $I_1 = I \cap (-\infty, b)$ and $I_2 = I \cap (b, +\infty)$. Let $g_1 \in C^1(I_1)$ and $g_2 \in C^1(I_2)$ be strictly convex functions on I_1 and I_2 respectively such that $g_1(b^-)$, $g_1'(b^-)$, $g_2(b^+)$ and $g_2'(b^+)$ exist, and $g_1(b^-) = g_2(b^+)$ as well as $g_1'(b^-) = g_2'(b^+)$. Then the function $g : I \to R$ defined by

$$g(x) = \begin{cases} g_1(x) & \text{if } x \in I_1 \\ g_2(x) & \text{if } x \in I_2 \\ g_1(b^-) & \text{if } x = b \end{cases},$$

in view of Lemma 3.1, is $C^1(I)$ and strictly convex.

Let g be a function defined on an interval I with $c = \inf I$ and $d = \sup I$. Note that c or d may be infinite, or may be outside the interval I, and that $g(c^+), g(d^-)$, $g'(c^+)$ or $g'(d^-)$ may not exist. In case c is finite and $g(c^+), g'(c^+)$ exist, we may define a straight line

$$L_{g|c^+}(x) = g'(c^+)(x - c) + g(c^+), \ x \in R.$$

Similarly, if d is finite and $g(d^-), g'(d^-)$ exist, we may define another straight line

$$L_{g|d^-}(x) = g'(d^-)(x - d) + g(d^-), \ x \in R.$$

Together with the tangent lines of g (see (1.3)), they form the set of *quasi-tangent lines* of g. The collection of all the quasi-tangent lines of g will be denoted by $\mathbf{F}(g)$.

Example 3.2. Suppose $c, d \in R$ and $g \in C^1[c, d)$. Then $L_{g|c^+} = L_{g|c}$.

Example 3.3. Suppose $c, d \in R$ and $g : (c, d) \to R$ is a smooth function such that $g(c^+), g(d^-), g'(c^+), g'(d^-)$ exist. Then an element of $\mathbf{F}(g)$ is either of the form $L_{g|\lambda}$ where $\lambda \in (c, d)$, or $L_{g|c^+}$ or $L_{g|d^-}$.

In view of the previous Example 3.3, the description of an element of $\mathbf{F}(g)$ may be quite inconvenient. For this reason, we will simply write $L_{g|c}$ for $L_{g|c^+}$ provided that c is finite and $g(c^+)$, $g'(c^+)$ exist, and $L_{g|d}$ for $L_{g|d^-}$ provided that d is finite and $g(d^-)$, $g'(d^-)$ exist. Therefore, the notation $L_{g|\delta}$ stands for a quasi-tangent line of g and it is a true tangent line of g if δ is an interior point in I such that $g'(\delta)$ exists, or is the left boundary point $c \in I$ such that $g'(c)$ exists, $g(c^+) = g(c)$ and $g'(c^+) = g'(c)$, or is the right boundary point $d \in I$ such that $g'(d)$ exists, $g(d^-) = g(d)$ and $g'(d^-) = g'(d)$.

Example 3.4. Suppose $c, d \in \mathbf{R}$ and $g : (c, d) \to \mathbf{R}$ is a smooth function. Then $\mathbf{F}(g) = \{L_{g|\lambda} : \lambda \in \Lambda\}$, where $\Lambda = [c, d]$ if $g(c^+)$, $g(d^-)$, $g'(c^+)$, $g'(d^-)$ exist; and $\Lambda = [c, d)$ if $g(c^+)$, $g'(c^+)$ exist but one of $g(d^-)$ or $g'(d^-)$ fails to exist.

The convenient notations stated above are also used in some of the figures to be presented in the sequel. For instance, in Figure 3.1, the function g is defined over (c, d), yet we write $L_{g|c}$ instead of the more precise notation $L_{g|c^+}$.

3.2 Asymptotes

In case g is a function defined for $x > d \in \mathbf{R}$, then g is said to have the asymptote $y(x) = \alpha x + \beta$ at $x = +\infty$ if

$$\lim_{x \to +\infty} \{g(x) - (\alpha x + \beta)\} = 0.$$

The asymptote of g at $+\infty$ will be denoted by $L_{g|+\infty}$. In case $\alpha = 0$, we say that g has a horizontal asymptote at $+\infty$. Recall that if

$$\overline{\alpha} = \lim_{x \to \infty} \frac{g(x)}{x},$$

and

$$\overline{\beta} = \lim_{x \to \infty} \{g(x) - \alpha x\}$$

exist, then the asymptote $L_{g|+\infty}$ exists and is given by

$$L_{g|+\infty}(x) = \overline{\alpha}x + \overline{\beta}.$$

The converse is also true.

Note further that if the graph of g is also described by the parametric functions $x = \psi(t)$ and $y = \phi(t)$ over the interval $I = (\sigma, \tau)$ such that $\lim_{t \to \tau^-} x(t) = +\infty$ (or $\lim_{t \to \tau^-} x(t) = -\infty$), then $L_{g|+\infty}(x) = \alpha x + \beta$ (respectively $L_{g|-\infty}(x) = \alpha x + \beta$) where

$$\alpha = \lim_{t \to \tau^-} \frac{\phi(t)}{\psi(t)} \quad \text{and} \quad \beta = \lim_{\lambda \to \tau^-} \{\phi(t) - \alpha\psi(t)\}.$$

The asymptote $L_{g|-\infty}$ at $-\infty$ is similarly defined, and its properties can similarly be stated.

Lemma 3.2. *Let* $\Phi = \{L_\lambda | \ \lambda \in (\sigma, \tau)\}$, *where* $\tau \in \mathbf{R}$, *be a family of straight lines* L_λ *defined by*

$$L_\lambda : a(\lambda)x + b(\lambda)y + c(\lambda) = 0, \ \lambda \in (\sigma, \tau),$$

where a, b, c *are smooth functions defined on* (σ, τ) *such that* $\acute{a}(\tau^-), b(\tau^-), c(\tau^-)$ *exist and* $b(\tau^-) \neq 0$. *Let the envelope* G *of the family* Φ *be described by the continuous parametric functions* $x = \psi(\lambda)$ *and* $y = \phi(\lambda)$ *for* $\lambda \in (\sigma, \tau)$. *Suppose* $\lim_{\lambda \to \tau^-} \psi(\lambda) = +\infty$ *(or* $\lim_{\lambda \to \tau^-} \psi(\lambda) = -\infty$*),* $\psi'(\lambda) \neq 0$ *for* $\lambda \in (\sigma, \tau)$ *and that* G *is also the graph of a function* $y = G(x)$ *defined on* $(\psi(\sigma^+), +\infty)$ *(respectively* $(-\infty, \psi(\sigma^+))$*). Then the asymptote* $L_{G|+\infty}$ *(respectively* $L_{G|-\infty}$*) of* G *is given by*

$$y = -\frac{a(\tau^-)}{b(\tau^-)}x + \frac{c(\tau^-)}{b(\tau^-)}, \ x \in \mathbf{R}. \tag{3.1}$$

Proof. For all λ sufficiently close to τ, say, for $\lambda \in (\tau - \delta, \tau)$ where δ is some positive number, $\psi(\lambda) > 0$ and $\psi'(t) > 0$. The tangent of the envelope G at the point $(\phi(\lambda), \psi(\lambda))$, where $\lambda \in (\tau - \delta, \tau)$, is given by both functions

$$y = \frac{\phi'(\lambda)}{\psi'(\lambda)}x + \phi(\lambda) - \frac{\phi'(\lambda)}{\psi'(\lambda)}\psi(\lambda),$$

and

$$y = -\frac{a(\lambda)}{b(\lambda)}x - \frac{c(\lambda)}{b(\lambda)}.$$

Hence

$$\frac{\phi'(\lambda)}{\psi'(\lambda)} = -\frac{a(\lambda)}{b(\lambda)}$$

and

$$\phi(\lambda) - \frac{\phi'(\lambda)}{\psi'(\lambda)}\psi(\lambda) = -\frac{c(\lambda)}{b(\lambda)}$$

for all $\lambda \in (\tau - \delta, \tau)$. Thus

$$\lim_{\lambda \to \tau^-} \frac{\phi'(\lambda)}{\psi'(\lambda)} = -\frac{a(\tau^-)}{b(\tau^-)}$$

and

$$\lim_{\lambda \to \tau^-} \left\{ \phi(\lambda) - \frac{\phi'(\lambda)}{\psi'(\lambda)}\psi(\lambda) \right\} = -\frac{c(\tau^-)}{b(\tau^-)}.$$

These show that the asymptote of G at $+\infty$ is given by (3.1). The proof is complete.

Example 3.5. Let

$$f(\lambda|x, y) = \lambda^3 - \lambda^2 + \lambda(\lambda - 1)x + y, \ \lambda > 0.$$

Let $a(\lambda) = \lambda(\lambda - 1)$, $b(\lambda) = 1$ and $c(\lambda) = \lambda^3 - \lambda^2$ for $\lambda > 0$. Then

$$f'_\lambda(\lambda|x, y) = 3\lambda^2 - 2\lambda + (2\lambda - 1)x, \ \lambda > 0.$$

Hence

$$D(\lambda) = a(\lambda)b'(\lambda) - a'(\lambda)b(\lambda) = 1 - 2\lambda,$$

which is not zero for $\lambda \in (0, 1/2)$. Let $\Phi = \{L_\lambda|\ \lambda \in (0, 1/2)\}$ be the family of straight lines defined by

$$L_\lambda : f(\lambda|x, y) = 0, \ \lambda \in (0, 1/2).$$

Then the envelope G of Φ is given by

$$\phi(\lambda) = \frac{3}{2} \frac{\lambda\left(\frac{2}{3} - \lambda\right)}{\lambda - \frac{1}{2}},$$

$$\psi(\lambda) = \frac{1}{2} \frac{\lambda^2(\lambda - 1)^2}{\lambda - \frac{1}{2}},$$

for $\lambda \in (0, 1/2)$. Note that

$$\psi'(\lambda) = -2\frac{-3\lambda + 1 + 3\lambda^2}{(2\lambda - 1)^2},$$

$$\phi'(\lambda) = 2\lambda\,(\lambda - 1)\,\frac{3\lambda^2 - 3\lambda + 1}{(2\lambda - 1)^2},$$

for $\lambda \in (0, 1/2)$. Since $\psi(0) = 0$, $\lim_{\lambda \to (1/2)-} \psi(\lambda) = -\infty$, $\psi(\lambda) < 0$ and $\psi'(\lambda) < 0$ and

$$\frac{\phi'(\lambda)}{\psi'(\lambda)} = -\lambda(\lambda - 1) > 0$$

for $\lambda \in (0, 1/2)$, we see that G is also the graph of a function $y = G(x)$ defined on $(-\infty, 0)$. By Lemma 3.2, the asymptote $L_{G|-\infty}$ is given by

$$L_{G|-\infty}(x) = \frac{1}{4}x + \frac{1}{8}.$$

The same conclusion can be arrived at by definition. Indeed, as can be checked easily,

$$\lim_{\lambda \to (1/2)-} \frac{\phi(\lambda)}{\psi(\lambda)} = \frac{1}{4} \text{ and } \lim_{\lambda \to (1/2)-} \left\{\phi(\lambda) - \frac{1}{4}\psi(\lambda)\right\} = \frac{1}{8}.$$

3.3   Intersections of Quasi-Tangent Lines and Vertical Lines

Let g be a strictly convex and smooth function defined on a real interval I with $c = \inf I$ and $d = \sup I$ (so that c may be $-\infty$ and d may be $+\infty$). Let $\mathbf{F}(g) = \{L_{g|\lambda} : \lambda \in \Lambda\}$ be the set of all quasi-tangent lines of g defined in the last Section.

Let the y-coordinate of the point of intersection of the vertical straight line $x = \alpha$ with $L_{g|\lambda}$ be denoted by $\hbar_g(\lambda|\alpha)$, that is,

$$\hbar_g(\lambda|\alpha) = L_{g|\lambda}(\alpha), \text{ where } \lambda \in \Lambda \text{ and } \alpha \in \mathbf{R}. \tag{3.2}$$

We will also write $\hbar(\lambda|\alpha)$ instead of $\hbar_g(\lambda|\alpha)$ in case no confusion is caused. It is easily seen that for each fixed $\alpha \in \mathbf{R}$, $\hbar$ as a function[1] of λ in Λ is continuous (since g is smooth).

It also appears that in case $d < +\infty$ and g has a 'vertical tangent at d', then the y-coordinate of the point of intersection will tend to $-\infty$. We will prove this statement and its converse as follows.

Lemma 3.3. *Suppose $c, d \in \mathbf{R}$ and $g : (c,d) \to \mathbf{R}$ is a smooth and strictly convex function. Then $g(d^-) > -\infty$. Furthermore, $g'(d^-) = +\infty$ if, and only if,*

$$\lim_{\lambda \to d^-} \hbar_g(\lambda|\alpha) = -\infty \text{ for any } \alpha < d. \tag{3.3}$$

Proof. Since $g'(x)$ is increasing in (c,d), if we pick any $a \in (c,d)$, then

$$g(x) = g(a) + \int_a^x g'(t)dt \geq g(a) + g'(a)(x - a), \ x \in (a, d).$$

Thus

$$g(d^-) \geq g(a) + g'(a)(d - a) > -\infty$$

as required. Consequently, either $g(d^-) < +\infty$ or $g(d^-) = +\infty$.

Suppose $g'(d^-) = +\infty$. If $g(d^-) < \infty$, then

$$\lim_{\lambda \to d^-} \{g'(\lambda)(\alpha - \lambda) + g(\lambda)\} = -\infty$$

for any $\alpha < d$. If $g(d^-) = +\infty$, then (3.3) also holds. Indeed, if $\hbar(\lambda|\alpha) = L_{g|\lambda}(\alpha)$ is bounded below, say by β. Fix $u \in \mathbf{R}$ with $\max\{c, \alpha\} < u < d$ and $g(u) > \beta$. Since $g(d^-) = g'(d^-) = +\infty$, there is $\lambda_1 \in (u, d)$ such that $0 < (g(u) - \beta)/(u - \alpha) < g'(\lambda_1)$. Let $L_{g|\lambda_1}$ be the tangent line of g passing through $(\lambda_1, g(\lambda_1))$. Then

$$L_{g|\lambda_1}(x) = g'(\lambda_1)(x - u) + L_{g|\lambda_1}(u).$$

Since g is convex, we see that $g(u) \geq L_{g|\lambda_1}(u)$. Therefore,

$$\hbar(\lambda_1|\alpha) = g'(\lambda_1)(\alpha - \lambda_1) + g(\lambda_1) = g'(\lambda_1)(\alpha - u + u - \lambda_1) + g(\lambda_1)$$

$$= g'(\lambda_1)(x - u) + L_{g|\lambda_1}(u) < \frac{g(u) - \beta}{u - \alpha}(\alpha - u) + g(u) = \beta,$$

which is a contradiction. Hence $\hbar(\lambda|\alpha)$ cannot be bounded below. Conversely, if (3.3) holds, then $g(d^-) > -\infty$ clearly implies $g'(d^-) = \infty$. The proof is complete.

We remark that the condition $c \in \mathbf{R}$ is not needed in the proof. Thus the interval (c,d) in the above result can be replaced by $(-\infty, d)$ or $[c, d)$.

Example 3.6. Suppose $c, d \in \mathbf{R}$ and $g : (c,d) \to \mathbf{R}$ is a strictly convex and smooth function such that $g(c^+), g'(c^+), g(d^-)$ exist and $\lim_{\lambda \to d^-} \hbar_g(\lambda|\alpha) = -\infty$ for any $\alpha < d$. Then the set of quasi-tangent lines of g is $\{L_{g|\lambda} : \lambda \in [c, d)\}$ but not $\{L_{g|\lambda} : \lambda \in [c, d]\}$ since $g'(d^-) = +\infty$ by Lemma 3.3.

---

[1]The function $\hbar(\lambda|\alpha)$, since it is equal to $L_{g|\lambda}(\alpha)$, may be avoided. However, it is used to remind us of the different roles played by λ and α. This function may be called the **sweeping function**.

Lemma 3.4. *Suppose $g : (c, \infty) \to \mathbf{R}$ is a strictly convex and smooth function such that $g'(+\infty) = \infty$. Then $\lim_{\lambda \to \infty} \hbar_g(\lambda | \alpha) = -\infty$ for any $\alpha \in \mathbf{R}$.*

Proof. Since $g'(+\infty) = +\infty$ and g is strictly convex, then there is $r > c$ such that $g'(r) > 0$ and $g(x) \geq L_{g|r}(x)$ for $x > c$. Hence $g(+\infty) = +\infty$. The rest of the proof is similar to that of Lemma 3.3.

It is important to note that the converse in Lemma 3.4 is not true in general as can be seen from the following result.

Lemma 3.5. *Assume $g : (c, \infty) \to \mathbf{R}$ is a strictly convex and smooth function such that $g(+\infty) = -\infty$ and $g'(+\infty) = 0$. Then $\lim_{\lambda \to \infty} \hbar_g(\lambda | \alpha) = -\infty$ for any $\alpha \in \mathbf{R}$.*

Proof. Since $g'(+\infty) = 0$, it suffices to show that $\lim_{\lambda \to \infty} \{-\lambda g'(\lambda) + g(\lambda)\} = -\infty$. For this purpose, take $t < \lambda$, then for some $\xi \in (t, \lambda)$,

$$g(\lambda) - g(t) = g'(\xi)(\lambda - t) < g'(\lambda)(\lambda - t).$$

Hence

$$\limsup_{\lambda \to \infty} \{g'(\lambda)(\omega - \lambda) + g(\lambda)\} \leq \limsup_{\lambda \to \infty} \{tg'(\lambda) + g(t)\} = g(t).$$

Since t is arbitrary, and since $g(+\infty) = -\infty$, we see that the left hand side is $-\infty$ as required. The proof is complete.

There are some obvious variants of the above results. For example, if $g : (-\infty, c) \to \mathbf{R}$ is a strictly convex and smooth function such that $g'(-\infty) = -\infty$, then $\lim_{\lambda \to \infty} \hbar_g(\lambda | \alpha) = -\infty$ for any $\alpha \in \mathbf{R}$.

The statement, that

$$\lim_{\lambda \to \infty} \hbar_g(\lambda | \alpha) = -\infty \text{ for any } \alpha \in \mathbf{R},$$

(where it is understood that $\hbar$ is defined for all large λ) is cumbersome. For this reason, we will adopt the equivalent notation $g \sim H_{+\infty}$ instead. Similarly, we set

$$g \sim H_{-\infty} \text{ if, and only if, } \lim_{\lambda \to -\infty} \hbar_g(\lambda | \alpha) = -\infty \text{ for any } \alpha \in \mathbf{R};$$

$$g \sim H_{d^-} \text{ if, and only if, } \lim_{\lambda \to d^-} \hbar_g(\lambda | \alpha) = -\infty \text{ for any } \alpha < d$$

and

$$g \sim H_{d^+} \text{ if, and only if, } \lim_{\lambda \to d^+} \hbar_g(\lambda | \alpha) = -\infty \text{ for any } \alpha > d.$$

Therefore, in case g is smooth and strictly convex on (c, d), in view of Lemma 3.3, $g'(d^-) = +\infty$ if, and only if, $g \sim H_{d^-}$. In case g is smooth and strictly convex on (c, ∞), then in view of Lemma 3.4, $g'(+\infty) = +\infty$ implies $g \sim H_{+\infty}$, etc.

Now suppose J is any subinterval of Λ. If $\alpha \leq \inf J$, then for any λ_1 and λ_2 that satisfy $\inf J \leq \lambda_1 < \lambda_2 \leq \sup J$, since g is strictly convex, the slope of g at $x = \lambda_1$ is less than the slope of the chord joining the points $(\lambda_1, g(\lambda_1))$ and $(\lambda_2, g(\lambda_2))$ which in turn is less than the slope of g at λ_2, hence

$$\hbar(\lambda_1|\alpha) = g'(\lambda_1)(\alpha - \lambda_1) + g(\lambda_1) > \frac{g(\lambda_1) - g(\lambda_2)}{\lambda_1 - \lambda_2}(\alpha - \lambda_1) + g(\lambda_1)$$

$$= \frac{g(\lambda_1) - g(\lambda_2)}{\lambda_1 - \lambda_2}(\alpha - \lambda_2) + g(\lambda_2) > g'(\lambda_2)(\alpha - \lambda_2) + g(\lambda_2) = \hbar(\lambda_2|\alpha).$$

Thus $\hbar$ is strictly decreasing on J. Similarly, if $\alpha \geq \sup J$, then $\hbar$ is strictly increasing on J.

Theorem 3.1. *Let g be a strictly convex and smooth function defined on a real interval I with $c = \inf I$ and $d = \sup I$. Let $\mathbf{F}(g) = \{L_{g|\lambda} : \lambda \in \Lambda\}$ be the set of all quasi-tangent lines of g. Let $\hbar$ be defined by (3.2). Then $\hbar$ is continuous on Λ. Furthermore, given any subinterval J of Λ, if $\alpha \leq \inf J$, then $\hbar$ is strictly decreasing on J, while if $\alpha \geq \sup J$, then $\hbar$ is strictly increasing on J. In particular, if α is an interior point of I, then $\hbar(\lambda|\alpha)$ has a local maximum $g(\alpha)$ at $\lambda = \alpha$.*

Example 3.7. If g is a strictly convex and smooth function defined on (c, d) and $g(c^+)$, $g(d^-)$, $g'(c^+)$ and $g'(d^-)$ exist. Then $\mathbf{F}(g) = \{L_{g|\lambda} : \lambda \in [c, d]\}$. If $\alpha \leq c$, then $\hbar$ is strictly decreasing on $[c, d]$, and hence

$$L_{g|d}(\alpha) = \hbar(d|\alpha) = \inf_{\mu \in [c,d]} \hbar(\mu|\alpha) < \sup_{\mu \in [c,d]} \hbar(\mu|\alpha) = \hbar(c|\alpha) = L_{g|c}(\alpha). \qquad (3.4)$$

If $\alpha \geq d$, then $\hbar$ is strictly increasing on $[c, d]$, and hence

$$L_{g|c}(\alpha) = \inf_{\mu \in [c,d]} \hbar(\mu|\alpha) < \sup_{\mu \in [c,d]} \hbar(\mu|\alpha) = L_{g|d}(\alpha). \qquad (3.5)$$

Next, suppose $\alpha \in (c, d)$. Such a case is slightly more complicated. First, note that g is strictly convex on (c, d), $g'(c^+) < g'(d^-)$ so that the lines $L_{g|c}$ and $L_{g|d}$ intersect at some unique point (a, b). Furthermore, $a > c$ for otherwise the line $L_{g|c}$, being the line that joins (a, b) and $(c, g(c^+))$, will have a slope which is greater than the slope of the line $L_{g|d}$ (which is the line joining (a, b) and $(d, g(d^-))$). This is contrary to the fact that $g'(c^+) < g'(d^-)$. Similarly, $a < d$. Thus $a \in (c, d)$. Now that $a \in (c, d)$, we may see further that $b < g(a)$ since the point $(a, g(a))$ lies above $L_{g|c}(a)$ and $L_{g|d}(a)$. If $\alpha \in (c, d)$, then either $\alpha \in (c, a]$ or $\alpha \in (a, d)$. Suppose $\alpha \in (c, a]$. Then $\hbar$ is strictly decreasing on $[\alpha, d]$ and

$$g(\alpha) = \sup_{\mu \in [\alpha,d]} \hbar(\mu|\alpha) > \hbar(\lambda|\alpha) > \inf_{\mu \in [\alpha,d]} \hbar(\mu|\alpha) = L_{g|d}(\alpha), \ \alpha < \lambda < d, \qquad (3.6)$$

as well as $\hbar$ is strictly increasing on $[c, \alpha]$ and

$$g(\alpha) = \sup_{\mu \in [c,\alpha]} \hbar(\mu|\alpha) > \hbar(\lambda|\alpha) > \inf_{\mu \in [c,\alpha]} \hbar(\mu|\alpha) = L_{g|d}(\alpha), \ c < \lambda < \alpha. \qquad (3.7)$$

Next suppose $\alpha \in (a, d)$. Then $\hbar$ is strictly increasing on $[c, \alpha]$ and

$$g(\alpha) = \sup_{\mu \in [c,\alpha]} \hbar(\mu|\alpha) > \hbar(\lambda|\alpha) > \inf_{\mu \in [c,\alpha]} \hbar(\mu|\alpha) = L_{g|c}(\alpha), \ c < \lambda < \alpha, \qquad (3.8)$$

as well as $\hbar$ is strictly decreasing on $[\alpha, d]$ and

$$g(\alpha) = \sup_{\mu \in [\alpha,d]} \hbar(\mu|\alpha) > \hbar(\lambda|\alpha) > \inf_{\mu \in [\alpha,d]} \hbar(\mu|\alpha) = L_{g|c}(\alpha), \ c < \lambda < d. \qquad (3.9)$$

Example 3.8. Let g be a strictly convex and smooth function defined on a real interval with $c = \inf I$ and $d = \sup I$. We now assume that $d < +\infty$ and $d \notin I$. Let $\mathbf{F}(g) = \{L_{g|\lambda} : \lambda \in \Lambda\}$ be the set of all quasi-tangent lines of g, and let J be a subinterval of I such that $\sup J = d$. If $\alpha \le \inf J$, then $\hbar$ is strictly decreasing on J, so that $\inf_{\mu \in J} \hbar(\mu|\alpha)$ occurs at $\mu = d$ and $\sup_{\mu \in J} \hbar(\mu|\alpha)$ occurs at $\mu = \inf J$. In case $g'(d^-) = +\infty$, then by Lemma 3.3, we may know further that $\inf_{\mu \in J} \hbar(\mu|\alpha) = \lim_{\lambda \to d^-} L_{g|\lambda}(\alpha) = -\infty$. If $\alpha > d$, then $\hbar$ is strictly increasing on J so that $\sup_{\mu \in J} \hbar(\mu|\alpha)$ occurs at $\mu = d$. In case $g'(d^-) = +\infty$, by Lemma 3.3, we may know further that

$$\sup_{\mu \in J} \hbar(\mu|\alpha) = \lim_{\lambda \to d^-} \hbar(\mu|\alpha) = \lim_{\lambda \to d^-} \{g'(\lambda)(\alpha - \lambda) + g(\lambda)\} = +\infty.$$

Finally, suppose $\alpha = d$. If $g'(d^-) = +\infty$, we assert that

$$\sup_{\mu \in J} \hbar(\mu|\alpha) = \lim_{\lambda \to d^-} \hbar(\lambda|\alpha) = g(d^-),$$

where we recall from Lemma 3.3 that $g(d^-)$ is either finite or equal to $+\infty$. Indeed, suppose $g(d^-)$ exists. Since for any $c < x < y < d$, $g(y) - g(x) = g'(z)(y - x)$ for some $z \in (x, y)$ and since g is strictly convex,

$$g(y) - g(x) > g'(x)(y - x).$$

If $x \to d^-$, then $y \to d^-$ and

$$0 \ge \lim_{x \to d^-} g'(x)(d - x) \ge 0.$$

So $\lim_{x \to d^-} g'(x)(d-x) = 0$. Hence $\lim_{\lambda \to d^-} \hbar(\lambda|\alpha) = g(d^-)$. Next, suppose $g(d^-) = +\infty$. Since $\lim_{x \to d^-} g'(x)(d - x) \ge 0$, we see that

$$\lim_{\lambda \to d^-} \hbar(\lambda|\alpha) = g(d^-) = \infty.$$

In the above example, we have assumed that $g'(d^-) = +\infty$. In view of Lemma 3.3, we may assume the equivalent condition $g \sim H_{d^-}$. We remark also that if we replace the condition that $d < +\infty$ with $\sup I = +\infty$, then similar conclusions hold if we also replace the condition $g'(d^-) = +\infty$ by $g \sim H_{+\infty}$.

Example 3.9. Suppose $c \in \mathbf{R}$ and $g : (c, \infty) \to \mathbf{R}$ is a strictly convex and smooth function such that $g(c^+)$ and $g'(c^+)$ exist, and the asymptote $L_{g|+\infty}$ of the graph g as $x \to \infty$ exists. Then $\{L_{g|\lambda} : \lambda \in [c, +\infty)\}$ is the set of quasi-tangent lines of g. We assert that $\lim_{\lambda \to \infty} \hbar(\lambda|\alpha) = \inf_{\lambda \in [\alpha, +\infty)} \hbar(\lambda|\alpha) = L_{g|+\infty}(\alpha)$ for any α. Indeed, we may assume without loss of generality that $L_{g|+\infty}(x) = 0$ and (hence) $g(+\infty) = 0$. Note first that $g(x) > 0$ for $x \in (c, \infty)$. Furthermore, $g'(x) < 0$ for $x \in (c, \infty)$ for otherwise if $g'(\xi) \ge 0$ for some $\xi \in (c, \infty)$, then $g(x) \ge g(\xi) > 0$ for $x \in (\xi, \infty)$, which is contrary to our assumption $g(+\infty) = 0$. We now assert that $\lim_{x \to \infty} (-x)g'(x) = 0$. Indeed, this follows from

$$0 \le (\alpha - x)g'(x) = (\alpha - x)\frac{L_{g|x}(x/2) - g(x)}{x/2 - x}$$

$$\le 2\left(\frac{\alpha}{x} - 1\right)\left\{L_{g|x}\left(\frac{x}{2}\right) - g(x)\right\}$$

$$\le 2\left(\frac{\alpha}{x} - 1\right)L_{g|x}\left(\frac{x}{2}\right) \le 2\left(\frac{\alpha}{x} - 1\right)g\left(\frac{x}{2}\right)$$

for $x > \alpha$ and $g(x/2) \to 0$ as $x \to \infty$. Hence

$$\lim_{u \to \infty} \hbar(u|\alpha) = 0 = L_{g|+\infty}(\alpha).$$

In view of the above examples, we may consider a strictly convex and smooth function g defined on I satisfying the following special cases and make similar conclusions:

C1: $I = (c, d)$, where $c, d \in \mathbf{R}$, and $g(c^+), g(d^-), g'(c^+)$ as well as $g'(d^-)$ exist.
C2: $I = (c, d)$, where $c, d \in \mathbf{R}$, $g(c^+)$ and $g'(c^+)$ exist, and $g \sim H_{d-}$.
C3: $I = (c, \infty)$, where $c \in \mathbf{R}$, $g(c^+)$ and $g'(c^+)$ exist, and $g \sim H_{+\infty}$.
C4: $I = (c, \infty)$, where $c \in \mathbf{R}$, $g(c^+)$ and $g'(c^+)$ exist, and $L_{g|+\infty}$ exists.
C5: $I = (-\infty, \infty)$, and $L_{g|+\infty}$ and $L_{g|-\infty}$ exist.
C6: $I = (-\infty, \infty)$, and $L_{g|-\infty}$ exists and $g \sim H_{+\infty}$.
C7: $I = (-\infty, d)$, and $L_{g|-\infty}$ exists and $g \sim H_{d-}$.
C8: $I = (-\infty, d)$, and $g \sim H_{-\infty}$ and $g \sim H_{d-}$.
C9: $I = [c, d)$, where $c, d \in \mathbf{R}$, and $g(d^-)$ as well as $g'(d^-)$ exist.
C10: $I = [c, d)$, where $c, d \in \mathbf{R}$, and $g \sim H_{d-}$.
C11: $I = [c, \infty)$, where $c \in \mathbf{R}$, and $g \sim H_{+\infty}$.
C12: $I = [c, \infty)$, where $c \in \mathbf{R}$, and $L_{g|+\infty}$ exists.

Indeed, case (C1) has already been discussed. As for case (C2), in view of the explanations in Examples 3.7 and 3.8, we may make the following conclusions: $\mathbf{F}(g) = \{L_{g|\lambda} : \lambda \in \Lambda\}$ where $\Lambda = [c, d)$. Furthermore, the following statements hold.

- When $\alpha \leq c$, $\hbar$ is strictly decreasing on $[c, d)$, and

$$L_{g|c}(\alpha) = \sup_{\mu \in [c,d)} \hbar(\mu|\alpha) > \hbar(\lambda|\alpha) > -\infty, \ \lambda \in (c, d). \quad (3.10)$$

- When $c < \alpha < d$, $\hbar$ is strictly increasing on $[c, \alpha]$ and

$$g(\alpha) = \sup_{\mu \in [c,\alpha]} \hbar(\mu|\alpha) > \hbar(\lambda|\alpha) > L_{g|c}(\alpha), \ \lambda \in (c, \alpha), \quad (3.11)$$

as well as $\hbar$ strictly decreasing on $[\alpha, d)$ and

$$g(\alpha) = \sup_{\mu \in [\alpha,d)} \hbar(\mu|\alpha) > \hbar(\lambda|\alpha) > -\infty = \lim_{\mu \to d^-} \hbar(\mu|\alpha), \ \lambda \in (\alpha, d). \quad (3.12)$$

- When $\alpha = d$, $\hbar$ is strictly increasing on $[c, d)$ and

$$g(d^-) = \sup_{\mu \in [c,d)} \hbar(\mu|\alpha) > \hbar(\lambda|\alpha) > L_{g|c}(\alpha), \ \lambda \in (c, d). \quad (3.13)$$

- When $\alpha > d$, $\hbar$ is strictly increasing on $[c, d)$ and

$$+\infty = \lim_{\mu \to d^-} \hbar(\mu|\alpha) > \hbar(\lambda|\alpha) > L_{g|c}(\alpha), \ \lambda \in (c, d). \quad (3.14)$$

Next, suppose (C3) holds. Then $\mathbf{F}(g) = \{L_{g|\lambda} : \lambda \in \Lambda\}$ where $\Lambda = [c, +\infty)$. Furthermore, the following statements hold.

- When $\alpha \leq c$, the function $\hbar$ is strictly decreasing on $[c, \infty)$ and
$$\sup_{\mu \in [c,\infty)} \hbar(\mu|\alpha) = L_{g|c}(\alpha) > \hbar(\lambda|\alpha) > -\infty = \inf_{\mu \in [c,\infty)} \hbar(\mu|\alpha), \ \lambda \in (c, \infty).$$

- When $c < \alpha$, the function $\hbar$ is strictly decreasing on $[\alpha, \infty)$ and
$$\sup_{\mu \in [\alpha,\infty)} \hbar(\mu|\alpha) = g(\alpha) > \hbar(\lambda|\alpha) > -\infty = \inf_{\mu \in [\alpha,\infty)} \hbar(\mu|\alpha), \ \lambda \in (\alpha, \infty),$$
and strictly increasing on $[c, \alpha]$ and
$$\sup_{\mu \in [c,\alpha]} \hbar(\mu|\alpha) = g(\alpha) > \hbar(\lambda|\alpha) > L_{g|c}(\alpha) = \inf_{\mu \in [c,\alpha]} \hbar(\mu|\alpha), \ \lambda \in (c, \alpha).$$

Next, suppose (C4) holds. Then $\mathbf{F}(g) = \{L_{g|\lambda} : \lambda \in \Lambda\}$ where $\Lambda = [c, +\infty)$. The lines $L_{g|c}$ and $L_{g|+\infty}$ intersect at exactly one point (a, b) with $a \in (c, +\infty)$ and $b < g(a)$. Furthermore, the following statements hold.

- When $\alpha \leq c$, $\hbar$ is strictly decreasing on $[c, \infty)$ and
$$\sup_{\mu \in [c,\infty)} \hbar(\mu|\alpha) = L_{g|c}(\alpha) > \hbar(\lambda|\alpha) > \inf_{\mu \in [c,\infty)} \hbar(\mu|\alpha) = L_{g|+\infty}(\alpha), \ \lambda \in (c, \infty).$$

- When $\alpha \in (c, a]$, $\hbar$ is strictly decreasing on $[\alpha, \infty)$ and
$$\sup_{\mu \in [\alpha,\infty)} \hbar(\mu|\alpha) = g(\alpha) > \hbar(\lambda|\alpha) > \inf_{\mu \in [\alpha,\infty)} \hbar(\mu|\alpha) = L_{g|+\infty}(\alpha), \ \lambda \in (\alpha, \infty),$$
and $\hbar$ is strictly increasing on $[c, \alpha]$ and
$$\sup_{\mu \in [c,\alpha]} \hbar(\mu|\alpha) = g(\alpha) > \hbar(\lambda|\alpha) > L_{g|c}(\alpha) = \inf_{\mu \in [c,\alpha]} \hbar(\mu|\alpha), \ \lambda \in (c, \alpha).$$

- When $\alpha \in [a, \infty)$, $\hbar$ is strictly decreasing on $[\alpha, \infty)$ and
$$\sup_{\mu \in [\alpha,\infty)} \hbar(\mu|\alpha) = g(\alpha) > \hbar(\lambda|\alpha) > L_{g|+\infty}(\alpha) = \inf_{\mu \in [\alpha,\infty)} \hbar(\mu|\alpha), \ \lambda \in (\alpha, \infty),$$
and $\hbar$ is strictly increasing on $[c, \alpha]$ and
$$\sup_{\mu \in [c,\alpha]} \hbar(\mu|\alpha) = L_{g|c}(\alpha) > \hbar(\lambda|\alpha) > L_{g|c}(\alpha) = \inf_{\mu \in [c,\alpha]} \hbar(\mu|\alpha), \ \lambda \in (c, \alpha).$$

Next suppose (C5) holds. Then $\mathbf{F}(g) = \{L_{g|\lambda} : \lambda \in \mathbf{R}\}$. The lines $L_{g|-\infty}$ and $L_{g|+\infty}$ intersect at exactly one point (a, b) with $b < g(a)$. Furthermore, the following statements hold.

- When $\alpha \leq a$, $\hbar$ is strictly decreasing on $[\alpha, \infty)$ and
$$g(\alpha) = \sup_{\mu \in [\alpha,\infty)} \hbar(\mu|\alpha) > \hbar(\lambda|\alpha) > \inf_{\mu \in [\alpha,\infty)} \hbar(\mu|\alpha) = L_{g|+\infty}(\alpha), \ \lambda \in (\alpha, \infty),$$
and $\hbar$ is strictly increasing on $(-\infty, \alpha]$ and
$$L_{g|-\infty}(\alpha) = \inf_{\mu \in (-\infty,\alpha]} \hbar(\mu|\alpha) < \hbar(\lambda|\alpha) < \sup_{\mu \in (-\infty,\alpha]} \hbar(\mu|\alpha) = g(\alpha), \ \lambda \in (-\infty, \alpha].$$

- When $\alpha > a$, $\hbar$ is strictly decreasing on $[\alpha, \infty)$ and
$$g(\alpha) = \sup_{\mu \in [\alpha, \infty)} \hbar(\mu|\alpha) > \hbar(\lambda|\alpha) > \inf_{\mu \in [\alpha, \infty)} \hbar(\mu|\alpha) = L_{g|+\infty}(\alpha), \ \lambda \in (\alpha, \infty),$$
and $\hbar$ is strictly increasing on $(-\infty, \alpha]$ and
$$L_{g|-\infty}(\alpha) = \inf_{\mu \in (-\infty, \alpha]} \hbar(\mu|\alpha) < \hbar(\lambda|\alpha) < \sup_{\mu \in (-\infty, \alpha]} \hbar(\mu|\alpha) = g(\alpha), \ \lambda \in (-\infty, \alpha).$$

Next suppose (C6) holds. Then $\mathbf{F}(g) = \{L_{g|\lambda} : \lambda \in \mathbf{R}\}$. For any $\alpha \in \mathbf{R}$, $\hbar$ is strictly decreasing on $[\alpha, \infty)$ and
$$g(\alpha) = \sup_{\mu \in [\alpha, \infty)} \hbar(\mu|\alpha) > \hbar(\lambda|\alpha) > \inf_{\mu \in [\alpha, \infty)} \hbar(\mu|\alpha) = -\infty, \ \lambda \in (\alpha, \infty),$$
and $\hbar$ is strictly increasing on $(-\infty, \alpha]$ and
$$g(\alpha) = \sup_{\mu \in (-\infty, \alpha]} \hbar(\mu|\alpha) > \hbar(\lambda|\alpha) > \inf_{\mu \in (-\infty, \alpha]} \hbar(\mu|\alpha) = L_{g|-\infty}(\alpha), \ \lambda \in (-\infty, \alpha).$$

The cases (C7) is similar to the case (C2). Now we have $\mathbf{F}(g) = \{L_{g|\lambda} : \lambda \in (-\infty, d)\}$. Furthermore, the following statements hold.

- When $\alpha < d$, $\hbar$ is strictly increasing on $(-\infty, \alpha]$ and
$$g(\alpha) = \sup_{\mu \in (-\infty, \alpha]} \hbar(\mu|\alpha) > \hbar(\lambda|\alpha) > L_{g|-\infty}(\alpha), \ \lambda \in (-\infty, \alpha),$$
as well as $\hbar$ is strictly decreasing on $[\alpha, d)$ and
$$g(\alpha) = \sup_{\mu \in [\alpha, d)} \hbar(\mu|\alpha) > \hbar(\lambda|\alpha) > -\infty = \lim_{\mu \to d^-} \hbar(\mu|\alpha), \ \lambda \in (\alpha, d).$$
- When $\alpha = d$, $\hbar$ is strictly increasing on $(-\infty, d)$ and
$$g(d^-) = \sup_{\mu \in (-\infty, d)} \hbar(\mu|\alpha) > \hbar(\lambda|\alpha) > L_{g|-\infty}(\alpha), \ \lambda \in (-\infty, d).$$
- When $\alpha > d$, $\hbar$ is strictly increasing on $(-\infty, d)$ and
$$+\infty = \lim_{\mu \to d^-} \hbar(\mu|\alpha) > \hbar(\lambda|\alpha) > L_{g|-\infty}(\alpha), \ \lambda \in (-\infty, d).$$

The cases (C8), (C9), (C10), (C11) and (C12) are handled in manners similar to the cases (C7), (C1), (C2), (C3) and (C4) respectively.

3.4   Distribution Maps for Dual Points

To count the number of tangent lines of a plane curve S that also pass through a point (α, β) in the plane, it suffices to consider the sweeping function $\hbar$ defined by (3.2).

Theorem 3.2. *Let g be a strictly convex and smooth function defined on a real interval I with $c = \inf I$ and $d = \sup I$. Let $\mathbf{F}(g) = \{L_{g|\lambda} : \lambda \in \Lambda\}$ be the set of all quasi-tangent lines of g. Let $\hbar$ be defined by (3.2). Then a point (α, β) in the plane is a dual point of order m of the graph g if, and only if, there are exactly m mutually distinct $\lambda_1, \lambda_2, ..., \lambda_m$ in I such that $\hbar(\lambda_1|\alpha) = \cdots = \hbar(\lambda_m|\alpha) = \beta$.*

Indeed, if there is a tangent line $L_{g|u}$ of the graph of g that passes through the point (α, β), then $\hbar(u|\alpha) = \beta$. Conversely, if $\hbar(u|\alpha) = \beta$ for some $u \in I$, then there is a tangent line $L_{g|u}$ of the graph of g that passes through the point (α, β).

In view of Theorem 3.2 and the properties of $\hbar$ described in the previous Section, we may count the number of tangents of smooth and strictly convex functions. There are several cases. Let us begin with $g : (c, d) \to \mathbf{R}$ that is strictly convex and smooth such that $g(c^+), g(d^-), g'(c^+)$ and $g'(d^-)$ exist (see Figure 3.1). First of all, the set of quasi-tangent lines of g is $\{L_{g|c} : c \in [c, d]\}$. Let (a, b) be the unique point of intersection of the lines $L_{g|c}$ and $L_{g|d}$ described in Example 3.7. Let $\hbar(\lambda|\alpha)$ (defined by (3.2)) be the y-coordinate of the intersection of the quasi-tangent line $L_{g|\lambda}$ with the vertical line $x = \alpha$. By the conclusions observed in Example 3.7, if (α, β) is a point in the plane such that $\alpha \leq c$, then $\hbar$ is strictly decreasing on $[c, d]$ and (3.4) holds. Thus

- if $\beta \in (-\infty, L_{g|d}(\alpha)]$, the equation $\hbar(\lambda|\alpha) = \beta$ does not have any solution in (c, d),
- if $\beta \in (L_{g|d}(\alpha), L_{g|c}(\alpha))$, the equation $\hbar(\lambda|\alpha) = \beta$ has exactly one solution in (c, d), and
- if $\beta \in [L_{g|c}(\alpha), +\infty)$, the equation $\hbar(\lambda|\alpha) = \beta$ has no solution in (c, d).

By Theorem 3.2, when $\alpha \leq c$, the point (α, β) which satisfies $L_{g|c}(\alpha) \leq \beta$ or $L_d(\alpha) \geq \beta$ is a dual point of order 0, and the point (α, β) which satisfies $L_{g|d}(\alpha) < \beta < L_{g|c}(\alpha)$ is a dual point of order 1. See Figure 3.1.

Similarly, if (α, β) is a point in the plane such that $\alpha \in (c, a]$, then

- if $\beta \in (-\infty, L_{g|d}(\alpha)]$, the equation $\hbar(\lambda|\alpha) = \beta$ does not have any solution in (c, d),
- if $\beta \in (L_{g|d}(\alpha), L_{g|c}(\alpha)]$, the equation $\hbar(\lambda|\alpha) = \beta$ has exactly one solution in (α, d),
- if $\beta \in (L_{g|c}(\alpha), g(\alpha))$, the equation $\hbar(\lambda|\alpha) = \beta$ has exactly one solution in (c, α) and exactly one solution in (α, d),
- if $\beta = g(\alpha)$, the equation $\hbar(\lambda|\alpha) = \beta$ has the exact solution $\lambda = \alpha$,
- if $\beta > g(\alpha)$, the equation $\hbar(\lambda|\alpha) = \beta$ does not have any solutions in (c, d).

By Theorem 3.2, when $\alpha \in (c, a]$, the point (α, β) which satisfies $\beta > g(\alpha)$ or $\beta < L_{g|d}(\alpha)$ is a dual point of order 0, the point (α, β) which satisfies $L_{g|d}(\alpha) < \beta \leq L_{g|c}(\alpha)$ is a dual point of order 1, and the point (α, β) which satisfies $L_{g|c}(\alpha) < \beta < g(\alpha)$ is a dual point of order 2. Similar conclusions hold for the cases $\alpha \geq d$ and $\alpha \in (a, d)$.

In view of the above discussions, we see that each point (α, β) in the plane may be regarded as a dual point of S and marked with its order and the corresponding 'distribution map' is shown in Figure 3.1. This map can be described in different manners. One convenient description is as follows.

Theorem 3.3. *Let* $g : (c, d) \to \mathbf{R}$ *be a strictly convex and smooth function such that* $g(c^+), g(d^-), g'(c^+)$ *and* $g'(d^-)$ *exist. Then the following statements hold.*

(1) (α, β) *in the plane is a dual point of order* 0 *of* g *if, and only if,*
  (i) $\alpha \le c$ *and* $\beta \ge L_{g|c}(\alpha)$, *or,*
  (ii) $\alpha \in (c, d)$ *and* $\beta > g(\alpha)$, *or,*
  (iii) $\alpha \ge d$ *and* $\beta \ge L_{g|d}(\alpha)$, *or,*
  (iv) $\beta \le \min \{ L_{g|c}(\alpha), L_{g|d}(\alpha) \}$.
(2) (α, β) *in the plane is a dual point of order* 1 *of* g *if, and only if,*
  (i) $\alpha \in (c, d)$ *and* $\beta = g(\alpha)$, *or,*
  (ii) $\alpha \in (c, d)$ *and* $\min \{ L_{g|c}(\alpha), L_{g|d}(\alpha) \} < \beta \le \max \{ L_{g|c}(\alpha), L_{g|d}(\alpha) \}$, *or,*
  (iii) $\alpha \in \mathbf{R} \backslash (c, d)$ *and* $\min \{ L_{g|c}(\alpha), L_{g|d}(\alpha) \} < \beta < \max \{ L_{g|c}(\alpha), L_{g|d}(\alpha) \}$.
(3) (α, β) *in the plane is a dual point of order* 2 *of* g *if, and only if,*
  (i) $\alpha \in (c, d)$ *and* $\max \{ L_{g|c}(\alpha), L_{g|d}(\alpha) \} < \beta < g(\alpha)$.
(4) *Dual points of order greater than 2 of* g *do not exist.*

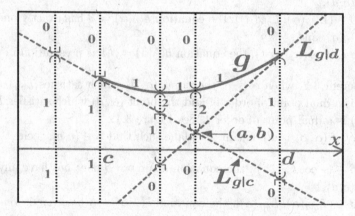

Fig. 3.1 $g \in C^1(c, d)$; $g(c^+), g(d^-), g'(c^+), g'(d^-)$ exist

We remark, as explained in the statements following Example 3.3, that the notations $L_{g|c}$ and $L_{g|d}$ are used instead of the more precise notations $L_{g|c^+}$ and $L_{g|d^-}$, since we have assumed the existence of $g(c^+), g(d^-), g'(c^+)$ and $g'(d^-)$ and therefore $L_{g|c}$ and $L_{g|d}$ can be treated as elements in the set of quasi-tangent lines of g. Similar practice will be used in the rest of this book.

We also remark that there are different ways to express our dual sets. For instance, the condition (iv) $\beta \le \min \{ L_{g|c}(\alpha), L_{g|d}(\alpha) \}$ in the conclusion (a) above is equivalent to $\alpha \le a$ and $\beta \le L_{g|c}(\alpha)$, or $\alpha > a$ and $\beta \le L_{g|d}(\alpha)$. The right choice depends on the specific problem at hand. Yet another possible way to describe the above result is by means of the ordering relations introduced for points and graphs. For example, the conclusion (a) is the same as saying the dual set of order 0 of g is

$$\left(\nabla(L_{g|c}) \oplus \vee(g) \oplus \nabla(L_{g|d}) \right) \cup \left(\triangle(L_{g|d}) \oplus \triangle(L_{g|c}) \right),$$

or

$$\left(\nabla(L_{g|c}\chi_{(-\infty,c]}) \oplus \vee(g) \oplus \nabla(L_{g|d}\chi_{[d,\infty)})\right) \cup \left(\triangle(L_{g|d}) \oplus \triangle(L_{g|c})\right),$$

and the conclusion (c) is the same as saying that the dual set of order 2 of g is

$$(\wedge(g)) \cup \left(\vee(L_{g|d}) \oplus \vee(L_{g|c})\right).$$

The conclusion (b) is more difficult to express but it is the complement of the dual sets in (a) and (b).

In view of the cumbersome descriptions of the dual sets, we will refer from time to time to the more appealing distribution maps in later discussions. Therefore a full understanding of the distribution map in Figure 3.1 is important. In particular, we remark that in Figure 3.1, we use 'closed' and 'open' brackets as we usually do in describing open and closed intervals in $\mathbf{R}$. For instance, on the vertical line $x = c$, the set of points lying above and *including* the point $(c, L_{g|c}(c))$ (which cannot be confused with $(c, g(c))$ since g is not defined at c) are dual points of order 0, and the set of points lying immediately below are dual points of order 1, and hence at the point $(c, L_{g|c}(c))$ we use a closed bracket and an open bracket to indicate these facts. This practice is convenient and self-explanatory and will be used in all subsequent discussions.

At least two variants of Theorem 3.3 are available. The first one is concerned with the same function g in Theorem 3.3 with the additional assumption that it is also strictly decreasing on (c, d). Although the same conclusions can be made, the corresponding distribution map, however, can be drawn differently, see Figure 3.2.

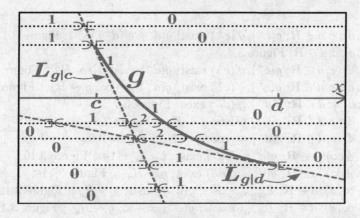

Fig. 3.2

The second has to do with a concave function. In view of Figure 3.2, if we now have a function $g : (c, d) \to \mathbf{R}$ which is strictly decreasing, *strictly concave* and smooth, then the corresponding distribution map can be depicted as in Figure 3.3.

Once we fully understand the 'sweeping method' used to obtain Theorem 3.3, we may apply the conclusions in the previous sections for various convex functions

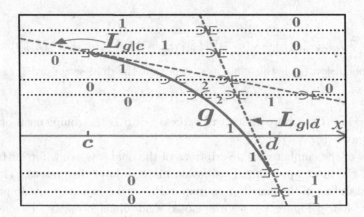

Fig. 3.3

to draw the corresponding distribution maps. Since the principles are the same and therefore most of the details will be skipped in the following discussion. Instead, we present the more important distribution maps and describe them in convenient manners. The variants of these distribution maps that are similar to those in Figures 3.2 and 3.3, however, will be assumed valid and skipped.

Before we proceed to describe the distribution maps, let us first classify our strictly convex and smooth function g defined on a real interval I in the following manners:

- $I = (c, d)$; $c, d \in \mathbf{R}$; $g(c^+), g'(c^+)$ exist; $g(d^-), g'(d^-)$ exist; Figure 3.1
- $I = [c, d]$; $c, d \in \mathbf{R}$; Figure 3.4
- $I = (c, d)$; $c, d \in \mathbf{R}$; $g(c^+), g'(c^+)$ exist; $g(d^-)$ exists; $g \sim H_{d^-}$; Figure 3.5
- $I = (c, d)$; $c, d \in \mathbf{R}$; $g(c^+), g'(c^+)$ exist; $g(d^-) = +\infty$; $g \sim H_{d^-}$; Figure 3.6
- $I = [c, d)$; $c, d \in \mathbf{R}$; $g(d^-), g'(d^-)$ exist; Figure 3.7
- $I = [c, d)$; $c, d \in \mathbf{R}$; $g(d^-)$ exists; $g \sim H_{d^-}$; Figure 3.8
- $I = [c, d)$; $c, d \in \mathbf{R}$; $g(d^-) = +\infty$; $g \sim H_{d^-}$; Figure 3.9
- $I = (c, +\infty)$; $c \in \mathbf{R}$; $g(c^+), g'(c^+)$ exist; $L_{g|+\infty}$ exists; Figure 3.10
- $I = (c, +\infty)$; $c \in \mathbf{R}$; $g(c^+), g'(c^+)$ exist; $g \sim H_{+\infty}$; Figure 3.11
- $I = (-\infty, d)$; $d \in \mathbf{R}$; $L_{g|-\infty}$ exists; $g(d^-)$ exists; $g \sim H_{d^-}$; Figure 3.12
- $I = (-\infty, d)$; $d \in \mathbf{R}$; $L_{g|-\infty}$ exists; $g(d^-) = +\infty$; $g \sim H_{d^-}$; Figure 3.13
- $I = (-\infty, d)$; $d \in \mathbf{R}$; $g \sim H_{-\infty}$; $g(d^-)$ exists; $g \sim H_{d^-}$; Figure 3.14
- $I = (-\infty, d)$; $d \in \mathbf{R}$; $g \sim H_{-\infty}$; $g(d^-) = +\infty$; $g \sim H_{d^-}$; Figure 3.15
- $I = [c, +\infty)$; $c \in \mathbf{R}$; $L_{g|+\infty}$ exists; Figure 3.16
- $I = [c, +\infty)$; $c \in \mathbf{R}$; $g \sim H_{+\infty}$; Figure 3.17
- $I = (-\infty, +\infty)$; $L_{g|-\infty}$ exists; $L_{g|+\infty}$ exists; Figure 3.18
- $I = (-\infty, +\infty)$; $L_{g|-\infty}$ exists; $g \sim H_{+\infty}$; Figure 3.19
- $I = (-\infty, +\infty)$; $g \sim H_{-\infty}$; $g \sim H_{+\infty}$; Figure 3.20

Theorem 3.4. *Let* $g : [c, d] \to \mathbf{R}$ *be a strictly convex and smooth function. Then the following statements hold.*

(1) (α, β) *in the plane is a dual point of order 0 of* g *if, and only if,*
  (i) $\beta > g(\alpha)$, *or*
  (ii) $\beta < \min\{L_{g|c}(\alpha), L_{g|d}(\alpha)\}$.
(2) (α, β) *in the plane is a dual point of order 1 of* g *if, and only if,*
  (i) $\alpha \in [c, d]$ *and* $\beta = g(\alpha)$, *or,*
  (ii) $\alpha \in (c, d)$ *and* $\min\{L_{g|c}(\alpha), L_{g|d}(\alpha)\} \leq \beta < \max\{L_{g|c}(\alpha), L_{g|d}(\alpha)\}$, *or,*
  (iii) $\alpha \in \mathbf{R}\backslash(c, d)$ *and* $\min\{L_{g|c}(\alpha), L_{g|d}(\alpha)\} \leq \beta \leq \max\{L_{g|c}(\alpha), L_{g|d}(\alpha)\}$.
(3) (α, β) *in the plane is a dual point of order 2 of* g *if, and only if,*
  (i) $\alpha \in (c, d)$ *and* $\max\{L_{g|c}(\alpha), L_{g|d}(\alpha)\} \leq \beta < g(\alpha)$.
(4) *Dual points of order greater than 2 of* g *do not exist.*

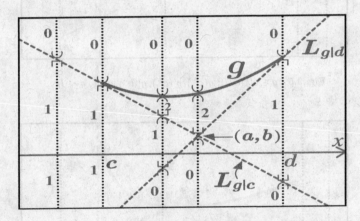

Fig. 3.4 $\quad g \in C^1[c, d]$

Theorem 3.5. *Let* $g : (c, d) \to \mathbf{R}$ *be a strictly convex and smooth function such that* $g(c^+), g'(c^+)$ *and* $g(d^-)$ *exist and* $g \sim H_{d^-}$. *Then the following statements hold.*

(1) (α, β) *in the plane is a dual point of order 0 of* g *if, and only if,*
  (i) $\alpha \leq c$ *and* $\beta \geq L_{g|c}(\alpha)$, *or,*
  (ii) $\alpha \in (c, d)$ *and* $\beta > g(\alpha)$, *or,*
  (iii) $\alpha = d$ *and* $\beta \geq g(d^-)$, *or,*
  (iv) $\alpha = d$ *and* $\beta \leq L_{g|c}(\alpha)$, *or*
  (v) $\alpha > d$ *and* $\beta \leq L_{g|c}(\alpha)$.
(2) (α, β) *in the plane is a dual point of order 1 of* g *if, and only if,*
  (i) $\alpha \in (c, d)$ *and* $\beta = g(\alpha)$, *or,*
  (ii) $\alpha \leq c$ *and* $\beta < L_{g|c}(\alpha)$, *or,*
  (iii) $\alpha \in (c, d)$ *and* $\beta \leq L_{g|c}(\alpha)$, *or,*
  (iv) $\alpha = d$ *and* $L_{g|c}(\alpha) < \beta < g(d^-)$, *or,*

    (v) $\alpha > d$ *and* $\beta > L_{g|c}(\alpha)$.
(3) (α, β) in the plane is a dual point of order 2 of g if, and only if,
    (i) $\alpha \in (c, d)$ *and* $L_{g|c}(\alpha) < \beta < g(\alpha)$.
(4) Dual points of order greater than 2 of g do not exist.

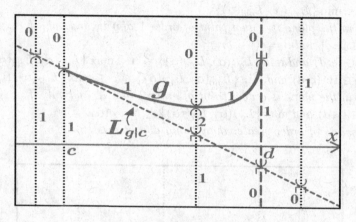

Fig. 3.5 $g \in C^1(c, d)$; $g(c^+), g'(c^+), g(d^-)$ exist; $g \sim H_{d^-}$

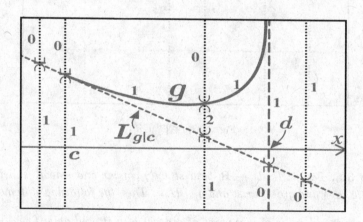

Fig. 3.6 $g \in C^1(c, d)$; $g(c^+), g'(c^+)$ exist; $g(d^-) = +\infty$; $g \sim H_{d^-}$

Theorem 3.6. *Let $g : (c, d) \to \mathbf{R}$ be a strictly convex and smooth function such that $g(c^+)$ and $g'(c^+)$ exist, $g(d^-) = +\infty$ and $g \sim H_{d^-}$. Then the following statements hold.*

(1) (α, β) in the plane is a dual point of order 0 of g if, and only if,
    (i) $\alpha \leq c$ *and* $\beta \geq L_{g|c}(\alpha)$, *or,*
    (ii) $\alpha \in (c, d)$ *and* $\beta > g(\alpha)$, *or,*
    (iii) $\alpha \geq d$ *and* $\beta \leq L_{g|c}(\alpha)$.

(2) (α, β) *in the plane is a dual point of order* 1 *of* g *if, and only if,*

    (i) $\alpha \in (c, d)$ *and* $\beta = g(\alpha)$, *or,*

    (ii) $\alpha \le c$ *and* $\beta < L_{g|c}(\alpha)$, *or,*

    (iii) $\alpha \in (c, d)$ *and* $\beta \le L_{g|c}(\alpha)$, *or,*

    (iv) $\alpha \ge d$ *and* $\beta > L_{g|c}(\alpha)$.

(3) (α, β) *in the plane is a dual point of order* 2 *of* g *if, and only if,*

    (i) $\alpha \in (c, d)$ *and* $L_{g|c}(\alpha) < \beta < g(\alpha)$.

(4) *Dual points of order greater than* 2 *of* g *do not exist.*

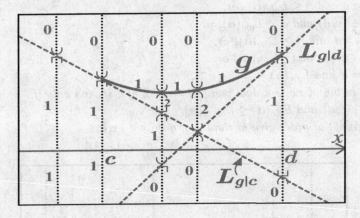

Fig. 3.7 $g \in C^1[c, d)$; $g(d^-), g'(d^-)$ exist

Theorem 3.7. *Let* $g : [c, d) \to \mathbf{R}$ *be a strictly convex and smooth function such that* $g(d^-)$ *and* $g'(d^-)$ *exist. Then the quasi-tangent lines* $L_{g|c}$ *and* $L_{g|d}$ *intersect at some unique point* (a, b) *where* $a \in (c, d)$. *Furthermore, the following statements hold.*

(1) (α, β) *in the plane is a dual point of order* 0 *of* g *if, and only if,*

    (i) $\alpha < c$ *and* $\beta > L_{g|c}(\alpha)$, *or,*

    (ii) $\alpha \in [c, d)$ *and* $\beta > g(\alpha)$, *or*

    (iii) $\alpha \ge d$ *and* $\beta \ge L_{g|d}(\alpha)$, *or,*

    (iv) $\beta \le L_{g|d}(\alpha)$ *and* $\beta < L_{g|c}(\alpha)$.

(2) (α, β) *in the plane is a dual point of order* 1 *of* g *if, and only if,*

    (i) $\alpha \in [c, d)$ *and* $\beta = g(\alpha)$, *or,*

    (ii) $\alpha \le c$ *and* $L_{g|d}(\alpha) < \beta \le L_{g|c}(\alpha)$, *or,*

    (iii) $\alpha \in (c, a)$ *and* $L_{g|d}(\alpha) < \beta < L_{g|c}(\alpha)$, *or,*

    (iv) $\alpha \in [a, d)$ *and* $L_{g|c}(\alpha) \le \beta \le L_{g|d}(\alpha)$, *or,*

    (v) $\alpha \in [d, +\infty)$ *and* $L_{g|c}(\alpha) \le \beta < L_{g|d}(\alpha)$.

(3) (c) (α, β) *in the plane is a dual point of order* 2 *of* g *if, and only if,*

    (i) $\alpha \in (c, d)$ *and* $L_{g|c}(\alpha) \le \beta$ *and* $L_{g|d}(\alpha) < \beta < g(\alpha)$.

(4) (d) *Dual points of order greater than* 2 *of* g *do not exist.*

Theorem 3.8. *Let $g : [c, d) \to \mathbf{R}$ be a strictly convex and smooth function such that $g(d^-)$ exists and $g \sim H_{d^-}$. Then the following statements hold.*

(1) (α, β) in the plane is a dual point of order 0 of g if, and only if,
 (i) $\alpha < c$ and $\beta > L_{g|c}(\alpha)$, or,
 (ii) $\alpha \in [c, d)$ and $\beta > g(\alpha)$, or,
 (iii) $\alpha \geq d$ and $\beta < L_{g|c}(\alpha)$, or,
 (iv) $\alpha = d$ and $\beta \geq g(d^-)$.
(2) (α, β) in the plane is a dual point of order 1 of g if, and only if,
 (i) $\alpha < c$ and $\beta \leq L_{g|c}(\alpha)$, or,
 (ii) $\alpha \in (c, d)$ and $\beta < L_{g|c}(\alpha)$, or,
 (iii) $\alpha > d$ and $\beta \geq L_{g|c}(\alpha)$, or,
 (iv) $\alpha \in [c, d)$ and $\beta = g(\alpha)$, or,
 (v) $\alpha = d$ and $L_{g|c}(\alpha) \leq \beta < g(d^-)$,
(3) (α, β) in the plane is a dual point of order 2 of g if, and only if,
 (i) $\alpha \in (c, d)$ and $L_{g|c}(\alpha) \leq \beta < g(\alpha)$.
(4) Dual points of order greater than 2 of g do not exist.

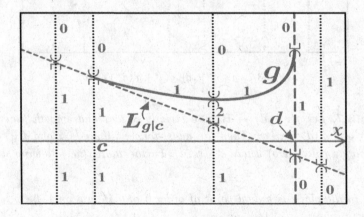

Fig. 3.8 $g \in C^1[c, d)$; $g(d^-)$ exists; $g \sim H_{d^-}$

Theorem 3.9. *Let $g : [c, d) \to \mathbf{R}$ be a strictly convex and smooth function such that $g(d^-) = +\infty$ and $g \sim H_{d^-}$. Then the following statements hold.*

(1) (α, β) in the plane is a dual point of order 0 of g if, and only if,
 (i) $\alpha < c$ and $\beta > L_{g|c}(\alpha)$, or,
 (ii) $\alpha \in [c, d)$ and $\beta > g(\alpha)$, or,
 (iii) $\alpha \geq d$ and $\beta < L_{g|c}(\alpha)$.
(2) (α, β) in the plane is a dual point of order 1 of g if, and only if,
 (i) $\alpha < c$ and $\beta \leq L_{g|c}(\alpha)$, or,
 (ii) $\alpha \in (c, d)$ and $\beta < L_{g|c}(\alpha)$, or,

    (iii) $\alpha \geq d$ and $\beta \geq L_{g|c}(\alpha)$, *or,*

    (v) $\alpha \in [c, d)$ and $\beta = g(\alpha)$.

(3) (α, β) *in the plane is a dual point of order 2 of g if, and only if,*

    (i) $\alpha \in (c, d)$ and $L_{g|c}(\alpha) \leq \beta < g(\alpha)$.

(4) Dual points of order greater than 2 of g do not exist.

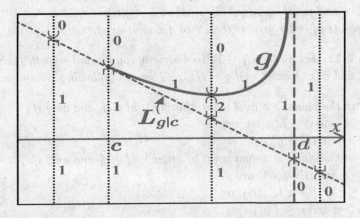

Fig. 3.9 $g \in C^1[c, d)$; $g(d^-) = +\infty$; $g \sim H_{d^-}$

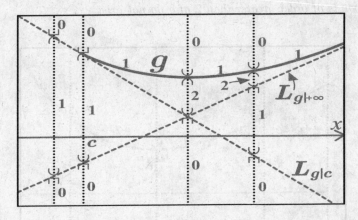

Fig. 3.10 $g \in C^1(c, \infty)$; $g(c^+), g'(c^+)$ exist; $L_{g|+\infty}$ exists

Theorem 3.10. *Let $g : (c, \infty) \to \mathbf{R}$ be a strictly convex and smooth function such that $g(c^+)$ and $g'(c^+)$ exist, and the asymptote $L_{g|+\infty}$ exists. Then the following statements hold.*

(1) (α, β) *in the plane is a dual point of order 0 of g if, and only if,*

    (i) $\alpha \leq c$ and $\beta \geq L_{g|c}(\alpha)$, *or,*

    (ii) $\alpha \in (c, \infty)$ and $\beta > g(\alpha)$, *or,*

(iii) $\beta \leq \min \left\{ L_{g|c}(\alpha), L_{g|+\infty}(\alpha) \right\}$.

(2) (α, β) in the plane is a dual point of order 1 of g if, and only if,
  (i) $\alpha > c$ and $\beta = g(\alpha)$, or,
  (ii) $\alpha \leq c$ and $L_{g|+\infty} < \beta < L_{g|c}$, or,
  (iii) $\alpha > c$ and $\min\{L_{g|+\infty}(\alpha), L_{g|c}(\alpha)\} < \beta \leq \max\{L_{g|+\infty}(\alpha), L_{g|c}(\alpha)\}$.

(3) (α, β) in the plane is a dual point of order 2 of g if, and only if,
  (i) $\alpha > c$ and $\max\{L_{g|+\infty}(\alpha), L_{g|c}(\alpha)\} < \beta < g(\alpha)$.

(4) Dual points of order greater than 2 of g do not exist.

Theorem 3.11. *Let $g : (c, \infty) \to \mathbf{R}$ be a strictly convex and smooth function such that $g(c^+)$ and $g'(c^+)$ exist, and $g \sim H_{+\infty}$. Then the following statements hold.*

(1) (α, β) in the plane is a dual point of order 0 of g if, and only if,
  (i) $\alpha \leq c$ and $\beta \geq L_{g|c}(\alpha)$, or,
  (ii) $\alpha > c$ and $\beta > g(\alpha)$.

(2) (α, β) in the plane is a dual point of order 1 of g if, and only if,
  (i) $\alpha > c$ and $\beta = g(\alpha)$, or,
  (ii) $\alpha \leq c$ and $\beta < L_{g|c}(\alpha)$, or,
  (iii) $\alpha > c$ and $\beta \leq L_{g|c}(\alpha)$.

(3) (α, β) in the plane is a dual point of order 2 of g if, and only if,
  (i) $\alpha > c$ and $L_{g|c}(\alpha) < \beta < g(\alpha)$.

(4) Dual points of order greater than 2 of g do not exist.

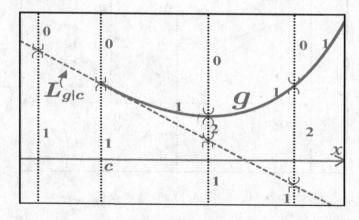

Fig. 3.11 $g \in C^1(c, \infty)$; $g(c^+), g'(c^+)$ exist; $g \sim H_{+\infty}$

Theorem 3.12. *Let $g : (-\infty, d) \to \mathbf{R}$ be a strictly convex and smooth function such that $g(d^-)$ exists, the asymptote $L_{g|-\infty}$ exists and $g \sim H_{d-}$. Then the following statements hold.*

(1) (α, β) in the plane is a dual point of order 0 of g if, and only if,

  (i) $\alpha < d$ and $\beta > g(\alpha)$, or,

  (ii) $\alpha \geq d$ and $\beta \leq L_{g|-\infty}(\alpha)$, or,

  (iii) $\alpha = d$ and $\beta \geq g(d^-)$.

(2) (α, β) in the plane is a dual point of order 1 of g if, and only if,

  (i) $\alpha < d$ and $\beta = g(\alpha)$, or,

  (ii) $\alpha < d$ and $\beta \leq L_{g|-\infty}(\alpha)$, or,

  (iii) $\alpha = d$ and $L_{g|-\infty}(\alpha) < \beta < g(d^-)$, or,

  (iv) $\alpha > d$ and $\beta > L_{g|-\infty}(\alpha)$.

(3) (α, β) in the plane is a dual point of order 2 of g if, and only if,

  (i) $\alpha < d$ and $L_{g|-\infty}(\alpha) < \beta < g(\alpha)$.

(4) Dual point of order greater than 2 of g do not exist.

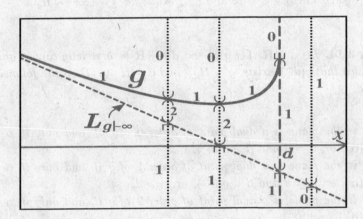

Fig. 3.12 $g \in C^1(-\infty, d)$; $L_{g|-\infty}$ exists; $g(d^-)$ exists; $g \sim H_{d-}$

Theorem 3.13. *Let $g : (-\infty, d) \to \mathbf{R}$ be a strictly convex and smooth function such that $g(d^-) = +\infty$, $g \sim H_{d-}$ and the asymptote $L_{g|-\infty}$ exists. Then the following statements hold.*

(1) *(α, β) in the plane is a dual point of order 0 of g if, and only if*

  (i) $\alpha < d$ and $\beta > g(\alpha)$, or

  (ii) $\alpha \geq d$ and $\beta \leq L_{g|-\infty}(\alpha)$.

(2) *(α, β) in the plane is a dual point of order 1 of g if, and only if,*

  (i) $\alpha < d$ and $\beta = g(\alpha)$, or,

  (ii) $\alpha \geq d$ and $\beta > L_{g|-\infty}(\alpha)$.

(3) *(α, β) in the plane is a dual point of order 2 of g if, and only if,*

  (i) $\alpha < d$ and $L_{g|-\infty}(\alpha) < \beta < g(\alpha)$.

(4) *Dual points of order greater than 2 of g do not exist.*

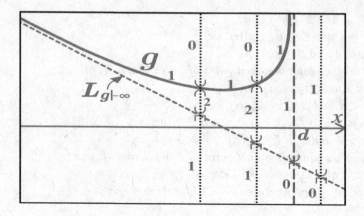

Fig. 3.13 $g \in C^1(-\infty, d)$; $L_{g|-\infty}$ exists; $g(d^-) = +\infty$; $g \sim H_{d-}$

Theorem 3.14. *Let $d \in \mathbf{R}$. Let $g : (-\infty, d) \to \mathbf{R}$ be a strictly convex and smooth function such that $g(d^-)$ exists, $g \sim H_{d-}$ and $g \sim H_{-\infty}$. Then the following statements hold.*

(1) (α, β) in the plane is a dual point of order 0 of g if, and only if, $\alpha < d$ and $\beta > g(\alpha)$, or, $\alpha = d$ and $\beta \geq g(d^-)$.

(2) (α, β) in the plane is a dual point of order 1 of g if, and only if, $\alpha < d$ and $\beta = g(\alpha)$, or, $\alpha = d$ and $\beta < g(d^-)$, or, $\alpha > d$.

(3) (α, β) in the plane is a dual point of order 2 of g if, and only if, $\alpha < d$ and $\beta < g(\alpha)$.

(4) Dual points of order greater than 2 of g do not exist.

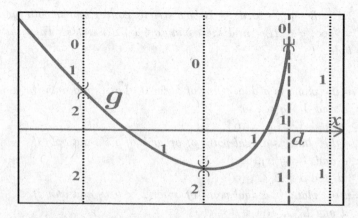

Fig. 3.14 $g \in C^1(-\infty, d)$; $g \sim H_{-\infty}$; $g(d^-)$ exists; $g \sim H_{d-}$

Theorem 3.15. *Let* $g : (-\infty, d) \to \mathbf{R}$ *be a strictly convex and smooth function such that* $g(d^-) = +\infty$, $g \sim H_{d^-}$ *and* $g \sim H_{-\infty}$. *Then the following statements hold.*

(1) (α, β) *in the plane is a dual point of order* 0 *of* g *if, and only if,* $\alpha < d$ *and* $\beta > g(\alpha)$.

(2) (α, β) *in the plane is a dual point of order* 1 *of* g *if, and only if,*
   (i) $\alpha < d$ *and* $\beta = g(\alpha)$, *or*
   (ii) $\alpha \geq d$.

(3) (α, β) *in the plane is a dual point of order* 2 *of* g *if, and only if,* $\alpha < d$ *and* $\beta < g(\alpha)$.

(4) *Dual points of order greater than* 2 *of* g *do not exist.*

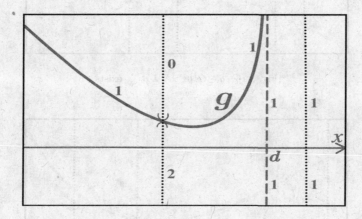

Fig. 3.15 $g \in C^1(-\infty, d)$; $g(d^-) = +\infty$; $g \sim H_{-\infty}$; $g \sim H_{d^-}$

Theorem 3.16. *Let* $g : [c, \infty) \to \mathbf{R}$ *be a strictly convex and smooth function such that the asymptote* $L_{g|+\infty}$ *exists. Then the quasi-tangent lines* $L_{g|c}$ *and* $L_{g|d}$ *intersect at the unique point* (a, b) *where* $a \in (c, d)$. *Furthermore, the following statements hold.*

(1) (α, β) *in the plane is a dual point of order* 0 *of* g *if, and only if,*
   (i) $\alpha < c$ *and* $\beta > L_{g|c}(\alpha)$, *or,*
   (ii) $\alpha \geq c$ *and* $\beta > g(\alpha)$, *or*
   (iii) $\beta \leq L_{g|+\infty}(\alpha)$ *and* $\beta < L_{g|c}(\alpha)$.

(2) (α, β) *in the plane is a dual point of order* 1 *of* g *if, and only if,*
   (i) $\alpha \in [c, d)$ *and* $\beta = g(\alpha)$, *or,*
   (ii) $\alpha \leq c$ *and* $L_{g|+\infty}(\alpha) < \beta \leq L_{g|c}(\alpha)$, *or,*
   (ii) $\alpha \in [c, a)$ *and* $L_{g|+\infty}(\alpha) < \beta < L_{g|c}(\alpha)$, *or,*
   (iii) $\alpha > a$ *and* $L_{g|+\infty}(\alpha) \geq \beta \geq L_{g|c}(\alpha)$, *or,*
   (iv) $\alpha = a$ *and* $\beta = L_{g|c}(a)$.

(3) (α, β) in the plane is a dual point of order 2 of g if, and only if,
   (i) $L_{g|c}(\alpha) \leq \beta$ and $L_{g|+\infty}(\alpha) < \beta < g(\alpha)$.
(4) Dual points of order greater than 2 of g do not exist.

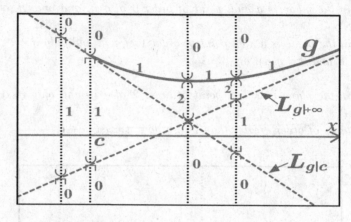

Fig. 3.16 $g \in C^1[c, \infty)$; $L_{g|+\infty}$ exists

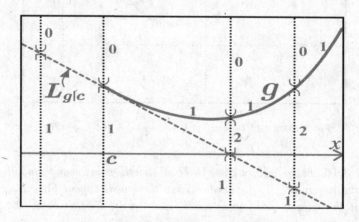

Fig. 3.17 $g \in C^1[c, \infty)$; $g \sim H_{+\infty}$

Theorem 3.17. *Let $g : [c, \infty) \to \mathbf{R}$ be a strictly convex and smooth function such that $g \sim H_{+\infty}$. Then the following statements hold.*

(1) (α, β) in the plane is a dual point of order 0 of g if, and only if,
   (i) $\alpha < c$ and $\beta > L_{g|c}(\alpha)$, or,
   (ii) $\alpha \geq c$ and $\beta > g(\alpha)$.
(2) (α, β) in the plane is a dual point of order 1 of g if, and only if,
   (i) $\alpha \in (c, \infty)$ and $\beta = g(\alpha)$, or,
   (ii) $\alpha \leq c$ and $\beta \leq L_{g|c}(\alpha)$, or,

(iii) $\alpha \in (c, \infty)$ and $\beta < L_{g|c}(\alpha)$.

(3) (α, β) in the plane is a dual point of order 2 of g if, and only if,

(i) $\alpha > c$ and $L_{g|c}(\alpha) \leq \beta < g(\alpha)$.

(4) Dual points of order greater than 2 of g do not exist.

Theorem 3.18. Let $g : \mathbf{R} \to \mathbf{R}$ be a strictly convex and smooth function such that the asymptotes $L_{g|-\infty}$ and $L_{g|+\infty}$ exist. Then the following statements hold.

(1) (α, β) in the plane is a dual point of order 0 of g if, and only if,
(i) $\beta > g(\alpha)$, or,
(ii) $\beta \leq \min \left\{ L_{g|-\infty}(\alpha), L_{g|+\infty}(\alpha) \right\}$.
(2) (α, β) in the plane is a dual point of order 1 of g if, and only if,
(i) $\beta = g(\alpha)$, or,
(ii) $\min \left\{ L_{g|-\infty}(\alpha), L_{g|+\infty}(\alpha) \right\} < \beta \leq \max \left\{ L_{g|-\infty}(\alpha), L_{g|+\infty}(\alpha) \right\}$.
(3) (α, β) in the plane is a dual point of order 2 of g if, and only if,
(i) $\max \left\{ L_{g|-\infty}(\alpha), L_{g|+\infty}(\alpha) \right\} < \beta < g(\alpha)$.
(4) Dual points of order greater than 2 of g do not exist.

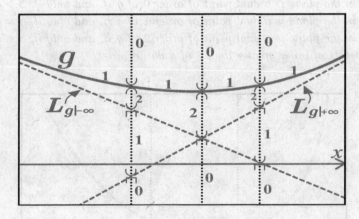

Fig. 3.18 $g \in C^1(\mathbf{R})$; $L_{g|-\infty}, L_{g|+\infty}$ exist

Theorem 3.19. Let $g : \mathbf{R} \to \mathbf{R}$ be a strictly convex and smooth function such that the asymptote $L_{g|-\infty}$ exists and $g \sim H_{+\infty}$. The the following statements hold.

(1) (α, β) in the plane is a dual point of order 0 of g if, and only if, $\beta > g(\alpha)$.
(2) (α, β) in the plane is a dual point of order 1 of g if, and only if,
(i) $\beta = g(\alpha)$, or,
(ii) $\beta \leq L_{g|-\infty}(\alpha)$.
(3) (α, β) in the plane is a dual point of order 2 of g if, and only if,
(i) $L_{g|-\infty}(\alpha) < \beta < g(\alpha)$.
(4) Dual points of order greater than 2 of g do not exist.

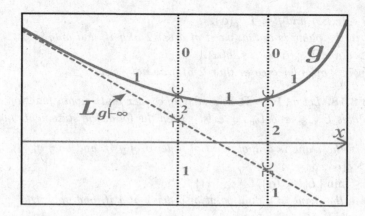

Fig. 3.19 $g \in C^1(\mathbf{R})$; $L_{g|-\infty}$ exists; $g \sim H_{+\infty}$

Theorem 3.20. *Let $g : \mathbf{R} \to \mathbf{R}$ be strictly convex and smooth function such that $g \sim H_{-\infty}$ and $g \sim H_{+\infty}$. Then the following statements hold.*

(1) (α, β) in the plane is a dual point of order 0 of g if, and only if, $\beta > g(\alpha)$.
(2) (α, β) in the plane is a dual point of order 1 of g if, and only if, $\beta = g(\alpha)$.
(3) (α, β) in the plane is a dual point of order 2 of g if, and only if, $\beta < g(\alpha)$.
(4) Dual points of order greater than 2 of g do not exist.

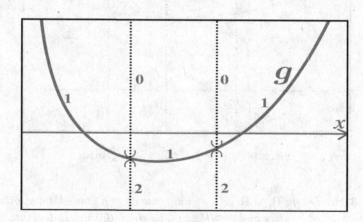

Fig. 3.20 $g \in C^1(\mathbf{R})$; $g \sim H_{-\infty}$, $g \sim H_{+\infty}$

A typical function g that satisfies the conditions in Theorem 3.20 is defined by $g(x) = x^2/4$ for $x \in \mathbf{R}$. Indeed, since $\lim_{x \to \pm\infty} g'(x) = \pm\infty$, by Lemma 3.4, we see that the conditions $g \sim H_{-\infty}$ and $g \sim H_{+\infty}$ are satisfied. Recall, however, that the condition $g'(+\infty) = +\infty$ is not necessary for $g \sim H_{+\infty}$. Indeed, in view of Lemma 3.5, we see that if $g : \mathbf{R} \to \mathbf{R}$ is strictly convex and smooth such that $g \sim H_{-\infty}$, $g(+\infty) = -\infty$ and $g'(+\infty) = 0$, then ($g \sim H_{+\infty}$ and hence) the conclusions in

Theorem 3.20 still hold (see Figure 3.21).

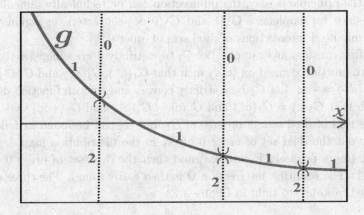

Fig. 3.21 $g \in C^1(\mathbf{R})$; $g \sim H_{-\infty}$; $g(+\infty) = -\infty$, $g'(+\infty) = 0$

For example, consider the curve S defined by the parametric functions

$$x(\lambda) = -2\lambda + \frac{1}{\lambda^2}, \ y(\lambda) = \lambda^2 - \frac{2}{\lambda},$$

defined for $\lambda \in (0, \infty)$. Since

$$(x(0^+), y(0^+)) = (+\infty, -\infty), \ (x(+\infty), y(+\infty)) = (-\infty, +\infty),$$

and

$$x'(\lambda) = -2 \left(\frac{\lambda^3 + 1}{\lambda^3} \right) < 0, \ \lambda \in (0, \infty),$$

we see that S is also the graph of a function $y = S(x)$ defined on $(-\infty, \infty)$. Furthermore, since

$$y'(\lambda) = 2 \left(\lambda + \frac{1}{\lambda^2} \right) = 2 \frac{\lambda^3 + 1}{\lambda^2},$$

we see that

$$\frac{dS}{dx} = -\lambda, \ \frac{d^2 S}{dx^2} = \frac{\lambda^3}{2(\lambda^3 + 1)}, \ \lambda > 0.$$

Hence S is a strictly decreasing, convex and smooth function which satisfies $S(-\infty) = +\infty$, $S'(-\infty) = -\infty$ $S(+\infty) = -\infty$ and $S'(+\infty) = 0$. Its dual set of order 0, in view of Theorem 3.20, is $\vee(S)$.

3.5 Intersections of Dual Sets of Order 0

Dual sets of some smooth and convex (or concave) functions are given in the previous section. Therefore, when we are facing with a curve that is composed of several smooth graphs of functions which are either convex or concave, its dual sets can then be obtained by unions and intersections of dual sets of the individual functions. Although the principle is easy, the intersection can be technically difficult to find. However, since we emphasize $\mathbf{C}\backslash\mathbf{R}$ and $\mathbf{C}\backslash(0,\infty)$-characteristic regions[2] in this book, we only need intersections of dual sets of order 0.

Let us first consider an example. Let G_1 be a strictly decreasing, strictly concave and smooth function defined on (c,d) such that $G_1(c^+), G_1'(c^+)$ and $G(d^-)$ exist as well as $G_1'(d^-) = -\infty$. Let G_2 be a strictly convex and smooth function defined on $[c,\infty)$ such that $G_2(c) = G_1(c^+)$ and $G_2'(c) = G_1'(c^+)$ and $G_2'(+\infty) = +\infty$. Then the intersection of dual sets of order 0 of G_1 and G_2 can be found as follows. We first single out the dual set of order 0 of G_1 in the distribution map depicted in Figure 3.1 (more precisely Figure 3.3) and then the dual set of order 0 of G_2 in Figure 3.11. The resulting intersection Ω is then easily found. The three dual sets are depicted from left to right in Figure 3.22.

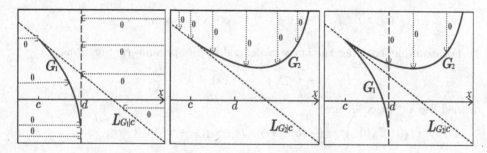

Fig. 3.22   Intersection of the dual sets of order 0 in Figures 3.3 and 3.11

The intersection Ω can also be described conveniently in the following manner: $(\alpha, \beta) \in \Omega$ if, and only if, $\alpha \geq d$ and $\beta > G_2(\alpha)$. By means of the chi-function χ_I defined in (1.2), we may also state our result as follows.

Theorem 3.21. *Let* $c, d \in \mathbf{R}$, $G_1 \in C^1(c,d)$ *and* $G_2 \in C^1[c,\infty)$. *Suppose the following conditions hold:*

(i) G_1 *is strictly decreasing, strictly concave on* (c,d) *such that* $G_1(c^+), G_1'(c^+)$ *and* $G(d^-)$ *exist as well as* $G_1'(d^-) = -\infty$;
(ii) G_2 *is strictly convex on* $[c,\infty)$ *such that* $G_2 \sim H_{+\infty}$;
(iii) $G_2(c) = G_1(c^+)$ *and* $G_2'(c) = G_1'(c^+)$.

Then the intersection of the dual sets of order 0 of G_1 *and* G_2 *is* $\vee(G_2\chi_{[d,\infty)})$.

As another example, recall the function g defined in Example 3.1. If we now let L be the straight line L defined by $y = g_1'(b^-)(x - b) + g_1(b^-)$ for $x \in \mathbf{R}$, then the dual set of order 0 of g is equal to the intersection of the dual sets of order 0 of g_1, g_2 and L. Indeed, this follows from the fact that a point (x, y) is in the dual

---

[2]Definitions will be given in Section 4.2.

set of order 0 of g if, and only if, no tangent lines of g passes through (x, y) and a tangent line of g is either a tangent line of g_1, g_2 or L. An example is illustrated in Figure 3.23 in which g_1 is the restriction of the quadratic function $g(x) = x^2/4$ over $(-\infty, 1)$ and g_2 the restriction of the same function over $(1, \infty)$.

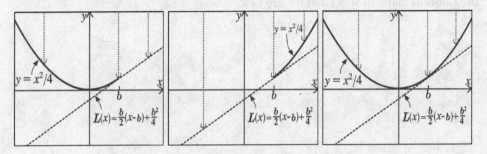

Fig. 3.23   Intersection of the dual sets of order 0 of g_1, g_2 and L

Theorem 3.22. *Let I be a real interval with $b \in I$. Let $I_1 = I \cap (-\infty, b)$ and $I_2 = I \cap (b, +\infty)$. Let $g_1 \in C^1(I_1)$ and $g_2 \in C^1(I_2)$ be strictly convex functions on I_1 and I_2 respectively such that $g_1(b^-)$, $g_1'(b^-)$, $g_2(b^+)$ and $g_2'(b^+)$ exist, and $g_1(b^-) = g_2(b^+)$ as well as $g_1'(b^-) = g_2'(b^+)$. Let the function $g : I \to R$ be defined*

$$g(x) = \begin{cases} g_1(x) & \text{if } x \in I_1 \\ g_2(x) & \text{if } x \in I_2 \\ g_1(b^-) & \text{if } x = b \end{cases},$$

and let

$$L(x) = g_1'(b^-)(x - b) + g_1(b^-), x \in \mathbf{R}.$$

Then $g \in C^1(I)$ and its dual set of order 0 is equal to the intersection of the dual sets of order 0 of g_1, g_2 and L.

As a variant, we have the following.

Theorem 3.23. *Let I be a real interval with $b \in I$. Let $I_1 = I \cap (-\infty, b]$ and $I_2 = I \cap (b, +\infty)$. Let $g_1 \in C^1(I_1)$ and $g_2 \in C^1(I_2)$ be strictly convex functions on I_1 and I_2 respectively such that $g_2(b^+)$ and $g_2'(b^+)$ exist, and $g_1(b) = g_2(b^+)$ as well as $g_1'(b) = g_2'(b^+)$. Let the function $g : I \to R$ be defined*

$$g(x) = \begin{cases} g_1(x) & \text{if } x \in I_1 \\ g_2(x) & \text{if } x \in I_2 \end{cases}.$$

Then $g \in C^1(I)$ and its dual set of order 0 is equal to the intersection of the dual sets of order 0 of g_1 and g_2.

The same principles can be used to derive intersections of dual sets of order 0 of various smooth convex or concave functions. We collect some of them in the Appendix for future references.

3.6 Notes

Elementary properties of convex functions are given in [32]. Several results in Section 3.3 are taken from [6]. Lemma 3.2 is provided by S. Y. Huang. The distribution maps in this Chapter are new.

Chapter 4

Quasi-Polynomials

4.1 Δ- and ∇-Polynomials

Real functions with several parameters arise in many situations. To see some examples, let us consider several population models as follows. First, let x_k denote the population size of a species in the time period k. The Malthus model asserts that the change $\Delta x_k = x_{k+1} - x_k$ is proportional to x_k, that is,

$$\Delta x_k = \gamma x_k, \ k \in \mathbf{N}, \tag{4.1}$$

where γ is a real proportionality constant (called the rate of growth). Assuming that it takes σ periods of time (naturally $\sigma \geq 1$) for the new born to mature and be ready for reproduction, a more general model may appear in the following form

$$\Delta x_k = \gamma x_{k-\sigma}, \ k \in \mathbf{N}. \tag{4.2}$$

When external factors (such as infectious diseases) cause depletion of the young people in the population, it is possible that the depleted population can be modeled in the form

$$x_{k-\tau+1} - x_{k-\tau} = \rho x_{k-\sigma}, \ k \in \mathbf{N},$$

where τ is a nonnegative integer such that $\tau \geq \sigma$. Therefore, the overall population now obeys the model

$$\Delta x_k - \Delta x_{k-\tau} = (\gamma - \rho) x_{k-\sigma}, \ k \in \mathbf{N},$$

which can be written as a 'neutral difference equation' with delay:

$$\Delta(x_k - x_{k-\tau}) + (\rho - \gamma) x_{k-\sigma} = 0, \ k \in \mathbf{N}. \tag{4.3}$$

Once a mathematical model is built, its implications will be needed to compare with realities. One way to deduce implications of our models is to find their solutions. For instance, in the case of (4.2), a solution is a (real or complex) sequence of the form $x = \{x_{-\sigma}, x_{-\sigma+1}, ...\}$ such that substitution of it into (4.2) renders an identity. A good guess may lead us to solutions which are geometric sequences, namely, $x = \left\{\lambda^k\right\}_{k=-\sigma}^{\infty}$. Substitution of x into (4.2) then leads us further to the equation

$$\lambda^k \left\{\lambda - 1 - \gamma \lambda^{-\sigma}\right\} = 0.$$

51

Therefore, for each root of the rational equation

$$\lambda - 1 - \gamma\lambda^{-\sigma} = 0,$$

we can find a corresponding solution of (4.2).

Similarly, by considering equation (4.3), we are led to the equation

$$\lambda^k \left\{ \lambda - 1 - \lambda^{-\tau}(\lambda - 1) + \lambda^{-\sigma}(\rho - \gamma) \right\} = 0$$

and then the rational equation

$$\lambda^{\tau+1} - \lambda^{\tau} - (\lambda - 1) + (\rho - \gamma)\lambda^{\tau-\sigma} = 0.$$

In general, given a recurrence relation of the form

$$a_n x_{k+n} + a_{n-1} x_{k+n-1} + a_{n-2} x_{k+n-2} + \cdots + a_0 x_k = 0, \; k \in \mathbf{Z}, \qquad (4.4)$$

where n is a fixed positive integer, the same consideration will lead us to the polynomial

$$f(\lambda | a_n, a_{n-1}, a_{n-2}, ..., a_0) = a_n \lambda^n + a_{n-1} \lambda^{n-1} + a_{n-2} \lambda^{n-2} + \cdots + a_0, \qquad (4.5)$$

while if we are given a difference equation more general than (4.3), to rational functions of the form

$$F(\lambda) = f_0(\lambda) + \lambda^{\tau_1} f_1(\lambda) + \cdots + \lambda^{\tau_m} f_m(\lambda) \qquad (4.6)$$

where $\tau_1, ..., \tau_m$ are integers, and $f_0(\lambda), ..., f_m(\lambda)$ are polynomials of the form (4.5).

Note that in case $\tau_1, ..., \tau_m$ are nonnegative integers, then the above function is just a polynomial. If one of $\tau_1, ..., \tau_m$ is negative, then the set of roots of (4.6) can be found by finding the roots of the polynomial

$$\lambda^{\max\{-\tau_1, ..., -\tau_m\}} F(\lambda).$$

In the sequel, we will therefore regard $F(\lambda)$ as a 'Δ-polynomial'. The integers $\tau_1, \tau_2, ..., \tau_m$ will be called the powers of $F(\lambda)$ and without loss of any generality, we may assume that they are *mutually distinct nonzero integers*. For instance, the Δ-polynomial $f_0(\lambda) + \lambda^{\tau_1} f_1(\lambda)$ is said to have one power if $\tau_1 \neq 0$, while $f_0(\lambda) + \lambda^{\tau_1} f_1(\lambda) + \lambda^{\tau_2} f_2(\lambda)$ is said to have two powers if τ_1 and τ_2 are distinct nonzero integers.

A further classification of Δ-polynomials is possible. To this end, we now treat (4.5) as a function with 'hidden parameters', that is, as a polynomial of $n + 2$ variables $\lambda, a_n, a_{n-1}, ...a_0$, where λ varies over a subset of $\mathbf{C}$ and each a_i varies over a subset of $\mathbf{R}$. Such a function is said to have degree[1] n if a_n varies over a subset of $\mathbf{R}$ containing points other than 0. We say that $F(\lambda)$ in (4.6) (now treated as a multivariate function) is a $\Delta(d_0, d_1, ..., d_m)$-polynomial if the degrees of $f_0, f_1, ..., f_m$ are $d_0, d_1, ..., d_m$ respectively. For instance,

$$\lambda - 1 + \lambda^{-3}p + \lambda^{-4}q, \; p, q \in \mathbf{R}$$

---

[1]This term is not to be confused with the degree of a polynomial with only one independent variable.

is a $\Delta(1,0,0)$-polynomial with powers -3 and -4, while the function

$$(a\lambda + b) + \lambda^{-1}c + \lambda^{-2}d + \lambda^{-3}p, \; a,b,c,d,p \in \mathbf{R},$$

is a $\Delta(1,0,0,0)$-polynomial with powers -1, -2 and -3. On the other hand, a $\Delta(1,1)$-polynomial must be of the form

$$a\lambda + b + (c\lambda + d)\lambda^{-\tau}$$

where $a \neq 0$ and $c \neq 0$.

So far we have considered discrete time population models. In some cases, it is better to consider continuous time models as well. A well known example is the decay process of a radioactive material. Let $x(t)$ be the mass of such a material at time t. One model assumes that the rate of change $x'(t)$ is proportional to its mass at t, that is,

$$x'(t) = \gamma x(t), \; t \geq 0. \tag{4.7}$$

While another model assumes that the rate of change $x'(t)$ not only depends on its mass at t, but also depends on its mass at a previous time, say $x(t - \tau)$. Then we are led to differential equations with 'delay':

$$x'(t) + ax(t) + bx(t - \tau) = 0, \; t \geq 0. \tag{4.8}$$

If we seek solutions of (4.7) in the form $e^{\lambda t}$, then we are led to

$$e^{\lambda t}(\lambda - \gamma) = 0.$$

Therefore, a root of the polynomial equation

$$\lambda - \gamma = 0$$

will lead us to the solution $e^{\gamma t}$. Similarly, if we seek solutions in the form $e^{\lambda t}$ of the linear differential equation (the damped oscillator equation)

$$x''(t) + ax'(t) + bx(t) = 0, \; t \in \mathbf{R}, \tag{4.9}$$

then we are led to the polynomial equation

$$\lambda^2 + a\lambda + b = 0 \tag{4.10}$$

obtained by substituting $x(t) = e^{\lambda t}$ into (4.9). In general, if we consider more general linear differential equations of the form

$$x^{(n)}(t) + a_{n-1}x^{(n-1)}(t) + a_{n-2}x^{(n-2)}(t) + \cdots + a_0 x(t) = 0, \; t \in \mathbf{R}, n \in \mathbf{Z}[1, \infty), \tag{4.11}$$

similar considerations will lead us to the polynomial (4.5) again.

However, if we substitute $e^{\lambda t}$ into (4.8), then instead of a polynomial equation, we will end up with

$$\lambda + a + be^{-\lambda\tau} = 0.$$

More general differential equations with delays and/or 'advancements' will further lead us to a more general class of '∇-polynomials'. More precisely, let

$f_0(\lambda), f_1(\lambda), ..., f_m(\lambda)$ be polynomials (with 'hidden parameters'). A function of the form

$$F(\lambda) = f_0(\lambda) + e^{-\lambda\tau_1}f_1(\lambda) + \cdots + e^{-\lambda\tau_m}f_m(\lambda)$$

where $\tau_1, ..., \tau_m$ are mutually distinct nonzero real numbers, is called a ∇-polynomial. The numbers $\tau_1, ..., \tau_m$ will be called the powers of F. We also say that F is a $\nabla(d_0, d_1, ..., d_m)$-polynomial if the degrees of $f_0, ..., f_m$ are $d_0, d_1, ..., d_m$ respectively.

In the sequel, a **quasi-polynomial** is either a polynomial, a Δ-polynomial or a ∇-polynomial[2].

Example 4.1. Δ-polynomials of the form

$$\lambda - 1 + \lambda^{-\tau}p + \lambda^{-\sigma}q = 0,$$

$$\lambda - 1 + \lambda^{\tau}p + \lambda^{\sigma}q = 0,$$

$$\lambda - 1 + \lambda^{-\tau}p + \lambda^{\sigma}q = 0,$$

$$\lambda - 1 + \lambda^{-\tau}(\lambda - 1)p + \lambda^{-\sigma}q = 0,$$

$$\lambda - 1 + \lambda^{-\tau}p + \lambda^{-\sigma}q + \lambda^{-\delta}r = 0,$$

$$\lambda - 1 + \lambda^{\tau}p + \lambda^{\sigma}q + \lambda^{\delta}r = 0,$$

$$\lambda - 1 + \lambda^{-\tau}(\lambda - 1)c + \lambda^{-\tau}p + \lambda^{-\delta}q = 0,$$

$$(\lambda - 1)^n + \lambda^{-\tau}(\lambda - 1)p + \lambda^{-\sigma}q = 0,$$

where $p, q, r, c \in \mathbf{R}$ and $\tau, \sigma, \delta \in \mathbf{Z}$ and $n \in \mathbf{Z}[1, \infty)$, arise from considering difference equations of the respective form

$$x_{k+1} - x_k + px_{k-\tau} + qx_{k-\sigma} = 0, \text{ where } \tau > \sigma \geq 0,$$

$$x_{k+1} - x_k + px_{k+\tau} + qx_{k+\sigma} = 0, \text{ where } \tau > \sigma \geq 0,$$

$$x_{k+1} - x_k + px_{k-\tau} + qx_{k+\sigma} = 0, \text{ where } \tau, \sigma \geq 0,$$

$$\Delta(x_k + px_{k-\tau}) + qx_{k-\sigma} = 0, \text{ where } \tau > 0; \sigma \geq 0,$$

$$x_{k+1} - x_k + px_{k-\tau} + qx_{k-\sigma} + rx_{k-\delta} = 0, \text{ where } 0 < \tau < \sigma < \delta,$$

$$x_{k+1} - x_k + px_{k+\tau} + qx_{k+\sigma} + rx_{k+\delta} = 0, \text{ where } \tau > \sigma > \delta > 0,$$

$$\Delta(x_k + cx_{k-\tau}) + px_{k-\tau} + qx_{k+\delta} = 0, \text{ where } \tau, \delta > 0,$$

and

$$\Delta^n(x_k + px_{k-\tau}) + qx_{k-\sigma} = 0, \text{ where } n, \tau > 0; \delta \geq 0.$$

---

[2]Pontryagyn in.[25] used the same term to indicate the class of functions of the form $G(\lambda, e^{\lambda})$ where $G(\lambda, x)$ is a polynomial in two variables. The same term has also been used for (4.6) in Gel'fond [14].

Example 4.2. ∇-polynomials of the form

$$\lambda + a + e^{-\tau\lambda}u + e^{-\sigma\lambda}v = 0,$$

$$\lambda + a + e^{\lambda\tau}u + e^{\lambda\sigma}v = 0,$$

$$\lambda + a + e^{\lambda\tau}b + e^{\lambda\sigma}u + e^{\lambda\delta}v = 0,$$

$$\lambda + a + e^{-\lambda\tau}u + e^{\lambda\sigma}v = 0,$$

$$\lambda + e^{-\lambda\tau}\lambda u + e^{\lambda\sigma}v = 0,$$

$$\lambda + a\lambda e^{-\lambda\tau} + b + e^{-\lambda\tau}u + e^{\lambda\sigma}v = 0,$$

$$\lambda^n + e^{-\lambda\tau}\lambda^n u + e^{-\lambda\sigma}v = 0$$

where $a, b, c, g, p, q, r, \tau, \sigma, \delta \in \mathbf{R}$ and $n \in \mathbf{Z}[1, \infty)$, arise from considering functional differential equations of the respective form

$$x'(t) + ax(t) + bx(t - \tau) + cx(t - \sigma) = 0, \text{ where } 0 < \tau < \sigma.$$

$$x'(t) + ax(t) + bx(t + \tau) + cx(t + \sigma) = 0, \text{ where } 0 < \tau < \sigma$$

$$x'(t) + ax(t) + bx(t + \tau) + cx(t + \sigma) + gx(t + \delta) = 0, \text{ where } 0 < \tau < \sigma < \delta$$

$$x'(t) + ax(t) + bx(t - \tau) + cx(t + \sigma) = 0, \text{ where } \tau, \sigma > 0$$

$$\frac{d}{d\lambda}[x(t) + px(t - \tau)] + qx(t + \sigma) = 0, \text{ where } \tau, \sigma > 0,$$

$$\frac{d}{d\lambda}[x(t) + cx(t - \tau)] + rx(t) + px(t - \tau) + qx(t + \sigma) = 0, \text{ where } \tau, \sigma > 0,$$

and

$$\frac{d^n}{d\lambda^n}[x(t) + px(t - \tau)] + qx(t - \sigma) = 0, \text{ where } \tau, \sigma > 0.$$

4.2   Characteristic Regions

For polynomials with order greater than 4, it is well known that there are no general formula for their roots. Therefore, finding roots of quasi-polynomials is a non-trivial problem. In some cases, however, we are not required to find the explicit roots of quasi-polynomials.

As an example, let us call a real solution of (4.11) oscillatory if it is neither eventually positive nor eventually negative. Then it is not difficult, using elementary theory of ordinary differential equations, to show that every real solution of the equation (4.9) oscillates if, and only if, the polynomial equation (4.10) does not have any real roots. But to determine the latter property of (4.10) is easy. Indeed, the roots of (4.10) are nonreal if, and only if, $a^2 - 4b > 0$.

More generally, a result in the theory of ordinary differential equations states that every real solution of the linear differential equation (4.11) with real constant coefficients $a_0, a_1, ..., a_{n-1}$ oscillates if, and only if, the polynomial (4.5) does not have any *real* roots. Then we need to decide when (4.5) does or does not have any *real* roots.

A similar result for (4.4) also holds. Let us call a real solution (sequence) of (4.4) oscillatory if it is neither eventually positive nor eventually negative. Then a result in the theory of ordinary difference equations (see e.g. [40]) states that every real solution of (4.4) oscillates if, and only if, the polynomial (4.5) does not have any *positive* roots.

As for differential equations with delays and advancements, we have the following result (see e.g. [1]): Consider the n-th order differential equation

$$\frac{d^n}{dt^n}\left[x(t) + \sum_{\vartheta} p_i x(t - \tau_i)\right] + \sum_{\varkappa} q_k x(t - \sigma_k) = 0, \tag{4.12}$$

where $n \geq 1$, ϑ, $\varkappa$ are initial segments of natural numbers, and $p_i, \tau_i, q_k, \delta_k \in \mathbf{R}$ for $i \in \vartheta$ and $k \in \varkappa$. A necessary and sufficient condition for the oscillation of all solutions of (4.12) is that the quasi-polynomial

$$f(\lambda) = \lambda^n + \lambda^n \sum_{\vartheta} p_i e^{-\lambda \tau_i} + \sum_{\varkappa} q_k e^{-\lambda \sigma_k}$$

has no real roots.

In another direction, it is desired to know when all the roots of a polynomial lie inside the open unit circle, or inside the open left-half plane (see e.g., [25, 33–35, 38]. Such problems attracted mechanical engineers who faced the problem of resolving the instability of steam engines. In general, given a function $f(\lambda | a_n, a_{n-1}, ..., a_0)$ with $n + 1$ real parameters, we are interested in finding the exact region Ψ for $(a_n, a_{n-1}, ..., a_0)$ so that for each $(a_n, a_{n-1}, ..., a_0) \in \Psi$, none, some or all of the roots of the function f lie in a specific subset Φ of the domain of f. It is natural to call Ψ the **characteristic region**. For example, the characteristic region for the function

$$f(\lambda | a, b) = \lambda^2 + b, \ b \in \mathbf{R},$$

to have all roots in $\mathbf{C}\backslash\mathbf{R}$ is $\{b \in \mathbf{R}|\, b > 0\}$, while the characteristic region for the function to have exactly one root in $(0, \infty)$ is $\{b \in \mathbf{R}|\, b < 0\}$.

In case Ψ is meant for all solutions to lie in Φ, we may also call Ψ the Φ-characteristic region for short.

In this book, we will emphasize $\mathbf{C}\backslash(0, \infty)$-characteristic regions for Δ-polynomials and $\mathbf{C}\backslash\mathbf{R}$-characteristic regions for ∇-polynomials.

Theorem 4.1. *Let I be a real interval which is the union of two disjoint intervals I_1 and I_2. Then the $\mathbf{C}\backslash I$-characteristic region of a quasi-polynomial $F(\lambda)$ is the intersection of the $\mathbf{C}\backslash I_1$- and the $\mathbf{C}\backslash I_2$-characteristic regions of $F(\lambda)$.*

The proof is quite easy which follows from the fact that $F(\lambda)$ does not have any roots in I if, and only if, it does not have any roots in I_1 nor I_2.

Example 4.3. As we have already seen in Chapter 1, the $\mathbf{C}\backslash\mathbf{R}$-characteristic region for $f(\lambda|a, b) = \lambda^2 + x\lambda + y$ in the x, y-plane is $\{(x, y) \in \mathbf{R}^2 : y < x^2/4\}$, and the $\mathbf{C}\backslash(0, \infty)$-characteristic region Ω for $f(\lambda|x, y) = \lambda^2 + x\lambda + y$ is defined by

$$x < 0 \text{ and } y > x^2/4, \text{ or, } x \geq 0 \text{ and } y \geq 0.$$

The $\mathbf{C}\backslash\{0\}$-characteristic region for $f(\lambda|x, y) = \lambda^2 + x\lambda + y$ is also easy to find. Indeed, $\lambda = 0$ is a root of $f(\lambda|x, y)$ if, and only if, $y = 0$. Thus the $\mathbf{C}\backslash\{0\}$-characteristic region is $\{(x, y) \in \mathbf{R}^2 : y = 0\}$. Once the $\mathbf{C}\backslash(0, \infty)$-characteristic region for $f(\lambda|x, y) = \lambda^2 + x\lambda + y$ is found, we can then find its $\mathbf{C}\backslash(-\infty, 0)$-characteristic region as well. Indeed, note that λ is a negative root of $f(\lambda|x, y)$ if, and only if, $-\lambda$ is a positive root of $f(\lambda|-x, y)$. Thus the $\mathbf{C}\backslash(-\infty, 0)$-characteristic region is defined by $x > 0$ and $y > x^2/4$, or, $x \leq 0$ and $y \geq 0$.

The next question is then to find the $\mathbf{C}\backslash(0, \infty)$-characteristic regions for more general quasi-polynomials. This, in general, is a difficult question.

In some special cases, however, we may find the desired $\mathbf{C}\backslash(0, \infty)$-characteristic regions by standard techniques. As an example, consider the quasi-polynomial

$$\lambda - 1 + \lambda^{-\sigma} p = 0,$$

where σ is an integer. In case σ is 0 or -1, the $\mathbf{C}\backslash(0, \infty)$-characteristic region is defined by $p \geq 1$ and $p \leq -1$ respectively. But in case σ is neither -1 nor 0, the corresponding $\mathbf{C}\backslash(0, \infty)$-characteristic region is not easy to find. We will see later (as a consequence of Theorems 6.1 and 6.2) that it is defined by $p(\sigma+1)^{\sigma+1}/\sigma^\sigma > 1$.

In this book, by means of finding dual sets of envelopes, we will show how characteristic regions for several important types of quasi-polynomials can be found. Before we study the general quasi-polynomials, let us illustrate the basic procedures by looking into two specific examples.

Example 4.4. Recall the function

$$f(\lambda|x, y) = \lambda^2 - \lambda + \lambda^{-1}(\lambda - 1)x + y, \quad \lambda > 0,$$

in Example 2.2. For each $\lambda \in (0, \infty)$, let L_λ be the straight line defined by $f(\lambda|x, y) = 0$. Let

$$x(\lambda) = \lambda^2 - 2\lambda^3, \quad y(\lambda) = 2\lambda(\lambda^2 - 2\lambda + 1),$$

for $\lambda > 0$. We have shown that the curves G_1 and G_2 described by $x(\lambda)$ and $y(\lambda)$ over the respective intervals $(0, 1/3)$ and $[1/3, \infty)$ are the envelopes of the families $\{L_\lambda : \lambda \in (0, 1/3)\}$ and $\{L_\lambda : \lambda \in [1/3, \infty)\}$ respectively, and that the curve G described by the parametric functions $x(\lambda)$ and $x(\lambda)$ over $(0, \infty)$ is the envelope of the family $\{L_\lambda : \lambda \in (0, \infty)\}$. The curves G_1 and G_2 are depicted in Figure 2.1 which shows that G_2 is the graph of a function $y = G_2(x)$ over $(-\infty, 1/27)$. By Theorem 2.6, the $\mathbf{C}\backslash(0, \infty)$-characteristic region of $f(\lambda|x, y)$ is just the dual set of order 0 of G. In view of the distribution of dual points exemplified by Figure A.1, we may see that the desired dual set is the set of points (x, y) that satisfies $x \leq 0$ and $y > G_2(x)$. An alternate derivation is as follows. By Theorem 4.1, the $\mathbf{C}\backslash(0, \infty)$-characteristic region of $f(\lambda|x, y)$ is the intersection of the $\mathbf{C}\backslash(0, 1/3)$- and the $\mathbf{C}\backslash[1/3, \infty)$-characteristic regions of $f(\lambda|x, y)$. By Theorem 2.6, the $\mathbf{C}\backslash(0, 1/3)$- and the $\mathbf{C}\backslash[1/3, \infty)$-characteristic regions of $f(\lambda|x, y)$ are the dual sets of order 0 of the envelopes G_1 and G_2 respectively. Again, in view of Figure A.1, we see that the same conclusion holds.

Example 4.5. Let

$$h(\lambda|x, y) = (\lambda - 1)^3 \lambda^2 + \lambda(\lambda - 1)^3 x + y, \quad \lambda > 0,$$

and let L_λ be the straight line defined by $h(\lambda|x, y) = 0$. Since

$$h'_\lambda(\lambda|x, y) = (\lambda - 1)^2(4\lambda - 1)x + \lambda(\lambda - 1)^2(5\lambda - 2), \quad \lambda > 0,$$

we see that the determinant of the linear system $h(\lambda|x, y) = 0 = h'_\lambda(\lambda|x, y)$ is

$$D(\lambda) = (\lambda - 1)^2(4\lambda - 1),$$

which has positive roots 1 and 1/4. By solving the linear system for x and y, we may obtain the parametric functions

$$x(\lambda) = -\frac{\lambda(5\lambda - 2)}{4\lambda - 1}, \quad y(\lambda) = \frac{(\lambda - 1)^4 \lambda^2}{4\lambda - 1}$$

for $\lambda \in (0, \infty) \backslash \{1, 1/4\}$. We have

$$x'(\lambda) = -2\frac{10\lambda^2 - 5\lambda + 1}{(4\lambda - 1)^2}, \quad y'(\lambda) = 2(\lambda - 1)^3 \lambda \frac{10\lambda^2 - 5\lambda + 1}{(4\lambda - 1)^2}$$

and

$$\frac{dy}{dx} = -(\lambda - 1)^3 \lambda, \quad \frac{d^2 y}{dx^2} = -\frac{1}{2}(\lambda - 1)^2 \frac{(4\lambda - 1)^3}{10\lambda^2 - 5\lambda + 1}$$

for $\lambda \in (0, \infty) \backslash \{1, 1/4\}$. Since $(x'(\lambda), y'(\lambda)) \neq 0$ for $\lambda \in (0, \infty) \backslash \{1/4, 1\}$, by Theorem 2.3, the curves G_1, G_2 and G_3 described by $x(\lambda)$ and $y(\lambda)$ over the respective intervals $(0, 1/4)$, $(1/4, 1)$ and $(1, \infty)$ are the envelopes of the families $\{L_\lambda : \lambda \in (0, 1/4)\}$, $\{L_\lambda : \lambda \in (1/4, 1)\}$ and $\{L_\lambda : \lambda \in (1, \infty)\}$ respectively.

By Theorem 4.1, the $\mathbf{C}\backslash(0, \infty)$-characteristic region of $h(\lambda|x, y)$ is the intersection of the $\mathbf{C}\backslash(0, 1/4)$-, $\mathbf{C}\backslash\{1/4\}$-, $\mathbf{C}\backslash(1/4, 1)$-, $\mathbf{C}\backslash\{1\}$- and $\mathbf{C}\backslash(1, \infty)$-characteristic regions of $h(\lambda|x, y)$. The $\mathbf{C}\backslash\{1\}$-characteristic region, since $h(1|x, y) = y$, is $\mathbf{C}\backslash\mathbf{R}$, which is also the dual set of order 0 of Θ_0. The $\mathbf{C}\backslash\{1/4\}$-characteristic region is the set of points that do not lie on the straight line $L_{1/4}$, that is, $\mathbf{C}\backslash L_{1/4}$, which is also the dual set of order 0 of $L_{1/4}$. As for the other three regions, by Theorem 2.6, the $\mathbf{C}\backslash(0, 1/4)$-characteristic region is the dual set of order 0 of the envelope G_1, the $\mathbf{C}\backslash(1/4, 1)$-characteristic region is the dual set of order 0 of the envelope G_2, and the $\mathbf{C}\backslash(1, \infty)$-characteristic region is the dual set of order 0 of the envelope G_3. In principle, we may thus find the desired $\mathbf{C}\backslash(0, \infty)$-characteristic region by taking intersection of the five different subsets just described.

There remains the computation of the intersection, however. To this end, we first observe that the information on $x(\lambda)$ and $y(\lambda)$ described above allow us to show that G_1 is the graph of a strictly increasing, strictly concave and smooth function $y = G_1(x)$ defined over the interval $(-\infty, 0)$ such that $G_1(0^-) = 0$ and $G_1'(0^-) = 0$, that G_2 is the graph of a strictly increasing, strictly convex and smooth function $y = G_2(x)$ defined over $(-1, \infty)$ such that $G_2((-1)^+) = 0$, $G_2'((-1)^+) = 0$ and $G_2(+\infty) = +\infty$, and that G_3 is the graph of a strictly decreasing and strictly convex function $y = G_3(x)$ defined over $(-\infty, -1)$ such that $G_3((-1)^-) = 0$, $G_3'((-1)^-) = 0$, $G_3(-\infty) = +\infty$ and $G_3'(-\infty) = -\infty$. Since $\lim_{\lambda \to (\lambda^*)^-} x(\lambda) = -\infty$ and $\lim_{\lambda \to (\lambda^*)^+} x(\lambda) = +\infty$, in view of Lemma 3.2, we also see that $L_{1/4}$ is the asymptote of G_1 at $-\infty$ and of G_2 at $+\infty$. See Figure 4.1

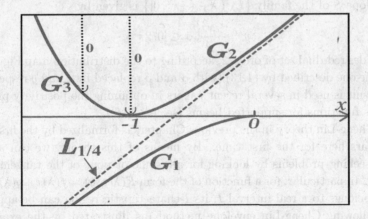

Fig. 4.1

Next, we apply Theorem 3.22 to conclude that the intersection of the dual sets of order 0 of G_2, G_3 and the straight line Θ_0 is equal to the dual set Ω of order 0 of the curve described by $x(\lambda)$ and $y(\lambda)$ over the interval $(1/4, \infty)$. Finally, in view of Figure A.12, the intersection of Ω with the dual sets of order 0 of $L_{1/4}$ and G_1 is equal to Ω itself. We have thus completed the computation part.

We remark that in the rest of this book, we will employ standard procedures similar to those just described for deriving more general characteristic regions. A good understanding of the previous two Examples is thus essential in the sequel.

4.3   Notes

Oscillation theory of difference and differential equations is a well developed subject. Several standard references are [1, 36, 37, 40]. In particular, necessary and sufficient conditions for oscillation in terms of characteristic functions can be found in these references. The term quasi-polynomial is used by different authors (see e.g. [14, 31]) and may have meaning different from ours.

Although we will emphasize $\mathbf{C}\backslash(0,\infty)$-characteristic regions for Δ-polynomials and $\mathbf{C}\backslash\mathbf{R}$-characteristic regions for ∇-polynomials, our method can be applied to obtain other characteristic regions as well. We demonstrate this by showing the following: If

$$\beta < 2\pi^2,\ 0 \geq \alpha \geq -\beta^2/4,\ \frac{\alpha}{\pi^4} + \frac{\beta}{\pi^2} < 1, \tag{4.13}$$

then the roots of the quadratic polynomial $\lambda^2 + \beta\lambda - \alpha$ fall inside the interval $(-\pi^2, 0]$. The proof is quite easy and follows the same principles in the previous examples. Indeed, for each $\lambda \in (-\pi^2, 0]$, consider the staright lines

$$L_\lambda : \lambda^2 + x\lambda - y = 0.$$

The envelope S of the family $\{L_\lambda : \lambda \in (-\pi^2, 0]\}$ is given by

$$y = -\frac{x^2}{4},\ x \in [0, 2\pi^2).$$

Then the desired dual set of order 2, according to the distribution map Figure 3.1 is exactly the one described by (4.13) (with α and β replaced by x and y respectively). Such a result is used in several recent papers in obtaining the positivity properties of Green's functions for supported beams (see e.g. [16]).

The Cheng-Lin theory in the previous Chapters is formalized by the first author and appears here for the first time. By means of this theory, we can approach the root seeking problems by looking for all, some or none of the tangent lines of envelopes. In particular, for a function of the form $F(\lambda|x,y) = f(\lambda)x + g(\lambda)y - h(\lambda)$ where λ belongs to a real interval I, its I-characteristic region can be approached by the following Cheng-Lin envelope method (as illustrated by the examples in the previous Section). More specifically, we first find $F'_\lambda(\lambda|x,y)$, then calculate the determinant $D(\lambda)$ of the linear system $F(\lambda|x,y) = 0 = F'_\lambda(\lambda|x,y)$. Then we break the interval I into several subintervals into two groups of subintervals, say, $I_1, I_2, ..., I_m$, and $I_{m+1}, ..., I_n$ according to whether $D(\lambda) \neq 0$ or $D(\lambda) = 0$. For each interval, we may try to find the corresponding characteristic regions by finding the dual sets of order 0 of envelopes or appropriate straight lines. Then the I-characteristic region is their intersection.

Chapter 5

$C\backslash(0, \infty)$-Characteristic Regions of Real Polynomials

5.1 Quadratic Polynomials

We have found the $C\backslash(0, \infty)$-characteristic regions of the quadratic polynomials in the first Chapter by means of the formula of the roots. Since explicit formulas of roots of general polynomials are not available, we need alternate means to find the corresponding regions. Dual sets of envelopes may be used to find some of these regions. Indeed, we will show that the $C\backslash(0, \infty)$-characteristic regions of real polynomials with degrees less than or equal to 5 can be found.

Before doing so, we will illustrate the principle behind by looking at the real quadratic polynomial again. Let

$$f(\lambda|x, y) = \lambda^2 + x\lambda + y, \ x, y \in \mathbf{R}, \tag{5.1}$$

and let us denote its $C\backslash(0, \infty)$-characteristic region by Ω.

Theorem 5.1. *A point $(x, y) \in \mathbf{R}^2$ belongs to the $C\backslash(0, \infty)$-characteristic region Ω of the quadratic polynomial $f(\lambda|x, y) = \lambda^2 + \lambda x + y$ if, and only if, $x \geq 0$ and $y \geq 0$, or, $x < 0$ and $y > x^2/4$.*

Proof: First of all, let L_λ be the straight line in the x, y-plane defined by $f(\lambda|x, y) = 0$, that is,

$$L_\lambda : \lambda x + y = -\lambda^2, \ \lambda > 0. \tag{5.2}$$

Since the determinant of the linear system

$$f(\lambda|x, y) = \lambda^2 + \lambda x + y = 0, \ \lambda > 0, \tag{5.3}$$

$$f'_\lambda(\lambda|x, y) = 2\lambda + x = 0, \ \lambda > 0, \tag{5.4}$$

is equal to -1, we may solve for x and y in terms of λ:

$$x(\lambda) = -2\lambda, \ y(\lambda) = \lambda^2, \ \lambda > 0.$$

Since

$$x'(\lambda) = -2, \ y'(\lambda) = 2\lambda, \ \lambda > 0,$$

by Theorem 2.3, the envelope G of the family $\{L_\lambda|\ \lambda \in (0, \infty)\}$ is described by the parametric functions $x(\lambda)$ and $y(\lambda)$

In view of Theorem 2.6, Ω is just the dual set of order 0 of the envelope G. To find the dual set, note that G can also be described by the graph of the function $y = G(x)$ defined on $(-\infty, 0)$, where

$$y = G(x) = \frac{x^2}{4}, \ x < 0.$$

Furthermore, $G(x)$ is a strictly convex and smooth function on $(-\infty, 0)$ such that $G(-\infty) = +\infty$, $G(0^-) = 0$, $G'(-\infty) = -\infty$ and $G'(0^-) = 0$. Thus $L_{G|0}$ is Θ_0, and $G \sim H_{-\infty}$ by Lemma 3.3. In view of the distribution of dual points exemplified by Figure 3.11, (a, b) is a dual point of order 0 of G if, and only if, $a < 0$ and $b > G(a)$, or, $a \geq 0$ and $b \geq 0$. The proof is complete.

We remark that the above result can also be stated as follows: a point $(x, y) \in \Omega$ if, and only if, $(x, y) \in \vee(G) \oplus \nabla(\Theta_0 \chi_{[0, +\infty)})$.

Note that the proof of the above result actually suggests more conclusions. Indeed, in view of the distribution of dual points exemplified by Figure 3.11, we may see that a point (x, y) is a dual point of order 1 of G if, and only if, $x < 0$ and $y = x^2/4$, or, $x < 0$ and $y \leq 0$, or, $x \geq 0$ and $y < 0$; while a point (x, y) is a dual point of order 2 of G if, and only if, $x < 0$ and $0 < y < x^2/4$.

Theorem 5.2. *The quadratic polynomial* $\lambda^2 + \lambda x + y$ *has exactly one positive root if, and only if, $x < 0$ and $y = x^2/4$, or, $x < 0$ and $y \leq 0$, or, $x \geq 0$ and $y < 0$; it has exactly two distinct positive roots if, and only if, $x < 0$ and $0 < y < x^2/4$.*

Since the roots of $\lambda^2 + \lambda x + y$ are $\left(-x \pm \sqrt{x^2 - 4y}\right)/2$, we may see further that when $\lambda^2 + \lambda x + y$ has exactly one positive root, then this root is a double root if, and only if, $y = x^2/4$.

We remark that the same principle can be used to handle other types of characteristic regions. For instance, let α be a fixed real number, and suppose we are concerned with the necessary and sufficient conditions that guarantee our quadratic polynomial does not have any roots in $(\alpha, +\infty)$. All we now need to do is to find the dual set of order 0 of the envelope G of the family $\{L_\lambda|\ \lambda \in (\alpha, +\infty)\}$. Then we will end up with the parametric functions

$$x(\lambda) = -2\lambda, \ y(\lambda) = \lambda^2 \text{ where } \lambda > \alpha.$$

Thus the envelope is now also the graph of the function $y = G(x)$ defined by

$$y = G(x) = \frac{x^2}{4}, \ x < -2\alpha.$$

In view of the distribution of dual points exemplified by Figure 3.11, the dual set of order 0 of G is then the set of points (a, b) that satisfies $a < -2\alpha$ and $b > a^2/4$, or, $a \geq -2\alpha$ and

$$b \geq G'(-2\alpha)\left(a - (-2\alpha)\right) + G(-2\alpha) = -a(x + 2\alpha) + \alpha^2.$$

We also remark that the proofs to be presented in the subsequent results will be based on similar principles, a clear understanding of the above explanations is therefore crucial at this point.

5.2   Cubic Polynomials

Let

$$f(\lambda|x, y, z) = \lambda^3 + x\lambda^2 + y\lambda + z, \ x, y, z \in \mathbf{R},$$

and let its $\mathbf{C}\backslash(0, \infty)$-characteristic region be Ω.

If $z < 0$, then $f(0|x, y, z) = z < 0$, and $\lim_{\lambda \to \infty, \lambda \in \mathbf{R}} f(\lambda|x, y, z) = +\infty$. Thus $f(\lambda|x, y, z)$ has at least one positive real root.

If $z = 0$, then

$$f(\lambda|x, y, 0) = \lambda^3 + x\lambda^2 + y\lambda = \lambda(\lambda^2 + x\lambda + y),$$

thus $f(\lambda|x, y, 0)$ does not have any positive roots if, and only if, $\lambda^2 + x\lambda + y$ does not have any positive roots. Theorem 5.1 can then be applied to handle this case.

We are now left with the case where $z > 0$. We will look for the level set

$$\Omega(z) = \{(x, y) \in \mathbf{R}^2 : (x, y, z) \in \Omega\} \tag{5.5}$$

of Ω for each $z > 0$.

Theorem 5.3. *Suppose z is a fixed positive number. Then the curve described by the parametric functions*

$$x(\lambda) = -2\lambda + \frac{z}{\lambda^2}, \ y(\lambda) = \lambda^2 - \frac{2z}{\lambda}, \ \lambda > 0,$$

is also the graph of a function $y = G(x)$ defined on $\mathbf{R}$. Furthermore, $\Omega(z)$ in (5.5) is equal to $\vee(G)$.

Proof. Let z be a fixed positive number c. Let L_λ be the straight line defined by

$$L_\lambda : \lambda^2 x + \lambda y = -\lambda^3 - c, \ \lambda \in (0, \infty). \tag{5.6}$$

Since the determinant of the linear system

$$f(\lambda|x, y, c) = \lambda^3 + x\lambda^2 + y\lambda + c = 0, \ \lambda > 0, \tag{5.7}$$

$$f'_\lambda(\lambda|x, y, c) = 3\lambda^2 + 2x\lambda + y, \ \lambda > 0, \tag{5.8}$$

is equal to $-\lambda^2$, we may solve x and y in terms of λ:

$$x(\lambda) = -2\lambda + \frac{c}{\lambda^2}, \ y(\lambda) = \lambda^2 - \frac{2c}{\lambda}, \ \lambda > 0.$$

Note that

$$x'(\lambda) = -2 - \frac{2c}{\lambda^3} < 0, \ y'(\lambda) = 2\lambda + \frac{2c}{\lambda^2} > 0, \ \lambda > 0.$$

By Theorem 2.3, the envelope G of the family $\{L_\lambda : \lambda \in (0, \infty)\}$ is described by $x(\lambda)$ and $y(\lambda)$ defined above. Also, by Theorem 2.6, $\Omega(z)$ is just the dual set of order 0 of the envelope G.

Note that

$$(x(0^+), y(0^+)) = (+\infty, -\infty), \quad (x(+\infty), y(+\infty)) = (-\infty, +\infty),$$

that $x(\lambda)$ is a strictly decreasing function on $(0, \infty)$, that $y(\lambda)$ is a strictly increasing function on $(0, \infty)$, and that

$$\frac{dy}{dx}(\lambda) = -\lambda < 0, \quad \frac{d^2y}{dx^2}(\lambda) = \frac{1}{2 + \frac{2c}{\lambda^3}} = \frac{\lambda^3}{2} \frac{1}{\lambda^3 + c} > 0, \quad \lambda > 0.$$

These properties together with other easily obtained information allow us to easily draw the envelope G. The curve G is the graph of a function $y = G(x)$ which is strictly decreasing, strictly convex and smooth on $(-\infty, +\infty)$ such that $G(-\infty) = +\infty$, $G'(-\infty) = -\infty$, $G(+\infty) = -\infty$ and $G'(+\infty) = 0$. Then in view of the distribution of dual points exemplified by Figure 3.21, (a, b) is a dual point of order 0 if, and only if, $b > G(a)$. The proof is complete.

We remark that since the function G is defined on $\mathbf{R}$, is strictly decreasing and strictly convex on $\mathbf{R}$, each $(a, b) \in \Omega(z)$ if and only if, $a = x(\lambda)$ for some (unique) $\lambda \in (0, \infty)$ and $b > y(\lambda)$.

Example 5.1. Consider the polynomial

$$f(\lambda| - 1, 1, 1) = \lambda^3 - \lambda^2 + \lambda + 1 = 0$$

The roots of $f(\lambda)$ are approximately $0.7718 + i1.1151, 0.7718 - i1.1151$ and -0.5437. Thus, $(a, b, z) = (-1, 1, 1)$ is a dual point in Ω. The same conclusion can be obtained by Theorem 5.3. Indeed, the curve G is now defined by the parametric functions

$$x(\lambda) = -2\lambda + \frac{1}{\lambda^2}, \ y(\lambda) = \lambda^2 - \frac{2}{\lambda}, \ \lambda > 0,$$

Since $x(1) = -1 = a$ and $y(1) = -1 < 1 = b$, we see that (a, b) is strictly above G.

5.3 Quartic Polynomials

In the previous Section, we have found $\mathbf{C}\backslash(0, \infty)$-characteristic regions of cubic polynomials by considering families of straight lines defined by (5.6). Alternatively, we may consider straight lines of the form

$$f(\lambda|x, y, c) = \lambda^3 + a\lambda^2 + y\lambda + z, \ a, y, z \in \mathbf{R},$$

in the y, z-plane. Since the parametric functions in Theorem 5.3 are relatively simple, it is probably not worth the effort to work out the details. In the case of quartic polynomials, different descriptions of the $\mathbf{C}\backslash(0, \infty)$-characteristic regions may make a difference in different applications. We will therefore provide two descriptions.

5.3.1   *First Description*

Let

$$f(\lambda|a,b,c,d) = \lambda^4 + a\lambda^3 + b\lambda^2 + c\lambda + d, \quad a,b,c,d \in \mathbf{R},$$

and let its $\mathbf{C}\backslash(0,\infty)$-characteristic region be denoted by Ω.

If $d < 0$, then $f(0|a,b,c,d) = d < 0$, and $\lim_{\lambda\to\infty,\lambda\in\mathbf{R}} f(\lambda|a,b,c,d) = +\infty$. Thus $f(\lambda|a,b,c,d)$ has a positive real root.

If $d = 0$, then

$$f(\lambda|a,b,c,0) = \lambda^4 + a\lambda^3 + b\lambda^2 + c\lambda = \lambda(\lambda^3 + a\lambda^2 + b\lambda + c),$$

thus $f(\lambda|a,b,c,0)$ does not have any positive roots if, and only if $\lambda^3 + a\lambda^2 + b\lambda + c$ does not have any positive roots. Theorem 5.3 can then be applied to handle this case.

We are left with the case where $d > 0$. We will look for the level set

$$\Omega(c,d) = \left\{(a,b) \in \mathbf{R}^2 : (a,b,c,d) \in \Omega\right\} \tag{5.9}$$

of Ω for each pair (c,d) in the upper half c,d-plane

$$\Gamma = \left\{(c,d) \in \mathbf{R}^2 : d > 0\right\}.$$

It turns out that we need to break Γ into three mutually disjoint parts Γ', Γ'' and Γ''' (see Figure 5.1) where

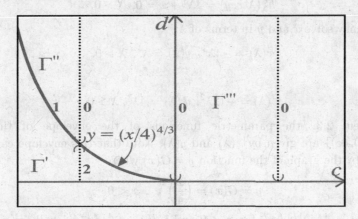

Fig. 5.1

$$\Gamma' = \left\{(c,d) \in \mathbf{R}^2 : c < 0, 0 < d < (c/4)^{4/3}\right\}, \tag{5.10}$$

$$\Gamma'' = \left\{(c,d) \in \mathbf{R}^2 : c < 0, d = (c/4)^{4/3}\right\}, \tag{5.11}$$

and

$$\Gamma''' = \left\{(c,d) \in \mathbf{R}^2 : c \geq 0, d > 0\right\} \cup \left\{(c,d) \in \mathbf{R}^2 : c < 0, d > (c/4)^{4/3}\right\}. \tag{5.12}$$

We also need the following preparatory result on the properties of the quartic polynomial

$$h(\lambda|c,d) = \lambda^4 + c\lambda + 3d, \ \lambda > 0, (c,d) \in \Gamma. \tag{5.13}$$

Lemma 5.1. *Let Γ', Γ'' and Γ''' be defined by (5.10), (5.11) and (5.12) respectively. The following results hold:*

(i) if $(c,d) \in \Gamma'''$, then $h(\lambda|c,d) > 0$ for $\lambda > 0$,
(ii) if $(c,d) \in \Gamma''$, then $h(\lambda|c,d) > 0$ for $\lambda \in (0, (-c/4)^{1/3}) \cup ((-c/4)^{1/3}, \infty)$, and
(iii) if $(c,d) \in \Gamma'$, then $h(\lambda|c,d)$ has two positive roots ξ_1 and ξ_2 such that $h(\lambda|c,d) > 0$ for $\lambda \in (0, \xi_1)$, $h(\lambda|c,d) < 0$ for $\lambda \in (\xi_1, \xi_2)$ and $h(\lambda|c,d) > 0$ for $\lambda \in (\xi_2, \infty)$.

Proof. Consider the quartic equation

$$h(\lambda|x,y) = \lambda^4 + x\lambda + 3y = 0. \tag{5.14}$$

We can interpret (5.14) as an equation describing a family $\{L_\lambda|\ \lambda \in (0,\infty)\}$ of straight lines defined by $L_\lambda : \lambda x + 3y = -\lambda^4$. Since the determinant of the linear system

$$h(\lambda|x,y) = \lambda^4 + x\lambda + 3y = 0, \ \lambda > 0,$$

$$h'_\lambda(\lambda|x,y) = 4\lambda^3 + x = 0, \ \lambda > 0, \tag{5.15}$$

is -3, we may solve x and y in terms of λ:

$$x(\lambda) = -4\lambda^3, \ y(\lambda) = \lambda^4, \ \lambda > 0. \tag{5.16}$$

Since

$$x'(\lambda) = -12\lambda^2, \ y'(\lambda) = 4\lambda^3, \ \lambda > 0,$$

by Theorem 2.3, the parametric functions of the envelope of the family $\{L_\lambda|\lambda \in (0,\infty)\}$ are given by $x(\lambda)$ and $y(\lambda)$. Note that this envelope can also be described by the graph of the function $y = G(x)$ where

$$y = G(x) = \left(\frac{x}{4}\right)^{4/3}, \ x < 0.$$

Since $G'(x) = \frac{4}{3}(\frac{x}{4})^{1/3} < 0$ for $x < 0$ and $G''(x) = \frac{4}{9}(\frac{x}{4})^{-2/3} > 0$ for $x < 0$, the function $G(x)$, as shown in Figure 5.1, is a strictly convex and smooth function on $(-\infty, 0)$ such that $G(-\infty) = +\infty$, $G'(-\infty) = -\infty$, $G(0^-) = 0$ and $G'(0^-) = 0$. In view of the distribution of dual points exemplified by Figure 3.11, a point $(c,d) \in \Gamma'''$ is a dual point of order 0 of G. In other words, there cannot be any positive λ such that $h(\lambda|c,d) = 0$. Since $\lim_{\lambda \to 0+} h(\lambda|c,d) = 3d > 0$, thus $h(\lambda|c,d) > 0$ for $\lambda > 0$. Next, a point $(c,d) \in \Gamma''$ is a dual point of order 1 of G. In other words, there exists a unique positive λ such that $h(\lambda|c,d) = 0$ and the unique root is $(-c/4)^{1/3}$. Since $\lim_{\lambda \to 0+} h(\lambda|c,d) = 3d > 0$ and $\lim_{\lambda \to +\infty} h(\lambda|c,d) \to \infty$, thus for $(c,d) \in \Gamma''$, $h(\lambda|c,d) > 0$ for $\lambda \in (0, (-c/4)^{1/3}) \cup ((-c/4)^{1/3}, \infty)$. Next, a point $(c,d) \in \Gamma'$ is a dual point of order 2 of G. In other words, $h(\lambda|c,d)$ has two positive roots ξ_1 and ξ_2 such that $h(\lambda|c,d) > 0$ for $\lambda \in (0,\xi_1)$, $h(\lambda|c,d) < 0$ for $\lambda \in (\xi_1,\xi_2)$ and $h(\lambda|c,d) > 0$ for $\lambda \in (\xi_2,\infty)$. The proof is complete.

Theorem 5.4. *Let Γ', Γ'' and Γ''' be defined by (5.10), (5.11) and (5.12) respectively. Let $\Omega(c,d)$ be defined in (5.9). Let G be the curve described by the parametric functions*

$$x(\lambda) = -2\lambda + \frac{c}{\lambda^2} + \frac{2d}{\lambda^3}, \ y(\lambda) = \lambda^2 - \frac{2c}{\lambda} - \frac{3d}{\lambda^2}, \ \lambda > 0.$$

(i) *If $(c,d) \in \Gamma'$, then the restriction G_1 of G over the interval $(0,\xi_1)$ and the restriction G_3 of G over (ξ_2,∞), where ξ_1 and ξ_2 are the positive roots of the function $h(\lambda|c,d)$ found in Lemma 5.1, are the graphs of smooth and strictly convex functions $y = G_1(x)$ and $y = G_3(x)$ defined over $(x_1(\xi_1),\infty)$ and $(-\infty, x(\xi_2))$ respectively; furthermore, $(a,b) \in \Omega(c,d)$ if, and only if, $a \le x(\xi_1)$ and $b > G_3(a)$, or, $a \in (x(\xi_1), x(\xi_2))$ and $b > \max\{G_1(a), G_3(a)\}$, or $a \ge x(\xi_2)$ and $b > G_1(a)$ (that is, $(a,b) \in \vee(G_1) \oplus \vee(G_3)$).*

(ii) *If $(c,d) \in \Gamma'' \cup \Gamma'''$, then the curve G is the graph of a function $y = G(x)$ defined on $\mathbf{R}$; furthermore, $\Omega(c,d)$ is the set of points strictly above the curve G.*

Proof. Let (c,d) be a fixed pair in Γ. Let L_λ be the straight line in the x,y-plane defined by

$$L_\lambda : \lambda^3 x + \lambda^2 y = -\lambda^4 - c\lambda - d, \ \lambda \in (0,\infty).$$

Since the determinant of the linear system

$$f(\lambda|x,y,c,d) = \lambda^4 + x\lambda^3 + y\lambda^2 + c\lambda + d = 0, \ \lambda > 0, \tag{5.17}$$

$$f'_\lambda(\lambda|x,y,c,d) = 4\lambda^3 + 3x\lambda^2 + 2y\lambda + c = 0, \ \lambda > 0, \tag{5.18}$$

is $-\lambda^4 < 0$, we may solve x and y in terms of λ:

$$x(\lambda) = -2\lambda + \frac{c}{\lambda^2} + \frac{2d}{\lambda^3}, \ y(\lambda) = \lambda^2 - \frac{2c}{\lambda} - \frac{3d}{\lambda^2}, \ \lambda > 0.$$

Let G be the curve described by $x(\lambda)$ and $y(\lambda)$ for $\lambda > 0$. Note that

$$x'(\lambda) = -2 - \frac{2c}{\lambda^3} - \frac{6d}{\lambda^4} = \frac{-2}{\lambda^4}h(\lambda|c,d),$$

$$y'(\lambda) = 2\lambda + \frac{2c}{\lambda^2} + \frac{6d}{\lambda^3} = \frac{2}{\lambda^3}h(\lambda|c,d),$$

for $\lambda > 0$, where $h(\lambda|c,d)$ is defined in (5.13), and

$$\frac{dy}{dx} = -\lambda, \ \frac{d^2y}{dx^2} = \frac{\lambda^4}{2}\frac{1}{h(\lambda|c,d)}$$

for $\lambda > 0$ except when $h(\lambda|c,d)$ is zero.

Suppose $(c,d) \in \Gamma'$. Then

$$(x(0^+), y(0^+)) = (+\infty, -\infty), \ (x(+\infty), y(+\infty)) = (-\infty, +\infty). \tag{5.19}$$

Furthermore, in view of Lemma 5.1,

$$x'(\lambda) = \frac{-2}{\lambda^4}h(\lambda|c,d) < 0, \ y'(\lambda) = \frac{2}{\lambda^3}h(\lambda|c,d) > 0, \ \lambda \in (0,\xi_1) \cup (\xi_2,\infty),$$

$$x'(\lambda) = \frac{-2}{\lambda^4}h(\lambda|c,d) > 0, \ y'(\lambda) = \frac{2}{\lambda^3}h(\lambda|c,d) < 0, \ \lambda \in (\xi_1,\xi_2)$$

$$\frac{dy}{dx} = -\lambda < 0, \ \lambda \in (0,\xi_1) \cup (\xi_1,\xi_2) \cup (\xi_2,\infty), \tag{5.20}$$

$$\frac{d^2y}{dx^2} = \frac{\lambda^4}{2}\frac{1}{h(\lambda|c,d)} > 0, \ \lambda \in (0,\xi_1) \cup (\xi_2,\infty),$$

and

$$\frac{d^2y}{dx^2} = \frac{\lambda^4}{2}\frac{1}{h(\lambda|c,d)} < 0, \ \lambda \in (\xi_1,\xi_2).$$

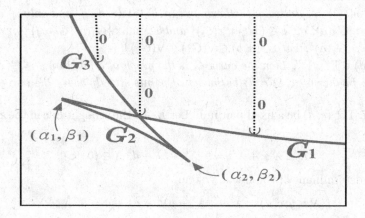

Fig. 5.2

These properties together with other easily obtained information allow us to depict the curve G as shown in Figure 5.2. It is composed of three pieces G_1, G_2 and G_3 and two turning points $(\alpha_1,\beta_1) = (x(\xi_1),y(\xi_1))$ and $(\alpha_2,\beta_2) = (x(\xi_2),y(\xi_2))$. The first piece G_1 corresponds to the case where $\lambda \in (0,\xi_1]$, the second G_2 to the case where $\lambda \in (\xi_1,\xi_2)$, and the third G_3 to the case where $\lambda \in [\xi_2,\infty)$. Furthermore, G_1 is the graph of a function $y = G_1(x)$ which is strictly decreasing, strictly convex, and smooth over $[\alpha_1,+\infty)$, G_2 is the graph of a function $y = G_2(x)$ which is strictly decreasing, strictly concave, and smooth over (α_1,α_2), and the curve G_3 is the graph of a function $y = G_3(x)$ which is strictly decreasing, strictly convex, and smooth over $(-\infty,\alpha_2]$ such that $G_3'(-\infty) = -\infty$. Hence, $G_1^{(v)}(\alpha_1) = G_2^{(v)}(\alpha_1^+)$ and $G_3^{(v)}(\alpha_2) = G_2^{(v)}(\alpha_2^-)$ for $v = 0,1$, furthermore, by Lemma 3.4, $G_3 \sim H_{-\infty}$.

In view of (5.20), we may infer from Theorem 2.4 that G_1 is the envelope of the family $\{L_\lambda | \lambda \in (0,\xi_1]\}$ (since $\lim_{\lambda \to \xi_1} y'(\lambda)/x'(\lambda) = \lim_{\lambda \to \xi_1}(-\lambda) = -\xi_1$). Similarly, by Theorems 2.3 and 2.5, we may show that G_2 is the envelope of the family $\{L_\lambda | \lambda \in (\xi_1,\xi_2)\}$, and that G_3 is the envelope of the family $\{L_\lambda | \lambda \in [\xi_2,\infty)\}$.

By Theorems 2.6 and 4.1, $\Omega(c, d)$ is just the intersection of the dual sets of order 0 of G_1, G_2 and G_3.

But in view of the distribution map in Figure A.14, (a, b) is in $\Omega(c, d)$ if, and only, if $a \leq \alpha_1$ and $b > G_3(a)$, or, $a \in (\alpha_1, \alpha_2)$ and $b > \max\{G_1(a), G_3(a)\}$, or $a \geq \alpha_2$ and $b > G_1(a)$. The proof of (i) is complete.

Next suppose $(c, d) \in \Gamma'''$. Then (5.19) holds again. Furthermore, in view of Lemma 5.1,

$$x'(\lambda) = \frac{-2}{\lambda^4} h(\lambda | c, d) < 0, \; y'(\lambda) = \frac{2}{\lambda^3} h(\lambda | c, d) > 0, \; \lambda \in (0, \infty),$$

$$\frac{dy}{dx} = -\lambda < 0, \; \lambda \in (0, \infty),$$

and

$$\frac{d^2 y}{dx^2} = \frac{\lambda^4}{2} \frac{1}{h(\lambda | c, d)} > 0, \; \lambda \in (0, \infty).$$

These properties together with other easily obtained information allow us to show that G is the envelope of the family $\{L_\lambda | \lambda \in (0, \infty)\}$ and that it is the graph of a function $y = G(x)$ which is strictly decreasing, strictly convex and smooth over $(-\infty, +\infty)$.

Next suppose $(c, d) \in \Gamma''$. Then (5.19) holds again. Furthermore, in view of Lemma 5.1,

$$x'(\lambda) = \frac{-2}{\lambda^4} h(\lambda | c, d) < 0, \; y'(\lambda) = \frac{2}{\lambda^3} h(\lambda | c, d) > 0,$$

$$\frac{dy}{dx} = -\lambda < 0, \tag{5.21}$$

and

$$\frac{d^2 y}{dx^2} = \frac{\lambda^4}{2} \frac{1}{h(\lambda | c, d)} > 0$$

for $\lambda \in (0, (-c/4)^{1/3}) \cup ((-c/4)^{1/3}, \infty)$. As in the first case (see also Example 2.2), we may consider the restrictions of G over the two intervals $(0, (-c/4)^{1/3}]$ and $((-c/4)^{1/3}, \infty)$ and then show that G is the envelope of the family $\{L_\lambda | \lambda \in (0, \infty)\}$ and that it essentially looks the same as the envelope in the previous case.

The envelope in both cases is the graph of a function $y = G(x)$ which is strictly decreasing, strictly convex and smooth (by Theorem 2.1) for $x \in (-\infty, +\infty)$. Furthermore, the function G satisfies $G'(-\infty) = -\infty$ and $G'(+\infty) = 0$. In view of the distribution of dual points exemplified by Figure A.26, the dual set of order 0 of G is the set of points (a, b) strictly above the curve G. The proof of (ii) is complete.

We remark that since the function G is defined on $\mathbf{R}$, is strictly decreasing and strictly convex on $\mathbf{R}$, $(a, b) \in \Omega(c, d)$ if, and only if, $a = x(\lambda)$ for some (unique) $\lambda \in (0, \infty)$ and $b > y(\lambda)$.

Example 5.2. Consider the polynomial

$$f(\lambda|0, 8, -4, 0.5) = \lambda^4 + 8\lambda^2 - 4\lambda + 0.5 = 0.$$

The roots of f are approximately $-0.2463 + i2.8497$, $-0.2463 - i2.8497$, $0.2463 + i0.0213$ and $0.2463 - i0.0213$. Thus, $(a, b, c, d) = (0, 8, -4, 0.5)$ is a dual point in Ω. The same conclusion can be obtained by Theorem 5.4. Indeed, $(c, d) = (-4, 0.5) \in \Gamma'$, and the roots of $h(\lambda| - 4, 0.5)$ are $\xi_1 = 0.3802...$ and $\xi_2 = 1.4349...$ The curves G_1, G_3 are now defined respectively by the parametric functions

$$G_1 : x(\lambda) = -2\lambda - \frac{4}{\lambda^2} + \frac{1}{\lambda^3}, \; y(\lambda) = \lambda^2 + \frac{8}{\lambda} - \frac{3}{2\lambda^2}, \; \lambda \in (0, 0.3802...),$$

and

$$G_3 : x(\lambda) = -2\lambda - \frac{4}{\lambda^2} + \frac{1}{\lambda^3}, \; y(\lambda) = \lambda^2 + \frac{8}{\lambda} - \frac{3}{2\lambda^2}, \; \lambda \in (1.4349..., \infty).$$

Since $(a, b) = (0, 8)$ satisfies

$$x(\lambda) = a = 0$$

for the unique number $\lambda = 0.2481... \in (0, \infty)$ and $x(\lambda) = a = 0 > x(1.4349...) = -4.4741...$, $b = 8 > G_1(0) = 7.9376...$. Thus $(a, b) \in \Omega(-4, 0.5)$.

Example 5.3. Consider the polynomial

$$f(\lambda| - 5, 10, -4, 0.5) = \lambda^4 - 5\lambda^3 + 10\lambda^2 - 4\lambda + 0.5 = 0.$$

The roots of f are approximately $2.2616 + i1.6322$, $2.2616 - i1.6322$, $0.2384 + i0.0862$ and $0.2384 - i0.0862$. Thus, $(a, b, c, d) = (-5, 10, -4, 0.5) \in \Omega$. The same conclusion can be obtained by Theorem 5.4. Indeed, $(c, d) = (-4, 0.5) \in \Gamma'$, and the roots of $h(\lambda| - 4, 0.5)$ are $\xi_1 \doteq 0.3802$ and $\xi_2 \doteq 1.4349$. The curves G_1, G_3 are now defined respectively by the parametric functions

$$G_1 : x(\lambda) = -2\lambda - \frac{4}{\lambda^2} + \frac{1}{\lambda^3}, \; y(\lambda) = \lambda^2 + \frac{8}{\lambda} - \frac{3}{2\lambda^2}, \; \lambda \in (0, 0.3802...),$$

$$G_3 : x(\lambda) = -2\lambda - \frac{4}{\lambda^2} + \frac{1}{\lambda^3}, \; y(\lambda) = \lambda^2 + \frac{8}{\lambda} - \frac{3}{2\lambda^2}, \; \lambda \in (1.4349..., \infty).$$

Since $(a, b) = (-5, 10)$ satisfies

$$x(\lambda) = a = -5$$

for the numbers $\lambda = 0.2725..., 1, 2.1007... \in (0, \infty)$ and $x(0.3802...) = -10.237... < x(\lambda) = a = -5 < x(1.4349...) = -4.4741...$, $b = 10 > \max\{G_1(-5) = 9.2317..., G_3(-5) = 7.8813...\}$. Thus $(a, b) \in \Omega(-4, 0.5)$.

Example 5.4. Consider the polynomial

$$f(\lambda| - 15, 57, -4, 0.5) = \lambda^4 - 15\lambda^3 + 57\lambda^2 - 4\lambda + 0.5 = 0$$

The roots of f are approximately $7.4655 + i0.4758$, $7.4655 - i0.4758$, $0.0345 + i0.0800$ and $0.0345 - i0.0800$. Thus, $(a, b, c, d) = (-15, 57, -4, 0.5) \in \Omega$. The same conclusion can be obtained by Theorem 5.4. Indeed, $(c, d) = (-4, 0.5) \in \Gamma'$, and the roots of $h(\lambda| - 4, 0.5)$ are $\xi_1 \doteq 0.3802$ and $\xi_2 \doteq 1.4349$. The curves G_1, G_3 are now respectively defined by the parametric functions

$$G_1 : x(\lambda) = -2\lambda - \frac{4}{\lambda^2} + \frac{1}{\lambda^3}, \ y(\lambda) = \lambda^2 + \frac{8}{\lambda} - \frac{3}{2\lambda^2}, \ \lambda \in (0, 0.3802...),$$

$$G_3 : x(\lambda) = -2\lambda - \frac{4}{\lambda^2} + \frac{1}{\lambda^3}, \ y(\lambda) = \lambda^2 + \frac{8}{\lambda} - \frac{3}{2\lambda^2}, \ \lambda \in (1.4349..., \infty).$$

Since $(a, b) = (-15, 57)$ satisfies

$$x(\lambda) = a = -15$$

for the unique number $\lambda \doteq 7.4653 \in (0, \infty)$ and $x(0.3802...) \doteq -10.237 > x(\lambda) = a = -15$, $b = 57 > G_3(-15) \doteq 56.775$, Thus $(a, b) \in \Omega(-4, 0.5)$.

Example 5.5. Consider the polynomial

$$f(\lambda|1, -2, 1, 1) = \lambda^4 + \lambda^3 - 2\lambda^2 + \lambda + 1 = 0.$$

The roots of f are approximately -2.0810, $0.7808 + i0.6248$, $0.7808 - i0.6248$ and -0.4805. Thus, $(a, b, c, d) = (1, -2, 1, 1) \in \Omega$. The same conclusion can be obtained by Theorem 5.4. Indeed, $(c, d) = (1, 1) \in \Gamma'''$ and

$$h(\lambda|1, 1) > 0, \ \lambda \in (0, \infty).$$

The curve G_4 is now defined by the parametric functions

$$G_4 : x(\lambda) = -2\lambda + \frac{1}{\lambda^2} + \frac{2}{\lambda^3}, \ y(\lambda) = \lambda^2 - \frac{2}{\lambda} - \frac{3}{\lambda^2}, \ \lambda \in (0, \infty).$$

Since $x(1) = 1 = a$ and $y(1) = -4 < -2 = b$, we see that $(a, b) \in \Omega(1, 1)$.

Example 5.6. Consider the polynomial

$$f(\lambda| - 4, 7, -4, 1) = \lambda^4 - 4\lambda^3 + 7\lambda^2 - 4\lambda + 1 = 0.$$

The roots of f are approximately $1.6248 + i1.3002$, $1.6248 - i1.3002$, $0.3752 + i0.3002$ and $0.3752 - i0.3002$. Thus, $(a, b, c, d) = (-4, 7, -4, 1) \in \Omega$. The same conclusion can be obtained by Theorem 5.4. Indeed, $(c, d) = (-4, 1) \in \Gamma''$ and

$$h(\lambda| - 4, 1) > 0, \ \lambda \in (0, 1) \cup (1, \infty).$$

The curve G_5 is now defined by the parametric functions

$$G_5 : x(\lambda) = -2\lambda - \frac{4}{\lambda^2} + \frac{2}{\lambda^3}, \ y(\lambda) = \lambda^2 + \frac{8}{\lambda} - \frac{3}{\lambda^2}, \ \lambda \in (0, \infty).$$

Since $x(1) = -4 = a$ and $y(1) = 6 < 7 = b$, we see that $(a, b) \in \Omega(-4, 1)$.

5.3.2   *Second Description*

Consider the real quartic polynomial

$$f(\lambda|a, b, x, y) = \lambda^4 + a\lambda^3 + b\lambda^2 + x\lambda + y, \ a, b, x, y \in \mathbf{R},$$

with $\mathbf{C}\backslash(0, \infty)$-characteristic region Ω. We will consider the level set of Ω at fixed $a, b \in \mathbf{R}$:

$$\Omega(a, b) = \{(x, y) \in \mathbf{R}^2 |\ (a, b, x, y) \in \Omega\}. \tag{5.22}$$

Let L_λ be the straight line in the x, y-plane defined by

$$L_\lambda : \lambda x + y = -\lambda^4 - a\lambda^3 - b\lambda^2. \tag{5.23}$$

Since the determinant of the linear system

$$f(\lambda|a, b, x, y) = \lambda^4 + a\lambda^3 + b\lambda^2 + x\lambda + y, \ \lambda > 0, \tag{5.24}$$

$$f'_\lambda(\lambda|a, b, x, y) = 4\lambda^3 + 3a\lambda^2 + 2b\lambda + x = 0, \ \lambda > 0, \tag{5.25}$$

is -1, by Theorem 2.6, $\Omega(a, b)$ is just the dual set of order 0 of the envelope of the family $\{L_\lambda|\ \lambda \in (0, \infty)\}$.

To find the envelope and its dual set of order 0, we first determine from (5.24) and (5.25) the parametric functions

$$x(\lambda) = -(4\lambda^3 + 3a\lambda^2 + 2b\lambda), \ y(\lambda) = 3\lambda^4 + 2a\lambda^3 + b\lambda^2, \ \lambda > 0.$$

Note that

$$(x(0^+), y(0^+)) = (0, 0), \ (x(+\infty), y(+\infty)) = (-\infty, +\infty).$$

Since

$$x'(\lambda) = -2(6\lambda^2 + 3a\lambda + b), \ y'(\lambda) = 2(6\lambda^3 + 3a\lambda^2 + b\lambda)$$

we have

$$y'(\lambda) = -\lambda x'(\lambda)$$

for $\lambda > 0$ and

$$\frac{dy}{dx} = -\lambda \text{ and } \frac{d^2y}{dx^2} = \frac{-1}{x'(\lambda)} \tag{5.26}$$

for $\lambda > 0$ except when $x'(\lambda) = 0$. Note that if $x(\lambda)$ is increasing, then $y(\lambda)$ is decreasing; and if $x(\lambda)$ is decreasing, then $y(\lambda)$ is increasing. In order to describe the curve G defined by $x(\lambda)$ and $y(\lambda)$, we consider the following four cases: (1) $x'(\lambda)$ has no positive roots, (2) $x'(\lambda)$ has a double positive root, (3) $x'(\lambda)$ has two distinct positive roots, and (4) $x'(\lambda)$ has a positive root (and a nonpositive root).

Case 1. Assume $x'(\lambda)$ has no positive roots. Then $x'(\lambda) < 0$ and $y'(\lambda) > 0$ for $\lambda > 0$. So

$$\frac{dy}{dx} = -\lambda < 0, \ \frac{d^2y}{dx^2} = \frac{-1}{x'(\lambda)} > 0, \ \lambda > 0.$$

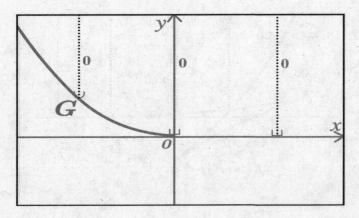

Fig. 5.3

These properties together with other easily obtained information allow us to depict the curve G in Figure 5.3. Furthermore, it is the graph of a function $y = G(x)$ which is a strictly decreasing, strictly convex and smooth function on $(-\infty, 0)$ such that $G(-\infty) = +\infty$, $G'(-\infty) = -\infty$, $G(0^-) = 0$ and $G'(0^-) = 0$. Hence $L_{G|0} = \Theta_0$ and $G \sim H_{-\infty}$. In view of the distribution map in Figure 3.11, (x, y) is a dual point of order 0 of G if, and only if, $x < 0$ and $y > G(x)$, or, $x \geq 0$ and $y \geq 0$.

Case 2. Assume $x'(\lambda)$ has a double positive root r. Then $x(\lambda)$ is strictly decreasing on $(0, r) \cup (r, \infty)$ and $y(\lambda)$ is strictly increasing on $(0, r) \cup (r, \infty)$. So we may see that the curve G is the graph of a function $y = G(x)$ which is smooth (by Theorem 2.1), strictly decreasing and strictly convex on $(-\infty, 0)$. The rest is the same as in the previous case.

Case 3. Assume $x'(\lambda)$ has two distinct positive roots r_1 and r_2 with $r_1 < r_2$. Then $x(\lambda)$ is strictly decreasing on $(0, r_1) \cup (r_2, \infty)$ and is strictly increasing on (r_1, r_2). The curve G is now made up of three curves G_1, G_2 and G_3 as well as two turning points $(\alpha_1, \beta_1) = (x(r_1), y(r_1))$ and $(\alpha_2, \beta_2) = (x(r_2), y(r_2))$. See Figures 5.4 and 5.5. The first curve G_1 corresponds to $\lambda \in (0, r_1]$ and is the graph of a function $y = G_1(x)$ which is smooth, strictly decreasing and strictly convex over $[\alpha_1, 0)$ such that $G_1(0^+) = 0 = G_1'(0^+)$. The second curve G_2 corresponds to $\lambda \in (r_1, r_2)$ and is the graph of a function $y = G_2(x)$ which is smooth, strictly decreasing and strictly concave over (α_1, α_2). The third curve G_3 corresponds to $\lambda \in [r_2, \infty)$ and is the graph of a function $y = G_3(x)$ which is smooth, strictly decreasing and strictly convex over $(-\infty, \alpha_2]$ such that $G_3'(-\infty) = -\infty$. Hence $G_1^{(v)}(\alpha_1) = G_2^{(v)}(\alpha_1^+)$ and $G_3^{(v)}(\alpha_2) = G_2^{(v)}(\alpha_2^-)$ for $v = 0, 1$, $G_3 \sim H_{-\infty}$ and $L_{G_1|0} = \Theta_0$. There are two possible cases: either G_1 and G_3 intersect, or they do not. In the first case, in view of Figure A.13, (α, β) is a dual point of order 0 of G if, and only if, $(a, b) \in \vee(G_1) \oplus \vee(G_3) \oplus \triangledown(\Theta_0)$. In the second case, in view of Figure A.13, (α, β) is a dual point of order 0 of G if, and only if, $(a, b) \in \vee(G_3) \oplus \triangledown(\Theta_0)$.

Case 4. Suppose $x'(\lambda)$ has a positive root (and a nonpositive root). Let the

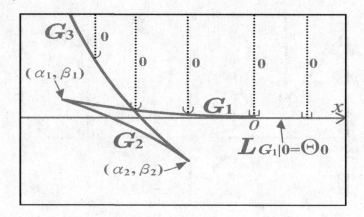

Fig. 5.4

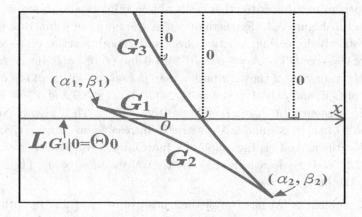

Fig. 5.5

positive root be r. Then $x(\lambda)$ is strictly increasing on $(0, r)$ and is strictly decreasing on (r, ∞). The curve G is now made up of two curves G_1 and G_2 as well as a turning point $(\alpha_3, \beta_3) = (x(r), y(r))$. See Figure 5.6. The first curve G_1 corresponds to $\lambda \in (0, r)$ and is the graph of a function $y = G_1(x)$ which is smooth, strictly decreasing and strictly concave over $(0, \alpha_3))$. The curve G_2 corresponds to $\lambda \in [r, \infty)$ and is the graph of a function $y = G_2(x)$ which is smooth, strictly decreasing and strictly convex over $(-\infty, \alpha_3]$ such that $G_2'(-\infty) = -\infty$. Hence $G_1^{(v)}(\alpha_3^+) = G_2^{(v)}(\alpha_3)$ for $v = 0, 1$, and $G_2 \sim H_{-\infty}$. In view of the distribution map in Figure A.3, (α, β) is a dual point of order 0 of G if, and only if, $(\alpha, \beta) \in \vee(G_2) \oplus \triangledown(\Theta_0)$.

As in Example 2.2, we may easily see that the curve G found above is the envelope of the family of straight lines defined by (5.23). In other words, we have found $\Omega(a, b)$ as well.

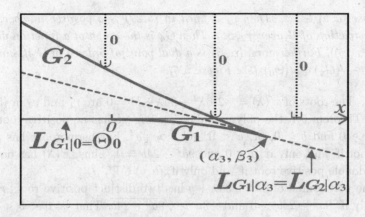

Fig. 5.6

Now we define a partition of the plane by

$$\Gamma' = \left\{(a,b) \in \mathbf{R}^2 : b \geq \frac{3a^2}{8}, \ a < 0\right\} \cup \left\{(a,b) \in \mathbf{R}^2 : a \geq 0, \ b \geq 0\right\},$$

$$\Gamma'' = \left\{(a,b) \in \mathbf{R}^2 : 0 < b < 3a^2/8 \text{ and } a < 0\right\},$$

$$\Gamma''' = \left\{(a,b) \in \mathbf{R}^2 : b < 0\right\} \cup \left\{(a,0) \in \mathbf{R}^2 : a < 0\right\}.$$

Theorem 5.5. *Let*

$$f(\lambda | a,b,x,y) = \lambda^4 + a\lambda^3 + b\lambda^2 + x\lambda + y = 0, \ a,b,x,y \in \mathbf{R},$$

let $\Omega(a,b)$ *be defined by (5.22) and let the parametric curve* G *be defined by*

$$x(\lambda) = -\lambda(4\lambda^2 + 3a\lambda + 2b), \ y(\lambda) = \lambda^2\left(3\lambda^2 + 2a\lambda + b\right), \ \lambda > 0.$$

(i) *Suppose* $(a,b) \in \Gamma'$. *Then* G *is the graph of a function defined on* $(-\infty, 0)$. *Furthermore,* $(x,y) \in \Omega(a,b)$ *if, and only if,* $x < 0$ *and* $y > G(x)$, *or,* $x \geq 0$ *and* $y \geq 0$ *(see Figure 5.3).*

(ii) *Suppose* $(a,b) \in \Gamma''$. *Then*

$$r_1 = \frac{-3a - \sqrt{9a^2 - 24b}}{12} \text{ and } r_2 = \frac{-3a + \sqrt{9a^2 - 24b}}{12} \qquad (5.27)$$

are positive numbers. Let G_1 *be the restriction of* G *over* $(0, r_1]$ *and* G_3 *be the restriction of* G *over* $[r_2, \infty)$. *Then* G_1 *is the graph of a function defined on* $[x(r_1), 0)$ *and* G_3 *the graph of a function on* $(-\infty, x(r_2)]$. *The curves* G_1 *and* G_3 *intersect with each other if, and only if,* $9a^2 - 32b > 0$ *and* $3\gamma_+^2 + 2a\gamma_+ + b \geq 0$, *where*

$$\gamma_+ = \frac{-3a + \sqrt{9a^2 - 32b}}{8}.$$

If intersection occurs, $(x,y) \in \Omega(a,b)$ *if, and only if,* $(x,y) \in \vee(G_1) \oplus \vee(G_3) \oplus \triangledown(\Theta_0)$ *(as in Figure 5.4). Otherwise,* (x,y) *is a dual point of order 0 of* G *if, and only if,* $(x,y) \in \vee(G_2) \oplus \vee(\Theta_0)$ *(as in Figure 5.5).*

(iii) Suppose $(a, b) \in \Gamma'''$. Then r_2 defined in (5.27) is a positive number. Let G_4 the restriction of G over $[r_2, \infty)$. Then G_4 is the graph of a function defined on $(-\infty, x(r_2)]$. Furthermore, (α, β) is a dual point of order 0 of G if, and only if, $(\alpha, \beta) \in \vee(G_4) \oplus \vee(\Theta_0)$ (see Figure 5.6).

Proof. The roots of $x'(\lambda) = -2(6\lambda^2 + 3a\lambda + b) = 0$ are r_1 and r_2 in (5.27).

(i) By Theorem 5.1, the polynomial $6\lambda^2 + 3a\lambda + b$ has no positive roots if, and only if, $a \geq 0$ and $b \geq 0$, or, $a < 0$ and $b > \frac{3}{8}a^2$. Furthermore, it has a double positive root if, and only if, $a < 0$ and $9a^2 - 24b = 0$. Thus, $x'(\lambda)$ has no positive root or a double positive root if, and only if, $(a, b) \in \Gamma'$.

(ii) The polynomial of $6\lambda^2 + 3a\lambda + b$ has two distinct positive roots r_1 and r_2 if, and only if, $9a^2 - 24b > 0$ and $-3a > \sqrt{9a^2 - 24b}$. That is, $0 < b < \frac{3}{8}a^2$ and $a < 0$. Thus, $x'(\lambda)$ has two positive roots r_1 and r_2 if, and only if, $(a, b) \in \Gamma''$. Let G_1 be the restriction of G over $(0, r_1]$, G_2 the restriction of G over (r_1, r_2) and G_3 be the restriction of G over $[r_2, \infty)$. We now examine when do the curves G_1 and G_3 intersect. First of all, the roots of $x(\lambda)$ are

$$\gamma_- = \frac{-3a - \sqrt{9a^2 - 32b}}{8} \quad \text{and} \quad \gamma_+ = \frac{-3a + \sqrt{9a^2 - 32b}}{8}.$$

In view of Figures 5.4 and 5.5, if $x(\lambda)$ has no positive roots, then clearly G_1 and G_3 intersect with each other. If $x(\lambda)$ has one positive root, then clearly G_1 and G_3 intersect with each other. If $x(\lambda)$ has two positive roots, then G_1 and G_3 intersect with each other if, and only if, $y(\gamma_+) \geq 0$, that is, $3\gamma_+^2 + 2a\gamma_+ + b \geq 0$. To summarize, when $(a, b) \in \Gamma''$, G_1 and G_3 intersect with each other if, and only if, $9a^2 - 32b > 0$ and $3\gamma_+^2 + 2a\gamma_+ + b \geq 0$.

(iii) When $b = 0$, the roots of the polynomial $6\lambda^2 + 3a\lambda + b$ are 0 and $-a/2$. Hence it has a positive root if, and only if, $a < 0$. When $b \neq 0$, since $\lim_{\lambda \to \pm\infty} 6\lambda^2 + 3a\lambda + b = +\infty$, we see that it has a positive root if, and only if, $b < 0$. Thus, $x'(\lambda)$ has a positive root and a nonpositive root if, and only if, $(a, b) \in \Gamma'''$.

We may now apply the previous discussions preceding our result and conclude our proof.

Example 5.7. Consider the polynomial

$$P(\lambda| - 3, 2, 0.4, 1) = \lambda^4 - 3\lambda^3 + 2\lambda^2 + 0.4\lambda + 1 = 0.$$

The roots of $P(\lambda)$ are approximately $1.7253 + \text{i}0.5197$, $1.7253 - \text{i}0.5197$, $-0.2253 + \text{i}0.5072$ and $-0.2253 - \text{i}0.5072$. Thus $(-3, 2, 0.4, 1) \in \Omega$. The same conclusion can be obtained by Theorem 5.5(ii). Indeed, let r_1 and r_2 be $(9 + \sqrt{33})/12$ and $(9 - \sqrt{33})/12$. Let the curve G be defined by

$$x(\lambda) = -(4\lambda^3 - 9\lambda^2 + 4\lambda), \quad y(\lambda) = 3\lambda^4 - 6\lambda^3 + 2\lambda^2, \quad \lambda > 0.$$

Let G_1 be the restriction of G over the interval $(0, r_1)$ and G_3 the restriction of the curve G over the interval (r_2, ∞). Then it is easily checked that $(0.4, 1) \in \Omega^{(4)}(-3, 2)$.

5.4   Quintic Polynomials

Let

$$f(\lambda|a,b,c,d,e) = \lambda^5 + a\lambda^4 + b\lambda^3 + c\lambda^2 + d\lambda + e, \quad a,b,c,d,e \in \mathbf{R},$$

and its $\mathbf{C}\backslash(0,\infty)$-characteristic region be Ω.

If $e < 0$, then $f(0|a,b,c,d,e) = e < 0$ and $\lim_{\lambda \to \infty, \lambda \in \mathbf{R}} f(\lambda|a,b,c,d,e) = +\infty$. Thus $f(\lambda|a,b,c,d,e)$ has a positive real root.

If $e = 0$, then

$$f(\lambda|a,b,c,d,0) = \lambda^5 + a\lambda^4 + b\lambda^3 + c\lambda^2 + d\lambda = \lambda(\lambda^4 + a\lambda^3 + b\lambda^2 + c\lambda + d),$$

so that $f(\lambda|a,b,c,d,e)$ does not have any positive roots if, and only if, the quartic polynomial $\lambda^4 + a\lambda^3 + b\lambda^2 + c\lambda + d$ does not have any positive roots. Theorem 5.4 can now be applied to handle this case.

We are now left with the case where $e > 0$. We will look for the level set

$$\Omega(c,d,e) = \{(a,b) \in \mathbf{R}^2 : (a,b,c,d,e) \in \Omega\} \tag{5.28}$$

of Ω where $e > 0$.

We will need the properties of the polynomial

$$H(\lambda|c,d,e) = \lambda^5 + c\lambda^2 + 3d\lambda + 6e, \quad \lambda > 0, \tag{5.29}$$

where e is a fixed positive number.

Lemma 5.2. *Let $e > 0$. Let Ψ be the curve in the plane defined by the parametric functions*

$$\Psi : x(\lambda) = -4\lambda^3 + \frac{6e}{\lambda^2}, \; y(\lambda) = \lambda^4 - \frac{4e}{\lambda}, \; \lambda > 0. \tag{5.30}$$

Then Ψ is the graph of a function defined on $\mathbf{R}$. Furthermore, the following statements hold.

(i) *For any point (c,d) which lies strictly above the curve Ψ, $H(\lambda|c,d,e) > 0$ for $\lambda > 0$.*

(ii) *For any (c,d) which lies on Ψ, there exists a unique λ^* such that $H(\lambda|c,d,e) > 0$ for $\lambda \in (0,\lambda^*) \cup (\lambda^*,\infty)$,*

(iii) *For any (c,d) which lies strictly below Ψ, there exists λ_1 and λ_2 such that $H(\lambda_1) = H(\lambda_2) = 0$, $H(\lambda|c,d,e) > 0$ for $\lambda \in (0,\lambda_1) \cup (\lambda_2,\infty)$ and $H(\lambda|c,d,e) < 0$ for $\lambda \in (\lambda_1,\lambda_2)$.*

Proof: Consider the family $\{L_\lambda | \lambda \in (0,\infty)\}$ of straight lines $L_\lambda : \lambda^2 x + 3\lambda y = -\lambda^5 - 6e$. Since the determinant of the linear system

$$H(\lambda|x,y,e) = \lambda^5 + x\lambda^2 + 3y\lambda + 6e = 0, \; \lambda > 0, . \tag{5.31}$$

$$H'_\lambda(\lambda|x,y,e) = 5\lambda^4 + 2x\lambda + 3y = 0, \; \lambda > 0, \tag{5.32}$$

is $D(\lambda) = -3\lambda^2$, which is not 0 for $\lambda > 0$, the corresponding parametric functions of the envelope Ψ of the family of straight lines are easily determined from (5.31) and (5.32) and given by

$$x(\lambda) = -4\lambda^3 + \frac{6e}{\lambda^2}, \quad y(\lambda) = \lambda^4 - \frac{4e}{\lambda}, \quad \lambda > 0.$$

Note that

$$(x(0^+), y(0^+)) = (+\infty, -\infty), \quad (x(+\infty), y(+\infty)) = (-\infty, +\infty),$$

$$x'(\lambda) = -12\lambda^2 - \frac{12e}{\lambda^3} < 0, \quad y'(\lambda) = 4\lambda^3 + \frac{4e}{\lambda^2} > 0, \quad \lambda > 0,$$

and

$$\frac{dy}{dx} = \frac{-\lambda}{3} < 0, \quad \frac{d^2y}{dx^2} = \frac{1}{3}\frac{\lambda^3}{12\lambda^5 + 12e} > 0, \quad \lambda > 0.$$

These are other easily obtained information allows us to easily draw the envelope of the family of straight lines defined by (5.31) for $\lambda > 0$. The envelope Ψ is the graph of a function $y = \Psi(x)$ which is strictly decreasing, strictly convex, and smooth on $(-\infty, +\infty)$ such that $\Psi(-\infty) = +\infty$, $\Psi'(-\infty) = -\infty$, $\Psi(+\infty) = +\infty$ and $\Psi'(+\infty) = 0$. Hence by the distribution map in Figure A.26, a point (c, d) which lies strictly above the curve Ψ is a dual point of order 0 of Ψ. In other words, there cannot be any positive λ such that $H(\lambda|c, d, e) = 0$. Since $\lim_{\lambda \to 0+} H(\lambda|c, d, e) = 6e > 0$, we see further that $H(\lambda|c, d, e) > 0$ for all $\lambda > 0$. Given a point (c, d) that lies on the curve Ψ, (c, d) is a dual point of order 1 of Ψ. In other words, there exists a unique positive root λ^* such that $H(\lambda^*|c, d, e) = 0$. Therefore $H(\lambda^*|c, d, e) = 0$ and $H(\lambda|c, d, e) > 0$ for $\lambda \in (0, \lambda^*) \cup (\lambda^*, \infty)$. Given a point (c, d) that lies strictly below the curve Ψ, (c, d) is a dual point of order 2 of Ψ. In other words, there exists two positive roots λ_1, λ_2 such that $H(\lambda_1|c, d, e) = H(\lambda_2|c, d, e) = 0$. Since $\lim_{\lambda \to 0+} H(\lambda|c, d, e) = 6e > 0$ and $\lim_{\lambda \to +\infty} H(\lambda|c, d, e) = +\infty$, thus $H(\lambda|c, d, e) > 0$ for $\lambda \in (0, \lambda_1) \cup (\lambda_2, \infty)$ and $H(\lambda|c, d, e) < 0$ for $\lambda \in (\lambda_1, \lambda_2)$. The proof is complete.

Theorem 5.6. *Let $e \in (0, \infty)$ and let Ψ be the curve defined by (5.30).*

(i) If (c, d) lies strictly below the curve Ψ, then the curves G_1 and G_3 defined by

$$G_1 : x(\lambda) = -2\lambda + \frac{c}{\lambda^2} + \frac{2d}{\lambda^3} + \frac{3e}{\lambda^4}, \quad y(\lambda) = \lambda^2 - \frac{2c}{\lambda} - \frac{3d}{\lambda^2} - \frac{4e}{\lambda^3}, \quad \lambda \in (0, \lambda_1)$$

and

$$G_3 : x(\lambda) = -2\lambda + \frac{c}{\lambda^2} + \frac{2d}{\lambda^3} + \frac{3e}{\lambda^4}, \quad y(\lambda) = \lambda^2 - \frac{2c}{\lambda} - \frac{3d}{\lambda^2} - \frac{4e}{\lambda^3}, \quad \lambda \in (\lambda_2, \infty),$$

where λ_1 and λ_2 are the roots of $H(\lambda|c, d, e)$ found in Lemma 5.2, are respectively the graphs of two functions defined on $(x(\lambda_1), +\infty)$ and $(-\infty, x(\lambda_2))$. Furthermore, $\Omega(c, d, e)$ is equal to $\vee(G_1) \oplus \vee(G_3)$.

(ii) If (c,d) lies strictly above or on the curve Ψ, then the curve G_4 defined by

$$G_4 : x(\lambda) = -2\lambda + \frac{c}{\lambda^2} + \frac{2d}{\lambda^3} + \frac{3e}{\lambda^4}, \ y(\lambda) = \lambda^2 - \frac{2c}{\lambda} - \frac{3d}{\lambda^2} - \frac{4e}{\lambda^3}, \ \lambda > 0,$$

is the graph of a function defined on $\mathbf{R}$. Furthermore, $\Omega(c,d,e)$ is equal to $\vee(G_4)$.

Proof: Let $e \in (0,\infty)$. Consider the quintic equation

$$f(\lambda|x,y,c,d,e) = \lambda^5 + x\lambda^4 + y\lambda^3 + c\lambda^2 + d\lambda + e = 0 \qquad (5.33)$$

We can interpret (5.33) as an equation describing a family $\{L_\lambda|\ \lambda \in (0,\infty)\}$ of a straight lines $L_\lambda : \lambda^4 x + \lambda^3 y = -\lambda^5 - c\lambda^2 - d\lambda - e$. Since the determinant of the linear system

$$f(\lambda|x,y,c,d,e) = \lambda^5 + x\lambda^4 + y\lambda^3 + c\lambda^2 + d\lambda + e = 0, \ \lambda > 0,$$

$$f'_\lambda(\lambda|x,y,c,d,e) = 5\lambda^4 + 4x\lambda^3 + 3y\lambda^2 + 2c\lambda + d = 0, \ \lambda > 0, \qquad (5.34)$$

is $-\lambda^6$, which is not 0 for $\lambda > 0$, the parametric functions of the envelope are easily determined from (5.33) and (5.34) and given by

$$x(\lambda) = -2\lambda + \frac{c}{\lambda^2} + \frac{2d}{\lambda^3} + \frac{3e}{\lambda^4}, \ y(\lambda) = \lambda^2 - \frac{2c}{\lambda} - \frac{3d}{\lambda^2} - \frac{4e}{\lambda^3}, \ \lambda > 0.$$

Hence,

$$x'(\lambda) = -2 - \frac{2c}{\lambda^3} - \frac{6d}{\lambda^4} - \frac{12e}{\lambda^5} = \frac{-2}{\lambda^5} H(\lambda|c,d,e),$$

$$y'(\lambda) = 2\lambda + \frac{2c}{\lambda^2} + \frac{6d}{\lambda^3} + \frac{12e}{\lambda^4} = \frac{2}{\lambda^4} H(\lambda|c,d,e),$$

for $\lambda > 0$, and

$$\frac{dy}{dx} = -\lambda, \ \frac{d^2 y}{dx^2} = \frac{\lambda^5}{2H(\lambda|c,d,e)}$$

for $\lambda > 0$ except when $H(\lambda|c,d,e) = 0$.

Suppose the point (c,d) lies strictly below the curve Ψ, Then

$$(x(0^+), y(0^+)) = (+\infty, -\infty), \ (x(+\infty), y(+\infty)) = (-\infty, +\infty). \qquad (5.35)$$

Furthermore, in view of Lemma 5.2,

$$x'(\lambda) = \frac{-2}{\lambda^5} H(\lambda|c,d,e) < 0, \ y'(\lambda) = \frac{2}{\lambda^4} H(\lambda|c,d,e) > 0, \ \lambda \in (0,\lambda_1) \cup (\lambda_2, \infty),$$

$$x'(\lambda) = \frac{-2}{\lambda^5} H(\lambda|c,d,e) > 0, \ y'(\lambda) = \frac{2}{\lambda^4} H(\lambda|c,d,e) < 0, \ \lambda \in (\lambda_1, \lambda_2),$$

$$\frac{dy}{dx} = -\lambda < 0, \ \frac{d^2 y}{dx^2} = \frac{\lambda^5}{2H(\lambda|c,d,e)} > 0, \ \lambda \in (0,\lambda_1) \cup (\lambda_2, \infty),$$

and

$$\frac{dy}{dx} = -\lambda < 0, \quad \frac{d^2y}{dx^2} = \frac{\lambda^5}{2H(\lambda|c,d,e)} < 0, \quad \lambda \in (\lambda_1, \lambda_2).$$

These properties together with other easily obtained information allow us to depict the curve defined by the parametric functions $x(\lambda)$ and $y(\lambda)$ in Figure 5.7. It is composed of three pieces G_1, G_2 and G_3 and two turning points $(\alpha_1, \beta_1) = (x(\lambda_1), y(\lambda_1))$ and $(\alpha_2, \beta_2) = (x(\lambda_2), y(\lambda_2))$. The first piece G_1 corresponds to the case where $\lambda \in (0, \lambda_1]$, the second G_2 to the case where $\lambda \in (\lambda_1, \lambda_2)$ and the third G_3 to the case where $\lambda \in [\lambda_2, \infty)$. Furthermore, G_1 is the graph of a function $y = G_1(x)$ which is strictly decreasing, strictly convex, and smooth over $[\alpha_1, +\infty)$. G_2 is the graph of a function $y = G_2(x)$ which is strictly decreasing, strictly concave, and smooth over (α_1, α_2), and the curve G_3 is the graph of a function $y = G_3(x)$ which is strictly decreasing, strictly convex, and smooth over $(-\infty, \alpha_2]$ such that $G_3'(-\infty) = -\infty$. Hence $G_1^{(v)}(\alpha_1) = G_2^{(v)}(\alpha_1^+)$ and $G_3^{(v)}(\alpha_2) = G_2^{(v)}(\alpha_2^-)$ for $v = 0, 1$, and $G_3 \sim H_{-\infty}$.

By Theorems 2.3 and 2.4, G_1 is the envelope of the family $\{L_\lambda : \lambda \in (0, \lambda_1]\}$, G_2 the envelope of $\{L_\lambda : \lambda \in (\lambda_1, \lambda_2)\}$ and G_3 the envelope of $\{L_\lambda : \lambda \in [\lambda_2, \infty)\}$. By Theorems 2.6 and 4.1, $\Omega(c, d, e)$ is the intersection of the dual sets of order 0 of G_1, G_2 and G_3. Then, in view of the distribution map in Figure A.14, $(a, b) \in \Omega(c, d, e)$ if, and only if, $(a, b) \in \vee(G_1) \oplus \vee(G_3)$.

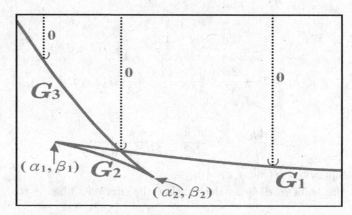

Fig. 5.7

Next suppose (c, d) lies strictly above the curve Ψ, Then (5.35) holds again. Furthermore, in view of Lemma 5.2,

$$x'(\lambda) = \frac{-2}{\lambda^5} H(\lambda|c, d, e) < 0, \quad y'(\lambda) = \frac{2}{\lambda^4} H(\lambda|c, d, e) > 0, \quad \lambda > 0,$$

and

$$\frac{dy}{dx} = -\lambda < 0, \quad \frac{d^2y}{dx^2} = \frac{\lambda^5}{2H(\lambda|c, d, e)} > 0, \quad \lambda > 0.$$

These properties together with other easily obtained information allow us to draw the curve G_4. It is the graph of a function $y = G_4(x)$ which is strictly decreasing, strictly convex, and smooth over $(-\infty, \infty)$.

Next, suppose (c, d) lie on the curve Ψ. Then (5.35) holds again. Furthermore, in view of Lemma 5.2,

$$x'(\lambda) = \frac{-2}{\lambda^5} H(\lambda|c, d, e) < 0, \; y'(\lambda) = \frac{2}{\lambda^4} H(\lambda|c, d, e) > 0, \; \lambda \in (0, \lambda^*) \cup (\lambda^*, \infty),$$

and

$$\frac{dy}{dx} = -\lambda < 0, \; \frac{d^2y}{dx^2} = \frac{\lambda^5}{2H(\lambda|c, d, e)} > 0, \; \lambda \in (0, \lambda^*) \cup (\lambda^*, \infty). \tag{5.36}$$

Also, in view of (5.36), These properties together with other easily obtained information allow us to draw the curve G_5 (which essentially looks the same as G_4). It is the graph of a function $y = G_5(x)$ which is strictly decreasing, smooth (by Theorem 2.1) over $(-\infty, \infty)$ and strictly convex on $(0, \infty)$.

The functions G_4 and G_5 satisfy $G_4(-\infty) = +\infty = G_5(-\infty)$, $G_4'(-\infty) = -\infty = G_5'(-\infty)$, $G_4(+\infty) = +\infty = G_5(+\infty)$ and $G_4'(+\infty) = 0 = G_5'(+\infty)$. As we have explained in the first case, $\Omega(c, d, e)$ is the dual set of order 0 of G_4 (or G_5). In view of the distribution map in Figure A.26, $\Omega(c, d, e)$ is the set of points (a, b) strictly above the curve G_4 (respectively G_5). The proof is complete.

We remark that since the function G_4 (or G_5) is defined on $\mathbf{R}$, is strictly decreasing and strictly convex on $\mathbf{R}$, each $(a, b) \in \Omega(c, d, e)$ if and only if, $a = x(\lambda)$ for some (unique) $\lambda \in (0, \infty)$ and $b > y(\lambda)$.

Example 5.8. Consider the polynomial

$$f(\lambda|0, 5, 2, -4, 1) = \lambda^5 + 5\lambda^3 + 2\lambda^2 - 4\lambda + 1 = 0.$$

The roots of f are approximately $0.1416 + i2.3970$, $0.1416 - i2.3970$, -1.0727, $0.3948 + i0.0764$, $0.3948 - i0.0764$. Thus, $(a, b, c, d, e) = (0, 5, 2, -4, 1)$ is a dual point in $\Omega^{(5)}$. The same conclusion can be obtained by Theorem 5.6. Indeed, since $e = 1$ and

$$\Psi : x(\lambda) = -4\lambda^3 + \frac{6}{\lambda^2}, \; y(\lambda) = \lambda^4 - \frac{4}{\lambda}, \; \lambda > 0,$$

$x(1) = 2$, $y(1) = -3$. Thus $(c, d) = (2, -4)$ lies strictly below the curve Ψ. Furthermore, the roots of $H(\lambda|2, -4, 1)$ are $\lambda_1 \doteq 0.5559$ and $\lambda_2 \doteq 1.4951$. The curves $G_1^{(5)}$, $G_3^{(5)}$ are now respectively defined by the parametric functions

$$G_1^{(5)} : x(\lambda) = -2\lambda + \frac{2}{\lambda^2} - \frac{8}{\lambda^3} + \frac{3}{\lambda^4}, \; y(\lambda) = \lambda^2 - \frac{4}{\lambda} + \frac{12}{\lambda^2} - \frac{4}{\lambda^3}, \; \lambda \in (0, 0.5559...),$$

$$G_3^{(5)} : x(\lambda) = -2\lambda + \frac{2}{\lambda^2} - \frac{8}{\lambda^3} + \frac{3}{\lambda^4}, \; y(\lambda) = \lambda^2 - \frac{4}{\lambda} + \frac{12}{\lambda^2} - \frac{4}{\lambda^3}, \; \lambda \in (1.4951..., \infty).$$

Since $(a, b) = (0, 5)$ satisfies

$$x(\lambda) = a = 0$$

for the unique number $\lambda \doteq 0.4150 \in (0, \infty)$ and $x(\lambda) = a = 0 > x(1.4951...) \doteq -3.8888$, $b = 5 > G_1^{(5)}(0) \doteq 4.2451$. Thus $(a, b) \in \Omega^{(5)}(2, -4, 1)$.

Example 5.9. Consider the polynomial

$$f(\lambda| - 5, 8, 2, -4, 1) = \lambda^5 - 5\lambda^4 + 8\lambda^3 + 2\lambda^2 - 4\lambda + 1 = 0.$$

The roots of f are approximately $2.4905 + \mathrm{i}1.4532$, $2.4905 - \mathrm{i}1.4532$, -0.7458, $0.3824 + \mathrm{i}0.1228$, $0.3824 - \mathrm{i}0.1228$. Thus, $(a, b, c, d, e) = (-5, 8, 2, -4, 1)$ is a dual point in $\Omega^{(5)}$. The same conclusion can be obtained by Theorem 5.6. Indeed, $(c, d) = (2, -4)$ lies strictly below the curve Ψ. Furthermore, the roots of $H(\lambda|2, -4, 1)$ are $\lambda_1 \doteq 0.5559$ and $\lambda_2 \doteq 1.4951$. The curves G_1, G_3 are now respectively defined by the parametric functions

$$G_1 : x(\lambda) = -2\lambda + \frac{2}{\lambda^2} - \frac{8}{\lambda^3} + \frac{3}{\lambda^4}, \; y(\lambda) = \lambda^2 - \frac{4}{\lambda} + \frac{12}{\lambda^2} - \frac{4}{\lambda^3}, \; \lambda \in (0, 0.5559...),$$

$$G_3 : x(\lambda) = -2\lambda + \frac{2}{\lambda^2} - \frac{8}{\lambda^3} + \frac{3}{\lambda^4}, \; y(\lambda) = \lambda^2 - \frac{4}{\lambda} + \frac{12}{\lambda^2} - \frac{4}{\lambda^3}, \; \lambda \in (1.4951..., \infty).$$

Since $(a, b) = (-5, 8)$ satisfies

$$x(\lambda) = a = -5$$

for the numbers $\lambda = 0.4444..., 1, 2.4342... \in (0, \infty)$ and $x(0.5559...) \doteq -9.7943 < x(\lambda) = a = -5 < x(1.4951...) \doteq -3.8888$, $b = 8 > \max\{G_1(-5) \doteq 6.3826, G_3(-5) \doteq 6.03\}$. Thus $(a, b) \in \Omega(2, -4, 1)$.

Example 5.10. Consider the polynomial

$$f(\lambda| - 10, 25, 2, -4, 1) = \lambda^5 - 10\lambda^4 + 25\lambda^3 + 2\lambda^2 - 4\lambda + 1 = 0.$$

The roots of f are approximately $5.0106 + \mathrm{i}0.4982$, $5.0106 - \mathrm{i}0.4982$, -0.4837, $0.2313 + \mathrm{i}0.1675$, $0.2313 - \mathrm{i}0.1675$. Thus, $(a, b, c, d, e) = (-10, 25, 2, -4, 1)$ is a dual point in Ω. The same conclusion can be obtained by Theorem 5.6. Indeed, $(c, d) = (2, -4)$ lies strictly below the curve Ψ. Furthermore, the roots of $H(\lambda|2, -4, 1)$ are $\lambda_1 \doteq 0.5559$ and $\lambda_2 \doteq 1.4951$. The curves G_1, G_3 are now respectively defined by the parametric functions

$$G_1 : x(\lambda) = -2\lambda + \frac{2}{\lambda^2} - \frac{8}{\lambda^3} + \frac{3}{\lambda^4}, \; y(\lambda) = \lambda^2 - \frac{4}{\lambda} + \frac{12}{\lambda^2} - \frac{4}{\lambda^3}, \; \lambda \in (0, 0.5559...),$$

$$G_3 : x(\lambda) = -2\lambda + \frac{2}{\lambda^2} - \frac{8}{\lambda^3} + \frac{3}{\lambda^4}, \; y(\lambda) = \lambda^2 - \frac{4}{\lambda} + \frac{12}{\lambda^2} - \frac{4}{\lambda^3}, \; \lambda \in (1.4951..., \infty).$$

Since $(a, b) = (-10, 25)$ satisfies

$$x(\lambda) = a = -10$$

for the unique number $\lambda \doteq 5.0104 \in (0, \infty)$ and $x(0.5559...) \doteq -9.7943 > x(\lambda) = a = -15$, $b = 25 > G_3(-10) \doteq 24.752$. Thus $(a, b) \in \Omega(2, -4, 1)$.

Example 5.11. Consider the polynomial

$$f(\lambda| - 1, 0, 2, -2, 1) = \lambda^5 - \lambda^4 + 2\lambda^2 - 2\lambda + 1 = 0.$$

The roots of f are approximately -1.3247, $0.50 + i0.866$, $0.50 - i0.866$, $0.6624 + i0.5623$, $0.6624 + i0.5623$. Thus, $(a, b, c, d, e) = (-1, 0, 2, -2, 1)$ is a dual point in Ω. The same conclusion can be obtained by Theorem 5.6. Indeed, $(c, d) = (2, -2)$ lies strictly above the curve Ψ. Furthermore,

$$H(\lambda|2, -2, 1) > 0, \ \lambda \in (0, \infty)$$

The curve G_4 is now defined by the parametric functions

$$G_4 : x(\lambda) = -2\lambda + \frac{2}{\lambda^2} - \frac{4}{\lambda^3} + \frac{3}{\lambda^4}, \ y(\lambda) = \lambda^2 - \frac{4}{\lambda} + \frac{6}{\lambda^2} - \frac{4}{\lambda^3}, \ \lambda \in (0, \infty).$$

Since $(a, b) = (-1, 0)$ satisfies

$$x(\lambda) = a = -1$$

for the unique number $\lambda = 1 \in (0, \infty)$ and $b = 0 > y(1) = -1$. Thus $(a, b) \in \Omega(2, -2, 1)$.

Example 5.12. Consider the polynomial

$$f(\lambda| - 3, 3, 2, -3, 1) = \lambda^5 - 3\lambda^4 + 3\lambda^3 + 2\lambda^2 - 3\lambda + 1 = 0.$$

The roots of f are approximately -0.9413, $0.4850 + i0.2570$, $0.4850 - i0.2570$, $1.4856 + i1.1486$, $1.4856 - i1.1486$. Thus, $(a, b, c, d, e) = (-3, 3, 2, -3, 1) \in \Omega$. The same conclusion can be obtained by Theorem 5.6. Indeed, $(c, d) = (2, -3)$ lies on the curve Ψ. Furthermore,

$$H(\lambda|2, -3, 1) > 0, \ \lambda \in (0, 1) \cup (1, \infty).$$

The curve G_5 is now defined by the parametric functions

$$G_5 : x(\lambda) = -2\lambda + \frac{2}{\lambda^2} - \frac{6}{\lambda^3} + \frac{3}{\lambda^4}, \ y(\lambda) = \lambda^2 - \frac{4}{\lambda} + \frac{9}{\lambda^2} - \frac{4}{\lambda^3}, \ \lambda \in (0, \infty).$$

Since $(a, b) = (-3, 3)$ satisfies

$$x(\lambda) = a = -3$$

for the unique number $\lambda = 1 \in (0, \infty)$ and $b = 3 > y(1) = 2$. Thus $(a, b) \in \Omega(2, -3, 1)$.

5.5 Notes

Most of the results in this Chapter are taken from [6]. Theorem 5.5, however, is from [12].

Chapter 6

$C\backslash(0, \infty)$-Characteristic Regions of Real Δ-Polynomials

6.1 Δ-Polynomials Involving One Power

There are good reasons to study real Δ-polynomials of the form

$$f_0(\lambda) + \lambda^{-\sigma} f_1(\lambda), \ \sigma \in \mathbf{Z}\backslash\{0\}, \tag{6.1}$$

involving one power. For instance, if we consider geometric sequences of the form $\{\lambda^k\}$ as solutions of the difference equation

$$\Delta x_k + p x_{k-\sigma} = 0, \ k \in \mathbf{N},$$

where $p \in \mathbf{R}$, then we are led to the Δ-polynomial

$$\lambda - 1 + \lambda^{-\sigma} p.$$

The $\mathbf{C}\backslash(0, \infty)$-characteristic region of the general real Δ-polynomial (6.1) is difficult to find. For this reason, we will restrict ourselves to the case when the degrees of f_0 and f_1 are less than or equal to 1.

The $\Delta(0, 0)$-polynomials are of the form $\alpha + \lambda^{-\sigma}\beta$, which can easily be handled and hence skipped.

6.1.1 $\Delta(1, 0)$-Polynomials

Let us consider $\Delta(1, 0)$-polynomials of the form

$$g(\lambda|a, b, c) = a\lambda - b + \lambda^{-\sigma} c, \ \sigma \in \mathbf{Z}\backslash\{0\}; a, c \in \mathbf{R}\backslash\{0\}.$$

When $\sigma = -1$,

$$g(\lambda|a, b, c) = (a + c)\lambda - b,$$

which is easy to handle and hence will be ignored. The case where $\sigma = -2$ is also easy since the resulting polynomial is quadratic, but we will include this case in the following general discussion: (i) $-\sigma \leq -1$, and (ii) $-\sigma = \tau \geq 2$.

6.1.1.1   *The Case* $-\sigma \le -1$

In this case,

$$g(\lambda|a,b,c) = a\lambda^{-\sigma}\left(\lambda^{\sigma+1} - \frac{b}{a}\lambda^{\sigma} + \frac{c}{a}\right), \ \sigma \in \mathbf{Z}[1,\infty), a \in \mathbf{R}\backslash\{0\}.$$

Therefore, we may consider the 'equivalent'

$$f(\lambda|x,y) = \lambda^{\sigma+1} - \lambda^{\sigma}x + y, \ \sigma \in \mathbf{Z}[1,\infty). \tag{6.2}$$

Since $c \ne 0$ and $a \ne 0$, we should require $y \ne 0$ in (6.2). There is no harm done, however, if we consider the more general case where $y \in \mathbf{R}$.

Theorem 6.1. *Suppose $\sigma \in \mathbf{Z}[1,\infty)$. Then the $\mathbf{C}\backslash(0,\infty)$-characteristic region for $f(\lambda|x,y)$ in (6.2) is the set of points (x,y) that satisfies $x \le 0$ and $y \ge 0$, or, $x > 0$ and $y > \frac{\sigma^{\sigma}}{a^{\sigma}(\sigma+1)^{\sigma+1}}x^{\sigma+1}$.*

Proof. Consider the family $\{L_{\lambda}| \ \lambda \in (0,\infty)\}$ of straight lines defined by L_{λ} : $f(\lambda|x,y) = 0$ where $\lambda \in (0,\infty)$. Since

$$f'_{\lambda}(\lambda|x,y) = (\sigma+1)\lambda^{\sigma} - \sigma\lambda^{\sigma-1}x,$$

the determinant of the system $f(\lambda|x,y) = 0 = f'_{\lambda}(\lambda|x,y)$ is $\sigma\lambda^{\sigma-1}$ which does not vanish for $\lambda > 0$. By Theorem 2.6, the $\mathbf{C}\backslash(0,\infty)$-characteristic region of $f(\lambda|x,y)$ is just the dual set of order 0 of the envelope G of the family $\{L_{\lambda}| \ \lambda \in (0,\infty)\}$. By Theorem 2.3, the parametric functions of G are given by

$$x(\lambda) = \frac{(\sigma+1)}{\sigma}\lambda, \ y(\lambda) = \frac{1}{\sigma}\lambda^{\sigma+1}, \ \lambda > 0.$$

But G can also be described by the graph of the function $y = G(x)$ where

$$y = G(x) = \frac{\sigma^{\sigma}}{(\sigma+1)^{\sigma+1}}x^{\sigma+1}, \ x > 0.$$

Since

$$G'(x) = \frac{\sigma^{\sigma}}{(\sigma+1)^{\sigma}}x^{\sigma}, \ G''(x) = \frac{\sigma^{\sigma+1}}{(\sigma+1)^{\sigma}}x^{\sigma-1}, \ x > 0,$$

$G(x)$ is a strictly increasing, strictly convex and smooth function on $(0,\infty)$ such that $G(0^{+}) = 0$, $G(+\infty) = +\infty$, $G'(0^{+}) = 0$ and $G'(+\infty) = +\infty$. In view of the distribution map in Figure 3.11, (p,q) is a dual point of G if, and only if, $p \le 0$ and $q \ge 0$, or, $p > 0$ and $q > G(p)$. The proof is complete.

As a corollary, if σ is a positive integer, then $h(\lambda) = \lambda^{\sigma+1} - \lambda^{\sigma} + p$ does not have any positive roots if, and only if, $p(\sigma+1)^{\sigma+1} > \sigma^{\sigma}$.

6.1.1.2 *The Case* $-\sigma = \tau \geq 2$

In this case,

$$g(\lambda|a,b,c) = a\lambda - b + \lambda^\tau c = a\left(\lambda - \frac{b}{a} + \lambda^\tau \frac{c}{a}\right).$$

Therefore it suffices to consider

$$f(\lambda|x,y) = \lambda - x + \lambda^\tau y, \ \tau \in \mathbf{Z}[2,\infty). \tag{6.3}$$

Theorem 6.2. *Suppose* $\tau \in \mathbf{Z}[2,\infty)$. *Then the* $\mathbf{C}\backslash(0,\infty)$-*characteristic region* Ω *for* $f(\lambda|x,y)$ *in (6.3) is the set of points* (x,y) *that satisfies* $x > 0$ *and* $y > \frac{\alpha^\tau(\tau+1)^{\tau-1}}{\tau^\tau}x^{1-\tau}$, *or,* $x \leq 0$ *and* $y \leq 0$.

Proof. Consider the family $\{L_\lambda | \lambda \in (0,\infty)\}$ of straight lines defined by L_λ : $f(\lambda|x,y) = 0$ where $\lambda \in (0,\infty)$. Since

$$f'_\lambda(\lambda|x,y) = 1 - \tau\lambda^{\tau-1}y,$$

the determinant of the linear system $f(\lambda|x,y) = 0 = f'_\lambda(\lambda|x,y)$ is $\tau\lambda^{\tau-1}$ which does not vanish for $\lambda > 0$. By Theorem 2.6, Ω is just the dual of the envelope G of the family $\{L_\lambda | \lambda \in (0,\infty)\}$. By Theorem 2.3, the parametric functions of G are given by

$$x(\lambda) = \frac{\tau+1}{\tau}\lambda, \ y(\lambda) = \frac{1}{\tau\lambda^{\tau-1}}, \ \lambda > 0.$$

Thus G can also be described by the graph of the function $y = G(x)$ where

$$G(x) = \frac{(\tau+1)^{\tau-1}}{\tau^\tau}x^{1-\tau}, \ x > 0.$$

Since

$$G'(x) = \frac{(\tau+1)^{\tau-1}}{\tau^\tau}(1-\tau)x^{-\tau}, \ G''(x) = -\frac{(\tau-1)^{\tau-1}}{\tau^\tau}(1-\tau)\tau x^{-\tau-1}, \ x > 0,$$

$G(x)$ is a strictly decreasing, strictly convex and smooth function on $(0,\infty)$ such that $G(0^+) = +\infty$, $G(+\infty) = 0$, $G'(0^+) = -\infty$ and $G'(+\infty) = 0$. In view of the distribution map in Figure 3.13, (p,q) is a dual point of G if, and only if, $p > 0$ and $q > G(p)$, or, $p \leq 0$ and $q \leq 0$.

As a corollary, if τ is a positive integer greater than or equal to 2, then $h(\lambda) = \lambda - 1 - \lambda^\tau p$ does not have any positive roots if, and only if, $p\tau^\tau > (\tau+1)^{\tau-1}$.

6.1.2 $\Delta(0,1)$-*Polynomials*

Let us consider $\Delta(0,1)$-polynomials of the form

$$g(\lambda|b,c,d) = b + \lambda^{-\sigma}(c\lambda + d), \ b,c \in \mathbf{R}\backslash\{0\}; \sigma \in \mathbf{Z}\backslash\{0\}.$$

When $\sigma = 1$, the corresponding case can easily be handled and hence skipped. Similarly, for the cases where $\sigma = -1, 2$, we will face with quadratic polynomials.

We will, however, include them in our general discussions: (i) $-\sigma = \tau \geq 1$, and (ii) $-\sigma \leq -2$.

In the former case,

$$g(\lambda|b, c, d) = b + \lambda^\tau(c\lambda + d) = b + \lambda^{\tau+1}c + \lambda^\tau d = c\left(\lambda^{\tau+1} + \frac{d}{c}\lambda^\tau + \frac{b}{c}\right).$$

Therefore, we may consider the 'equivalent'

$$f(\lambda|x, y) = \lambda^{\tau+1} - x\lambda^\tau + y,$$

where $\tau \in \mathbf{Z}[1, \infty)$. But this is the polynomial (6.2) with σ replaced by τ. Hence Theorem 6.1 can be applied.

In the case where $-\sigma \leq -2$,

$$g(\lambda|b, c, d) = b\lambda^{-\sigma}\left(\lambda^\sigma + \frac{c}{b}\lambda + \frac{d}{b}\right).$$

Therefore, we may consider the 'equivalent'

$$f(\lambda|x, y) = \lambda^\sigma - \lambda x + y, \tag{6.4}$$

where $\sigma \geq 2$.

Let $\{L_\lambda | \lambda \in (0, \infty)\}$ be the family of straight lines defined by $L_\lambda : f(\lambda|x, y) = 0$. Since the determinant of the linear system

$$f(\lambda|x, y) = \lambda^\sigma - \lambda x + y = 0, \ \lambda > 0,$$
$$f'_\lambda(\lambda|x, y) = \sigma\lambda^{\sigma-1} - x = 0, \ \lambda > 0,$$

is 1, hence we may solve for x and y to yield

$$x(\lambda) = \sigma\lambda^{\sigma-1}, y(\lambda) = (\sigma - 1)\lambda^\sigma,$$

for $\lambda > 0$. Since $(x'(\lambda), y'(\lambda)) \neq 0$ for $\lambda > 0$, G is the envelope of $\{L_\lambda : \lambda \in (0, \infty)\}$. The curve G described by $x(\lambda)$ and $y(\lambda)$ is also the graph of the function

$$y = G(x) = (\sigma - 1)\left(\frac{x}{\sigma}\right)^{\sigma/(\sigma-1)}, x > 0.$$

Since

$$G'(x) = \sigma^{1/(1-\sigma)}x^{1/(\sigma-1)}, \ G''(x) = \frac{\sigma^{1/(1-\sigma)}}{\sigma - 1}x^{(2-\sigma)/(\sigma-1)}, \ x > 0,$$

G is a strictly increasing, strictly convex and smooth function over $(0, \infty)$ such that $G(0^+) = 0$, $G'(0^+) = 0$, $G(+\infty) = +\infty$ and $G'(+\infty) = +\infty$. Hence $L_{G|0} = \Theta_0$ and $G \sim H_{+\infty}$ In view of the distribution map in Figure 3.11, we see that (x, y) is a dual point of order 0 of G if, and only if, $x \leq 0$ and $y \geq 0$, or, $x > 0$ and $y > G(x)$.

Theorem 6.3. *Suppose $\tau \in \mathbf{Z}[1, \infty)$. Then the $\mathbf{C}\backslash(0, \infty)$-characteristic region for $f(\lambda|x, y)$ in (6.4) is the set of points (x, y) that satisfies $x \leq 0$ and $y \geq 0$, or, $x > 0$ and $y > (\sigma - 1)\left(\frac{x}{\sigma}\right)^{\sigma/(\sigma-1)}$.*

6.1.3 $\Delta(1,1)$-*Polynomials*

Let us consider $\Delta(1,1)$-polynomials of the form

$$g(\lambda|a,b,c,d) = a\lambda + b + \lambda^{-\sigma}(c\lambda + d), \ a,c \in \mathbf{R}\backslash\{0\}; \sigma \in \mathbf{Z}\backslash\{0\}.$$

When $\sigma = 1$ or $\sigma = -1$,

$$g(\lambda|a,b,c,d) = \lambda^{-1}\left(a\lambda^2 + (c+1)\lambda + d\right)$$

and

$$g(\lambda|a,b,c,d) = c\lambda^2 + (a+d)\lambda + b$$

respectively. These quadratic polynomials have been handled before. Therefore we may assume (i) $-\sigma = \tau \geq 2$, or (ii) $-\sigma \leq -2$. In the former case,

$$g(\lambda|a,b,c,d) = c\lambda^{\tau+1} + d\lambda^\tau + a\lambda + b = c\left(\lambda^{\tau+1} + \frac{d}{c}\lambda^\tau + \frac{a}{c}\lambda + \frac{b}{c}\right).$$

Therefore it suffices to consider

$$f(\lambda|\alpha,\beta,\gamma) = \lambda^{\tau+1} + \alpha\lambda^\tau + \beta\lambda + \gamma,$$

where $\tau \geq 2$. In the latter case,

$$g(\lambda|a,b,c,d) = a\lambda^{-\sigma}\left(\lambda^{\sigma+1} + \frac{b}{a}\lambda^\sigma + \frac{c}{a}\lambda + \frac{d}{a}\right).$$

Therefore, it suffices to consider

$$f(\lambda|\alpha,\beta,\gamma) = \lambda^{\sigma+1} + \alpha\lambda^\sigma + \beta\lambda + \gamma,$$

where $\sigma \geq 2$.

Both cases can be combined as

$$f(\lambda|\alpha,x,y) = \lambda^{\sigma+1} + \alpha\lambda^\sigma - x\lambda + y, \tag{6.5}$$

where $\sigma \geq 2$.

Consider the family of straight lines $\{L_\lambda|\ \lambda \in (0,\infty)\}$ defined by L_λ : $f(\lambda|\alpha,x,y) = 0$. The determinant of the linear system

$$f(\lambda|\alpha,x,y) = \lambda^{\sigma+1} + \alpha\lambda^\sigma - x\lambda + y = 0, \ \lambda > 0,$$

$$f'_\lambda(\lambda|\alpha,x,y) = (\sigma+1)\lambda^\sigma + \alpha\sigma\lambda^{\sigma-1} - x = 0, \ \lambda > 0,$$

is 1, and hence we may solve for x and y to yield the parametric functions

$$x(\lambda) = (\sigma+1)\lambda^\sigma + \alpha\sigma\lambda^{\sigma-1}, \ y(\lambda) = \sigma\lambda^{\sigma+1} + \alpha(\sigma-1)\lambda^\sigma,$$

where $\lambda > 0$. We may compute that

$$x'(\lambda) = \sigma(\sigma+1)\lambda^{\sigma-1} + \alpha\sigma(\sigma-1)\lambda^{\sigma-2},$$

$$y'(\lambda) = \sigma(\sigma+1)\lambda^\sigma + \alpha\sigma(\sigma-1)\lambda^{\sigma-1} = \lambda x'(\lambda),$$

and

$$\frac{dy}{dx}(\lambda) = \lambda,$$

$$\frac{d^2y}{dx^2}(\lambda) = \frac{1}{\sigma(\sigma+1)\lambda^{\sigma-1} + \alpha\sigma(\sigma-1)\lambda^{\sigma-2}} \tag{6.6}$$

when $x'(\lambda) \neq 0$.

We consider three cases: (i) $\alpha = 0$, (ii) $\alpha > 0$ and (iii) $\alpha < 0$.

Suppose $\alpha = 0$. Then $x(\lambda) = (\sigma+1)\lambda^\sigma$, $x'(\lambda) = \sigma(\sigma+1)\lambda^{\sigma-1} \neq 0$ and $y(\lambda) = \sigma\lambda^{\sigma+1}$, so that G is the envelope of $\{L_\lambda : \lambda \in (0, \infty)\}$ which is also the graph of a function $y = G(x)$ defined by

$$G(x) = \sigma\left(\frac{x}{\sigma+1}\right)^{(\sigma+1)/\sigma}, \quad x > 0. \tag{6.7}$$

By computing G' and G'', we easily see that G is a strictly increasing, strictly convex and smooth function on $(0, \infty)$ such that $G(0^+) = 0$, $G'(0^+) = 0$, $G(+\infty) = +\infty$ and $G'(+\infty) = +\infty$. In view of the distribution map in Figure 3.11, we see that (x, y) is a dual point of order 0 of G if, and only if, $x \leq 0$ and $y \geq 0$, or, $x > 0$ and $y > G(x)$.

Suppose $\alpha > 0$. Then

$$(x(0^+), y(0^+)) = (0, 0), \quad (x(+\infty), y(+\infty)) = (+\infty, +\infty), \tag{6.8}$$

and

$$x'(\lambda) > 0, y'(\lambda) > 0, \frac{dy}{dx}(\lambda) > 0, \frac{d^2y}{dx^2}(\lambda) > 0$$

for $\lambda > 0$. Hence G is the desired envelope which is also the graph of a function $y = G(x)$ over $(0, \infty)$ which is strictly increasing, strictly convex and smooth on $(0, \infty)$ such that $G(0^+) = 0$, $G'(0^+) = 0$, $G(+\infty) = +\infty$ and $G'(+\infty) = +\infty$. In view of the distribution map in Figure 3.11, we see that (x, y) is a dual point of order 0 of G if, and only if, $x \leq 0$ and $y \geq 0$, or, $x > 0$ and $y > G(x)$.

Suppose $\alpha < 0$. Then (6.8) still holds. Since $x'(\lambda)$ as well as $y'(\lambda)$ have a unique common root

$$\lambda_* = \frac{-\alpha(\sigma-1)}{\sigma+1} > 0,$$

we see that $x(\lambda)$ and $y(\lambda)$ are strictly decreasing on $(0, \lambda_*)$ and strictly increasing on (λ_*, ∞). Furthermore, since the denominator on the right hand side of (6.6) has the unique root λ_* as well, we see that $d^2y/dx^2 < 0$ for $\lambda \in (0, \lambda_*)$ and $d^2y/dx^2 > 0$ for $\lambda \in (\lambda_*, \infty)$. As in Example 2.2, these information show that the envelope G is made up of two pieces G_1 and G_2 corresponding to $\lambda \in (0, \lambda_*)$ and $\lambda \in [\lambda_*, \infty)$ respectively, and the turning point $(x(\lambda_*), y(\lambda_*))$ corresponds to $\lambda_* = -\alpha(\sigma-1)/(\sigma+1)$ (see Figure 6.2). The piece G_1 is also the graph of a strictly increasing, strictly concave and smooth function on $(\alpha, 0)$, while G_2 is also the graph of a strictly increasing, strictly convex and smooth function on $[\alpha, +\infty)$ such that $G_2(+\infty) = +\infty$ and $G_2'(+\infty) = +\infty$. In view of the distribution map in Figure A.3, we see that the $\mathbf{C}\backslash(0, \infty)$-characteristic region of G is $\nabla(\Theta_0) \oplus \vee(G_2)$.

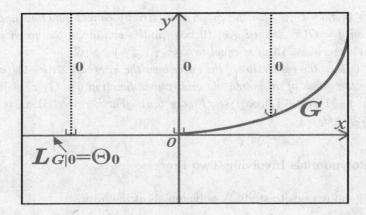

Fig. 6.1

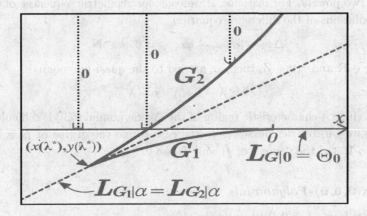

Fig. 6.2

We remark that G_2 and the x-axis have a unique point of intersection, say, $(x^*, 0)$. Thus, (x, y) is a dual point of order 0 if, and only if, $x \leq x^*$ and $y \geq 0$, or, $x > x^*$ and $y > G_2(x)$.

Theorem 6.4. *Suppose $\sigma \in \mathbf{Z}[2, \infty)$. Let*

$$f(\lambda|\alpha, x, y) = \lambda^{\sigma+1} + \alpha\lambda^\sigma - x\lambda + y, \ \alpha \in \mathbf{R}.$$

and let $\Omega(\alpha)$ be the set of points (x, y) such that $f(\lambda|\alpha, x, y)$ does not have any positive roots. Let G be the curve defined by the parametric functions

$$x(\lambda) = (\sigma+1)\lambda^\sigma + \alpha\sigma\lambda^{\sigma-1}, \ y(\lambda) = \sigma\lambda^{\sigma+1} + \alpha(\sigma-1)\lambda^\sigma, \ \lambda > 0.$$

The following results hold:

(i) *If $\alpha = 0$, then G is also the graph of a strictly convex and smooth function $y = G(x)$ defined in (6.7) (see Figure 6.1). Furthermore, $\Omega(\alpha)$ is equal to $\nabla(\Theta_0\chi_{(-\infty,0]}) \oplus \vee(G)$.*

(ii) *If $\alpha > 0$, then G is also the graph of a strictly convex function and smooth function $y = G(x)$ defined over $(0, \infty)$ (and is similar to the graph in Figure 6.1). Furthermore, $\Omega(\alpha)$ is equal to $\nabla(\Theta_0 \chi_{(-\infty, 0]}) \oplus \vee(G)$.*

(iii) *If $\alpha < 0$, then the restriction of the curve over the interval $(-\alpha(\sigma-1)/(\sigma+1), \infty)$ is also the graph of an increasing and convex function $y = G_2(x)$ defined over $(x(-\alpha(\sigma - 1)/(\sigma + 1)), \infty)$ (see Figure 6.2). Furthermore, $\Omega(\alpha)$ is equal to $\vee(G_2) \oplus \nabla(\Theta_0)$.*

6.2   Δ-Polynomials Involving Two Powers

There are good reasons to study Δ-polynomials of the form

$$f_0(\lambda) + \lambda^{-\tau} f_1(\lambda) + \lambda^{-\sigma} f_2(\lambda) \tag{6.9}$$

involving two powers. For instance, if we consider geometric sequences of the form $\{\lambda^k\}$ as solutions of the difference equation

$$\Delta x_k + p x_{k-\tau} + q x_{k-\sigma} = 0, \; k \in \mathbf{N},$$

where $p, q \in \mathbf{R}$ and $\tau, \sigma \in \mathbf{Z}$, then we are led to the quasi-polynomial

$$g(\lambda) = \lambda - 1 + \lambda^{-\tau} p + \lambda^{-\sigma} q.$$

The $\mathbf{C}\backslash(0, \infty)$-characteristic region of the Δ-polynomial (6.9) is difficult to find. Therefore we will restrict ourselves to the case where the degree of f_0 is less than or equal to 1, and the degrees of f_1 and f_2 are 0.

6.2.1   Δ(0, 0, 0)-*Polynomials*

The general form of $\Delta(0, 0, 0)$-polynomials is

$$F(\lambda) = -\beta + \lambda^{-\tau} x + \lambda^{-\sigma} y,$$

where $\beta \in \mathbf{R}\backslash\{0\}$ and τ, σ are distinct nonzero integers (if $\tau = \sigma$, the resulting Δ-polynomial is equivalent to a Δ-polynomial involving only one power which has already been discussed before). Since $\beta \neq 0$, by dividing $F(\lambda)$ by β if necessary, we may then assume without loss of generality that our Δ-polynomial is of the form

$$-1 + \lambda^{-\tau} x + \lambda^{-\sigma} y.$$

Thus we will consider this function in which $\tau \neq \sigma$ (and $\tau\sigma \neq 0$).

6.2.1.1   *The Case $\tau, \sigma > 0$*

In this case, we may assume without loss of generality that $\tau > \sigma > 0$. We therefore consider the following equivalent polynomial

$$f(\lambda | x, y) = -\lambda^\tau + x + \lambda^{\tau - \sigma} y \tag{6.10}$$

where $\tau > \sigma \geq 1$.

Theorem 6.5. *Suppose $\tau,\sigma \in \mathbf{Z}[1,\infty)$ such that $\tau > \sigma \geq 1$. Then (x,y) is in the $\mathbf{C}\backslash(0,\infty)$-characteristic region Ω of the polynomial (6.10) if, and only if, $x = 0$ and $y \leq 0$, or, $x > 0$ and*

$$y < \tau(\tau-\sigma)^{\sigma/\tau-1}\left(\frac{x}{\sigma}\right)^{\sigma/\tau}.$$

Proof. Consider the family $\{L_\lambda |\ \lambda \in (0,\infty)\}$ of straight lines defined by L_λ : $f(\lambda|x,y) = 0$ where $\lambda \in (0,\infty)$. Since

$$f'_\lambda(\lambda|x,y) = -\tau\lambda^{\tau-1} + (\tau-\sigma)\lambda^{\tau-\sigma-1}y,$$

the determinant of the linear system $f(\lambda|x,y) = 0 = f'_\lambda(\lambda|x,y)$ is $(\tau-\sigma)\lambda^{\tau-\sigma-1}$ which does not vanish for $\lambda > 0$. By Theorem 2.6, Ω is just the dual of the envelope G of the family $\{L_\lambda |\ \lambda \in (0,\infty)\}$. By Theorem 2.3, we may easily show that the parametric functions of G are given by

$$x(\lambda) = -\frac{\sigma}{\tau-\sigma}\lambda^\tau,\ y(\lambda) = \frac{\tau}{\tau-\sigma}\lambda^\sigma,$$

for $\lambda > 0$. Hence G is also the graph of the function $y = G(x)$ defined by

$$G(x) = \tau(\tau-\sigma)^{\sigma/\tau-1}\left(\frac{-x}{\sigma}\right)^{\sigma/\tau},\ x < 0$$

By computing G' and G'', we may easily see that $G(x)$ is strictly decreasing, strictly concave and smooth on $(-\infty,0)$ such that $G(0^-) = 0$, $G'(0^-) = -\infty$, $G(-\infty) = \infty$ and $G'(-\infty) = 0$. Hence $G \sim H_{-\infty}$ by Lemma 3.5. In view of the distribution map in Figure 3.14, (x,y) is a dual point of order 0 of G if, and only if, $x = 0$ and $y \leq 0$, or, $x < 0$ and $y < G(x)$. The proof is complete.

6.2.1.2   *The Case $\tau,\sigma < 0$*

In this case, we may consider the equivalent polynomial

$$f(\lambda|x,y) = -1 + \lambda^\tau x + \lambda^\sigma y \tag{6.11}$$

where now $\tau,\sigma \in \mathbf{Z}[1,\infty)$. We may further assume without loss of generality that $\tau > \sigma > 0$.

Theorem 6.6. *Suppose $\tau,\sigma \in \mathbf{Z}[1,\infty)$ such that $\tau > \sigma \geq 1$. Then (p,q) is in the $\mathbf{C}\backslash(0,\infty)$-characteristic region Ω of the polynomial (6.11) if, and only if, $p = 0$ and $q \leq 0$, or, $p < 0$ and*

$$q < \frac{\tau}{\sigma^{\sigma/\tau}}(\tau-\sigma)^{\sigma/\tau-1}(-p)^{\sigma/\tau}.$$

Proof. Consider the family $\{L_\lambda |\ \lambda \in (0,\infty)\}$ of straight lines defined by L_λ : $f(\lambda|x,y) = 0$ where $\lambda \in (0,\infty)$. Since

$$f'_\lambda(\lambda|x,y) = \tau\lambda^{\tau-1}x + \sigma\lambda^{\sigma-1}y,$$

the determinant of the linear system $f(\lambda|x,y) = 0 = f'_\lambda(\lambda|x,y)$ is $(\sigma - \tau)\lambda^{\sigma+\tau-1}$ which does not vanish for $\lambda > 0$. By Theorem 2.6, Ω is just the dual of the envelope G of the family $\{L_\lambda| \lambda \in (0,\infty)\}$. By Theorem 2.3, we may easily show that the parametric functions of G are given by

$$x(\lambda) = \frac{-\sigma}{\tau - \sigma}\lambda^{-\tau}, \; y(\lambda) = \frac{\tau}{\tau - \sigma}\lambda^{-\sigma},$$

for $\lambda > 0$. Thus G is the graph of the function $y = G(x)$ defined by

$$y = \frac{\tau}{\sigma^{\sigma/\tau}}(\tau - \sigma)^{\sigma/\tau-1}(-x)^{\sigma/\tau}, \; x < 0.$$

By computing G' and G'', we may easily see that G is strictly decreasing, strictly concave and smooth on $(-\infty, 0)$ such that $G(-\infty) = \infty$, $G'(-\infty) = 0$, $G(0^-) = 0$ and $G'(0^-) = -\infty$. In view of the distribution map in Figure 3.14, (x,y) is a dual point of G if, and only if, $x = 0$ and $y \le 0$, or $x < 0$ and $y < G(x)$. The proof is complete.

6.2.1.3 *The Case $\tau\sigma < 0$*

In this case, we may assume $\tau > 0$ and $\sigma < 0$. We may consider the equivalent Δ-polynomial

$$-1 + \lambda^{-\tau}x + \lambda^\sigma y,$$

or the equivalent polynomial

$$f(\lambda|x,y) = -\lambda^\tau + x + \lambda^{\tau+\sigma}y \tag{6.12}$$

where $\tau, \sigma \ge 1$.

Theorem 6.7. *Suppose τ and σ are positive integers. Then (p,q) is in the $C\backslash(0,\infty)$-characteristic region Ω of $f(\lambda|x,y)$ in (6.12) if, and only if, $p \le 0$ and $q \le 0$, or, $p > 0$ and*

$$q > \frac{\tau}{\tau+\sigma}\left(\frac{\sigma}{\tau+\sigma}\right)^{\sigma/\tau}\frac{1}{p^{\sigma/\tau}}.$$

Proof. Consider the family $\{L_\lambda| \lambda \in (0,\infty)\}$ of straight lines defined by L_λ : $f(\lambda|x,y) = 0$ where $\lambda \in (0,\infty)$. Since

$$f'_\lambda(\lambda|x,y) = -\tau\lambda^{\tau-1} + (\tau+\sigma)\lambda^{\tau+\sigma-1}y,$$

the determinant of the linear system $f(\lambda|x,y) = 0 = f'_\lambda(\lambda|x,y)$ is $(\tau + \sigma)\lambda^{\tau+\sigma-1}$ which does not vanish for $\lambda > 0$. By Theorem 2.6, Ω is just the dual of the envelope G of the family $\{L_\lambda| \lambda \in (0,\infty)\}$. By Theorem 2.3, we may easily show that the parametric functions of G are given by

$$x(\lambda) = \frac{\sigma}{\tau+\sigma}\lambda^\tau, \; y(\lambda) = \frac{\tau}{\tau+\sigma}\lambda^{-\sigma},$$

for $\lambda > 0$. Thus G is the graph of the function $y = G(x)$ defined by

$$G(x) = \frac{\tau}{\tau + \sigma} \left(\frac{\sigma}{\tau + \sigma} \right)^{-\sigma/\tau} \frac{1}{x^{\sigma/\tau}}, \, x > 0.$$

By computing G' and G'', we may easily see that $G(x)$ is strictly decreasing, strictly convex and smooth on $(0, \infty)$ such that $G(0^+) = \infty$, $G'(0^+) = -\infty$, $G(\infty) = 0$ and $G'(+\infty) = 0$. In view of the distribution map in Figure 3.13, (x, y) is a dual point of G if, and only if, $x \leq 0$ and $y \leq 0$, or, $x > 0$ and $y > G(x)$. The proof is complete.

6.2.2 $\Delta(1, 0, 0)$-*Polynomials*

The general $\Delta(1, 0, 0)$-polynomial is of the form

$$g(\lambda) = \alpha\lambda - \beta + \lambda^{-\tau}p + \lambda^{-\sigma}q,$$

where $\alpha \neq 0$ and τ, σ are distinct nonzero integers. If $\beta = 0$, then

$$g(\lambda) = \alpha\lambda + \lambda^{-\tau}p + \lambda^{-\sigma}q = \lambda \left\{ \alpha + \lambda^{-\tau-1}p + \lambda^{-\sigma-1}q \right\},$$

where

$$\alpha + \lambda^{-\tau-1}p + \lambda^{-\sigma-1}q$$

is a $\Delta(0, 0, 0)$-polynomial. We may therefore assume that $\beta \neq 0$. But then if we divide $g(\lambda)$ by β, we have

$$\frac{\alpha}{\beta}\lambda - 1 + \lambda^{-\tau}\frac{p}{\beta} + \lambda^{-\sigma}\frac{q}{\beta}$$

which can be written as

$$\tilde{\alpha}\lambda - 1 + \lambda^{-\tau}\tilde{p} + \lambda^{-\sigma}\tilde{q},$$

where $\tilde{\alpha} \neq 0$. If we now make the change of variable $\tilde{\alpha}\lambda = t$, then we get

$$t - 1 + t^{-\tau}\tilde{\alpha}^{\tau}\tilde{p} + t^{-\sigma}\tilde{\beta}^{\sigma}\tilde{q}$$

which can then be written as

$$t - 1 + t^{-\tau}\bar{p} + t^{-\sigma}\bar{q}.$$

Therefore we only need to find the **C\(0,∞)**-characteristic regions for Δ-polynomials of the form

$$\lambda - 1 + \lambda^{-\tau}p + \lambda^{-\sigma}q.$$

It is convenient to classify our quasi-polynomial into several different types neglecting the cases where $\tau = \sigma$ or $\tau\sigma = 0$.

6.2.2.1   *The Case $\tau, \sigma > 0$*

In this case, we may assume without loss of generality that $\tau > \sigma > 0$. Then to find the $\mathbf{C}\backslash(0, \infty)$-characteristic region of g, it suffices to consider the polynomial

$$f(\lambda|p, q) \equiv \lambda^{\tau+1} - \lambda^{\tau} + x + y\lambda^{\tau-\sigma} \tag{6.13}$$

where $\tau > \sigma \geq 1$.

Theorem 6.8. *Suppose $\tau, \sigma \in \mathbf{Z}[1, \infty)$ such that $\tau > \sigma \geq 1$. Then the curve defined by*

$$x(\lambda) = \frac{\sigma+1}{\tau-\sigma}\lambda^{\tau}\left(\lambda - \frac{\sigma}{\sigma+1}\right), \; y(\lambda) = -\frac{\tau+1}{\tau-\sigma}\lambda^{\sigma}\left(\lambda - \frac{\tau}{\tau+1}\right), \; \lambda \geq \frac{\sigma}{\sigma+1},$$

is also the graph of a strictly decreasing and smooth function $y = G(x)$ defined over $(0, \infty)$. Furthermore, the $\mathbf{C}\backslash(0, \infty)$-characteristic region Ω of $f(\lambda|x, y)$ in (6.13) is the set of points (x, y) that satisfies $x \geq 0$ and $y > G(x)$.

Proof. Consider the family $\{L_\lambda | \lambda \in (0, \infty)\}$ of straight lines defined by L_λ : $f(\lambda|x, y) = 0$ where $\lambda \in (0, \infty)$. Since

$$f'_\lambda(\lambda|x, y) = (\tau+1)\lambda^{\tau} - \tau\lambda^{\tau-1} + y(\tau-\sigma)\lambda^{\tau-\sigma-1},$$

the determinant of the linear system $f(\lambda|x, y) = 0 = f'_\lambda(\lambda|x, y)$ is $(\tau-\sigma)\lambda^{\tau-\sigma-1}$ which does not vanish for $\lambda > 0$. By Theorem 2.6, Ω is just the dual of the envelope G of the family $\{L_\lambda | \lambda \in (0, \infty)\}$. To find the envelope, we solve from $f(\lambda|x, y) = 0 = f'_\lambda(\lambda|x, y)$ for x and y to yield

$$x(\lambda) = \frac{\sigma+1}{\tau-\sigma}\lambda^{\tau}\left(\lambda - \frac{\sigma}{\sigma+1}\right), \; y(\lambda) = -\frac{\tau+1}{\tau-\sigma}\lambda^{\sigma}\left(\lambda - \frac{\tau}{\tau+1}\right),$$

where $\lambda > 0$. We have

$$(x(0^+), y(0^+)) = (0, 0), \; (x(+\infty), y(+\infty)) = (+\infty, -\infty),$$

$$x'(\lambda) = \frac{\sigma+1}{\tau-\sigma}\lambda^{\tau-1}\left((\tau+1)\lambda - \frac{\sigma\tau}{\sigma+1}\right),$$

and

$$y'(\lambda) = -\frac{\tau+1}{\tau-\sigma}\lambda^{\sigma-1}\left((\sigma+1)\lambda - \frac{\sigma\tau}{\tau+1}\right).$$

So $x'(\lambda)$ and $y'(\lambda)$ have the unique real root $\lambda^* = \tau\sigma/(\sigma+1)(\tau+1)$. Furthermore, $x(\lambda)$ is strictly decreasing and $y(\lambda)$ is strictly increasing on $(0, \lambda^*)$, and, $x(\lambda)$ is strictly increasing and $y(\lambda)$ is strictly decreasing on (λ^*, ∞). Thus

$$\frac{dy}{dx}(\lambda) = -\lambda^{-(\tau-\sigma)} < 0, \; \lambda \in (0, \lambda^*) \cup (\lambda^*, \infty),$$

and

$$\frac{d^2y}{dx^2}(\lambda) = \frac{(\tau-\sigma)^2\lambda^{\sigma-2\tau}}{(\sigma+1)(\tau+1)\lambda - \sigma\tau}, \; \lambda \in (0, \lambda^*) \cup (\lambda^*, \infty).$$

As in Example 2.2, the desired envelope is the curve described by $x(\lambda)$ and $y(\lambda)$, which is composed of two pieces G_1 and G_2 and one turning point $(x(\lambda^*), y(\lambda^*))$ (see Figure 6.3). The first piece G_1 corresponds to the case where $\lambda \in (0, \lambda^*)$ and the second G_2 to the case where $\lambda \in [\lambda^*, \infty)$. Furthermore, G_1 is the graph of a function $y = G_1(x)$ which is strictly decreasing, strictly concave, and smooth over $(x(\lambda^*), 0)$ such that $G_1(0^+) = 0$ and $G_1'(0^+) = -\infty$; and G_2 is the graph of a function $y = G_2(x)$ which is strictly decreasing, strictly convex, and smooth over $[x(\lambda^*), \infty)$ such that $G_2(+\infty) = -\infty$ and $G_2'(+\infty) = 0$. In view of the distribution map in Figure A.1, the dual set of order 0 of G is the set of points (x, y) that satisfies $x \geq 0$ and $y > G_2(x)$. The proof is complete.

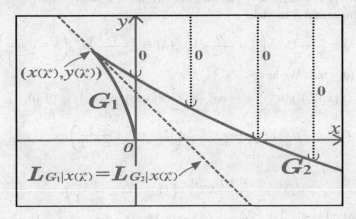

Fig. 6.3

6.2.2.2   *The Case* $\tau, \sigma < 0$

In this case, we will consider the equivalent polynomial

$$f(\lambda|p, q) = \lambda - 1 + \lambda^\tau p + \lambda^\sigma q \tag{6.14}$$

where now τ, σ are distinct nonzero integers. Since no additional restrictions are imposed on p or q, we may assume without loss of generality that $\tau > \sigma$. There are two cases to consider: (a) $\tau > \sigma > 1$ and (b) $\tau > \sigma = 1$. However, in the latter case,

$$f(\lambda|p, q) = \lambda(q + 1) - 1 + \lambda^\tau p$$

which is also a Δ-polynomial with one power. Therefore we only need to handle the first case.

Theorem 6.9. *Suppose* $\tau > \sigma > 1$. *Then the curve described by*

$$x(\lambda) = \frac{\sigma - 1}{\tau - \sigma}\lambda^{-\tau}\left(\lambda - \frac{\sigma}{\sigma - 1}\right), \quad y(\lambda) = -\frac{\tau - 1}{\tau - \sigma}\lambda^{-\sigma}\left(\lambda - \frac{\tau}{\tau - 1}\right),$$

over $(0, \sigma/(\sigma - 1)]$ is also the graph of a strictly decreasing function $y = G(x)$ defined over $(-\infty, 0)$. Furthermore, the $\mathbf{C}\backslash(0, \infty)$-characteristic region Ω of $f(\lambda|x, y)$ in (6.14) is the set of points (x, y) that satisfies $x \leq 0$ and $y < G(x)$.

Proof. Consider the family $\{L_\lambda | \lambda \in (0, \infty)\}$ of straight lines defined by L_λ : $f(\lambda|x, y) = 0$ where $\lambda \in (0, \infty)$. Since

$$f'_\lambda(\lambda|x, y) = 1 + \tau x \lambda^{\tau-1} + \sigma y \lambda^{\sigma-1},$$

the determinant of the linear system $f(\lambda|x, y) = 0 = f'_\lambda(\lambda|x, y)$ is $(\sigma - \tau)\lambda^{\tau+\sigma-1}$ which does not vanish for $\lambda > 0$. By Theorem 2.6, Ω is just the dual of the envelope of the family $\{L_\lambda | \lambda \in (0, \infty)\}$. To find the envelope, we solve from $f(\lambda|x, y) = 0 = f'_\lambda(\lambda|x, y)$ for x and y to yield

$$x(\lambda) = \frac{\sigma - 1}{\tau - \sigma} \lambda^{-\tau} \left(\lambda - \frac{\sigma}{\sigma - 1} \right), \quad y(\lambda) = -\frac{\tau - 1}{\tau - \sigma} \lambda^{-\sigma} \left(\lambda - \frac{\tau}{\tau - 1} \right)$$

where $\lambda \in (0, +\infty)$. We have

$$(x(0^+), y(0^+)) = (-\infty, +\infty), \quad (x(+\infty), y(+\infty)) = (0, 0),$$

$$x'(\lambda) = \frac{\sigma - 1}{\tau - \sigma} \lambda^{-\tau-1} \left((1 - \tau)\lambda + \frac{\sigma\tau}{\sigma - 1} \right),$$

and

$$y'(\lambda) = -\frac{\tau - 1}{\tau - \sigma} \lambda^{-\sigma-1} \left((1 - \sigma)\lambda + \frac{\sigma\tau}{\tau - 1} \right).$$

So $x'(\lambda)$ and $y'(\lambda)$ have the unique real root $\lambda^* = \tau\sigma/(\sigma - 1)(\tau - 1)$. Furthermore, $x(\lambda^*) > 0$, $y(\lambda^*) < 0$, $x(\lambda)$ is strictly increasing and $y(\lambda)$ is strictly decreasing on $(0, \lambda^*)$, and, $x(\lambda)$ is strictly decreasing and $y(\lambda)$ is strictly increasing on (λ^*, ∞). Thus

$$\frac{dy}{dx}(\lambda) = -\lambda^{\tau-\sigma} < 0, \quad \lambda \in (0, \lambda^*) \cup (\lambda^*, +\infty),$$

and

$$\frac{d^2y}{dx^2}(\lambda) = \frac{(\tau - \sigma)^2 \lambda^{2\tau-\sigma}}{(\sigma - 1)(\tau - 1)\lambda - \sigma\tau}, \quad \lambda \in (0, \lambda^*) \cup (\lambda^*, +\infty).$$

As in Example 2.2, the envelope is equal to the curve G described by $x(\lambda)$ and $y(\lambda)$ and is composed of two pieces G_1 and G_2 and one turning points $(x(\lambda^*), y(\lambda^*))$ (see Figure 6.4). The first piece G_1 corresponds to the case where $\lambda \in (0, \lambda^*)$ and the second G_2 to the case where $\lambda \in [\lambda^*, \infty)$. Furthermore, G_1 is the graph of a function $y = G_1(x)$ which is strictly decreasing, strictly concave, and smooth over $(-\infty, x(\lambda^*))$ such that $G_1(-\infty) = +\infty$ and $G'_1(-\infty) = 0$; and G_2 is the graph of a function $y = G_2(x)$ which is strictly decreasing, strictly convex, and smooth over $(0, x(\lambda^*)]$ such that $G_2(0^+) = 0$ and $G'_2(0^+) = -\infty$. In view of the distribution map in Figure A.1, the dual set of order 0 of G is the set of points (x, y) that satisfies $x \leq 0$ and $y < G_1(x)$. Note that $x(\lambda)$ has the unique root $\sigma/(\sigma - 1)$. Hence, we may restrict G_1 to the interval $(0, \sigma/(\sigma - 1)]$ and conclude our proof.

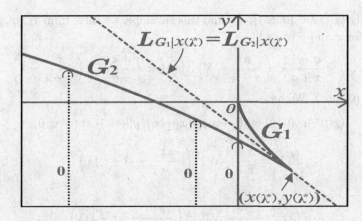

Fig. 6.4

6.2.2.3 *The Case* $\tau\sigma < 0$

In this case, we will consider the quasi-polynomial

$$g(\lambda) = \lambda - 1 + \lambda^{-\tau}p + \lambda^{\sigma}q$$

where $\tau, \sigma > 0$. In case $\sigma = 1$,

$$g(\lambda) = \lambda(q+1) - 1 + \lambda^{-\tau}p$$

which is a Δ-polynomial involving one power. Therefore we will consider the polynomial

$$f(\lambda|p,q) = \lambda^{\tau+1} - \lambda^{\tau} + p + q\lambda^{\tau+\sigma}, \tag{6.15}$$

where now $\tau \geq 1$ and $\sigma > 1$.

Theorem 6.10. *Suppose* $\tau \geq 1$ *and* $\sigma \geq 2$. *Then the curves* G_1 *and* G_2 *defined by*

$$x(\lambda) = \frac{\sigma - 1}{\tau + \sigma}\lambda^{\tau}\left(\frac{\sigma}{\sigma - 1} - \lambda\right), \; y(\lambda) = \frac{\tau + 1}{\tau + \sigma}\lambda^{-\sigma}\left(\frac{\tau}{\tau + 1} - \lambda\right)$$

over the intervals $(0, \tau/(\tau + 1))$ *and* $[\sigma/(\sigma - 1), \infty)$ *are also the graphs of strictly decreasing functions* $y = G_1(x)$ *and* $y = G_2(x)$ *defined over* $(0, \infty)$ *and* $(-\infty, 0]$ *respectively. Furthermore, the* C\(0, ∞)-*characteristic region* Ω *of* $f(\lambda|x, y)$ *in (6.15) is the set of points* (x, y) *that satisfies* $x \leq 0$ *and* $y < G_2(x)$, *or,* $x > 0$ *and* $y > G_1(x)$.

Proof. Consider the family $\{L_{\lambda}| \lambda \in (0, \infty)\}$ of straight lines defined by L_{λ} : $f(\lambda|x, y) = 0$ where $\lambda \in (0, \infty)$. Since

$$f'_{\lambda}(\lambda|x, y) = (\tau + 1)\lambda^{\tau} - \tau\lambda^{\tau-1} + (\tau + \sigma)\lambda^{\tau+\sigma-1}y,$$

the determinant of the linear system $f(\lambda|x, y) = 0 = f'_{\lambda}(\lambda|x, y)$ is $(\tau + \sigma)\lambda^{\tau+\sigma-1}$ which does not vanish for $\lambda > 0$. By Theorem 2.6, Ω is just the dual of the envelope of the family $\{L_\lambda | \lambda \in (0, \infty)\}$. To find this envelope, we solve from $f(\lambda | x, y) = 0 = f'_\lambda(\lambda | x, y)$ for x and y to yield

$$x(\lambda) = \frac{\sigma - 1}{\tau + \sigma} \lambda^\tau \left(\frac{\sigma}{\sigma - 1} - \lambda \right), \quad y(\lambda) = \frac{\tau + 1}{\tau + \sigma} \lambda^{-\sigma} \left(\frac{\tau}{\tau + 1} - \lambda \right)$$

where $\lambda \in (0, +\infty)$. We have

$$(x(0^+), y(0^+)) = (0, +\infty), \quad (x(+\infty), y(+\infty)) = (-\infty, 0),$$

$$x'(\lambda) = \frac{\sigma - 1}{\tau + \sigma} \lambda^{\tau - 1} \left(\frac{\sigma \tau}{\sigma - 1} - (1 + \tau)\lambda \right),$$

and

$$y'(\lambda) = -\frac{\tau + 1}{\tau + \sigma} \lambda^{-\sigma - 1} \left(\frac{\sigma \tau}{\tau + 1} - (\sigma - 1)\lambda \right).$$

So $x'(\lambda)$ and $y'(\lambda)$ have the unique positive real root $\lambda^* = \tau\sigma/(\sigma - 1)(\tau + 1)$. Furthermore, $x(\lambda^*) > 0$, $y(\lambda^*) < 0$, $x(\lambda)$ is strictly increasing and $y(\lambda)$ is strictly decreasing on $(0, \lambda^*)$, and, $x(\lambda)$ is strictly decreasing and $y(\lambda)$ is strictly increasing on (λ^*, ∞). Thus

$$\frac{dy}{dx}(\lambda) = -\lambda^{-\sigma - \tau} < 0, \quad \lambda \in (0, \lambda^*) \cup (\lambda^*, \infty).$$

and

$$\frac{d^2y}{dx^2}(\lambda) = \frac{(\sigma + \tau)^2 \lambda^{-\sigma - 2\tau}}{\sigma\tau - (\sigma - 1)(1 + \tau)\lambda}, \quad \lambda \in (0, \lambda^*) \cup (\lambda^*, \infty).$$

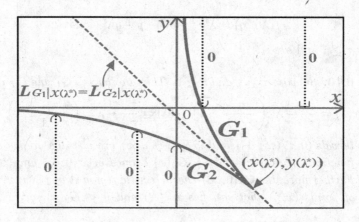

Fig. 6.5

As in Example 2.2, the desired envelope is the curve G described by $x(\lambda)$ and $y(\lambda)$, and is composed of two pieces G_1 and G_2 and one turning points $(x(\lambda^*), y(\lambda^*))$ (see Figure 6.5), The first piece G_1 corresponds to the case where $\lambda \in (0, \lambda^*)$ and the second G_2 to the case where $\lambda \in [\lambda^*, \infty)$. Furthermore, G_1 is the graph of a function $y = G_1(x)$ which is strictly decreasing, strictly convex, and smooth over

$(0, x(\lambda^*))$ such that $G_1(0^+) = +\infty$ and $G_1'(0^+) = -\infty$; and G_2 is the graph of a function $y = G_2(x)$ which is strictly decreasing, strictly concave, and smooth over $(-\infty, x(\lambda^*)]$ such that $G_2(-\infty) = 0$ and $G_2'(-\infty) = 0$. In view of the distribution map in Figure A.2, the dual of G is the set of points (x, y) that satisfies $x \le 0$ and $y < G_2(x)$, or, $x > 0$ and $y > G_1(x)$. The proof is complete.

6.2.3 $\Delta(1,1,0)$-*Polynomials I*

If we consider geometrical sequences of the form $\{\lambda^k\}$ as solutions of the neutral difference equation with two delays

$$\Delta(x_k + px_{k-\tau}) + qx_{k-\sigma} = 0, \ k \in \mathbf{N},$$

where $p, q \in \mathbf{R}$, τ and σ are *nonnegative integers*, then we are led to functions of the form

$$f(\lambda|p,q) = \lambda - 1 + \lambda^{-\tau}(\lambda - 1)p + \lambda^{-\sigma}q. \tag{6.16}$$

If $\tau\sigma = 0$, (6.16) is a Δ-polynomial with one power. More precisely, when $\tau = \sigma = 0$,

$$f(\lambda|p,q) = \lambda - 1 + (\lambda - 1)p + q. \tag{6.17}$$

When $\tau = 0$ but $\sigma \neq 0$,

$$f(\lambda|p,q) = \lambda - 1 + (\lambda - 1)p + \lambda^{-\sigma}q, \tag{6.18}$$

and when $\tau \neq 0$ but $\sigma = 0$,

$$f(\lambda|p,q) = \lambda - 1 + q + \lambda^{-\tau}(\lambda - 1)p. \tag{6.19}$$

Furthermore, if $\tau = \sigma$, then

$$f(\lambda|p,q) = \lambda - 1 + \lambda^{-\tau}\left\{(\lambda - 1)p + q\right\}. \tag{6.20}$$

The polynomial (6.17) is easy to handle. (6.18) is a $\Delta(1,0)$-polynomial, while (6.19) and (6.20) are $\Delta(1,1)$-polynomials. They have been handled in previous sections. In particular, as a corollary of Theorem 6.4, we have the following result: If $\tau \in \mathbf{Z}[1,\infty)$ and if the curve G is described by

$$x(\lambda) = (\tau+1)\lambda^{\tau-1}\left(\frac{\tau}{\tau+1} - \lambda\right), \ y(\lambda) = \tau(\lambda - 1)^2\lambda^{\tau-1}$$

for $\lambda > (\tau-1)/(\tau+1)$, then G is the graph of a strictly convex and smooth function $y = G(x)$ defined on $(-\infty, x((\tau - 1)/\tau + 1))$; furthermore, (p, q) is a point in the $\mathbf{C}\backslash(0,\infty)$-characteristic region of

$$f(\lambda|p,q) = \lambda^{\tau+1} - \lambda^\tau + (\lambda - 1)p + q,$$

if, and only if, $(p, q) \in \vee(G) \oplus \triangledown(\Upsilon)$.

We will therefore find the $\mathbf{C}\backslash(0,\infty)$-characteristic region of the Δ-polynomial (6.16) when τ and σ are *distinct positive integers*. For this purpose, we may consider the equivalent function

$$f(\lambda|p,q) \equiv \lambda^{\sigma+1} - \lambda^{\sigma} + \lambda^{\sigma-\tau}(\lambda-1)p + q. \tag{6.21}$$

The determinant of the linear system

$$f(\lambda|x,y) \equiv \lambda^{\sigma+1} - \lambda^{\sigma} + \lambda^{\sigma-\tau}(\lambda-1)x + y = 0, \ \lambda > 0,$$

$$f'_{\lambda}(\lambda|x,y) = (\sigma+1)\lambda^{\sigma} - \sigma\lambda^{\sigma-1} + ((\sigma-\tau+1)\lambda - (\sigma-\tau))\lambda^{\sigma-\tau-1}x = 0, \ \lambda > 0,$$

is

$$D(\lambda) = (\sigma - \tau + 1)\lambda - (\sigma - \tau), \tag{6.22}$$

which has the unique root

$$\bar{\lambda} = \frac{\sigma - \tau}{\sigma - \tau + 1}$$

if $\tau \neq \sigma + 1$.

When τ and σ are distinct positive integers, there are three possible cases: (1) $\tau = \sigma + 1$, (2) $\tau \neq \sigma + 1$ and $(\sigma - \tau)/(\sigma - \tau + 1) > 0$, and (3) $\tau \neq \sigma + 1$ and $\sigma - \tau)/(\sigma - \tau + 1) < 0$. The third case, however, does not arise since it is equivalent to $\sigma + 1 > \tau > \sigma$, which is not possible when τ is an integer.

6.2.3.1   *The Case $\tau = \sigma + 1$*

This case is relatively easy.

Theorem 6.11. *Suppose $\tau = \sigma + 1$. Let the curve G be defined by*

$$x(\lambda) = \sigma\lambda^{\sigma+1} - (\sigma+1)\lambda^{\sigma+2}, \ y(\lambda) = \lambda^{\sigma}\{(\sigma+1)\lambda^2 - 2(\sigma+1)\lambda + 1 + \sigma\}$$

for $\lambda > \sigma/(\sigma+2)$. Then G is also the graph of a strictly convex and smooth function $y = G(x)$ over $(-\infty, x(\sigma/(\sigma+2))$. Furthermore, (p,q) is a point in the $\mathbf{C}\backslash(0,\infty)$-characteristic region of (6.21) if, and only if, (p,q) satisfies $p \leq 0$ and $q > G(p)$.

Proof. In this case, since $D(\lambda) = \tau - \sigma = 1$, by Theorem 2.6, the $\mathbf{C}\backslash(0,\infty)$-characteristic region Ω for $f(\lambda|p,q)$ is just the dual of the envelope G of the family $\{L_{\lambda}| \lambda \in (0,\infty)\}$ of straight lines $L_{\lambda} : f(\lambda|x,y) = 0$. To find this envelope, we solve from $f(\lambda|x,y) = 0 = f'_{\lambda}(\lambda|x,y)$ for x and y to yield

$$x(\lambda) = \sigma\lambda^{\sigma+1} - (\sigma+1)\lambda^{\sigma+2}, \ y(\lambda) = \lambda^{\sigma}\{(\sigma+1)\lambda^2 - 2(\sigma+1)\lambda + 1 + \sigma\},$$

where $\lambda > 0$. We have

$$(x(0^+), y(0^+)) = (0,0), \ (x(+\infty), y(+\infty)) = (-\infty, +\infty),$$

$$x'(\lambda) = (\sigma+1)\lambda^{\sigma}(\sigma - (\sigma+2)\lambda),$$

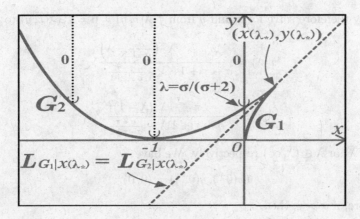

Fig. 6.6

and
$$y'(\lambda) = (\sigma + 1)(\lambda - 1)((\sigma + 2)\lambda - \sigma)\lambda^{\sigma-1}.$$
So $x'(\lambda)$ has the unique positive root
$$\lambda_* = \sigma/(\sigma + 2).$$
Furthermore, $x(\lambda)$ is strictly increasing on $(0, \lambda_*)$ and strictly decreasing on (λ_*, ∞).
Also,
$$\frac{dy}{dx} = \frac{1}{\lambda} - 1, \ \lambda \in (0, \lambda_*) \cup (\lambda_*, \infty),$$

$$\frac{d^2y}{dx^2} = \frac{1}{(\sigma + 1)\lambda^{\sigma+2}((\sigma + 2)\lambda - \sigma)}, \ \lambda \in (0, \lambda_*) \cup (\lambda_*, \infty).$$

As in Example 2.2, The envelope is just G which is composed of two pieces G_1 and G_2 as shown in Figure 6.6. The first piece G_1 corresponds to the case where $\lambda \in (0, \lambda^*)$ and the second G_2 to the case where $\lambda \in [\lambda^*, \infty)$. Furthermore, G_1 is the graph of a strictly increasing, strictly concave and smooth function $y = G_1(x)$ defined on $(0, x(\lambda^*))$ such that $G_1(0^+) = 0$ and $G_1'(0^+) = +\infty$; and G_2 is the graph of a function $y = G_2(x)$ which is strictly convex and smooth over $(-\infty, x(\lambda^*)]$ such that $G_2(-\infty) = +\infty$ and $G_2'(-\infty) = -\infty$. In view of the distribution map in Figure A.1, the dual of G is the set of points (p, q) that satisfies $p \leq 0$ and $q > G_2(p)$. Finally, since $x(\lambda)$ has the unique root $\sigma/(\sigma + 2)$ in $(0, \infty)$, we may restrict G_2 to the interval $(\sigma/(\sigma + 2), \infty)$ to conclude our proof.

6.2.3.2   *The Case* $(\sigma - \tau)/(\sigma - \tau + 1) > 0$ *and* $\sigma - \tau + 1 \neq 0$

The conditions $\tau \neq \sigma + 1$ and $(\sigma - \tau)/(\sigma - \tau + 1) > 0$ hold if, and only if, $\sigma > \tau$ or $\sigma + 1 < \tau$. Since the unique root $\overline{\lambda}$ of $D(\lambda)$ in (6.22) is $(\sigma - \tau)/(\sigma - \tau + 1)$, therefore when $\sigma > \tau$ or $\sigma + 1 < \tau$, we have
$$D(\lambda) \neq 0, \ \lambda \in (0, \overline{\lambda}) \cup (\overline{\lambda}, \infty).$$

We may therefore solve for x and y from $f(\lambda|x, y) = 0 = f'_\lambda(\lambda|x, y)$ to yield

$$x(\lambda) = \frac{\sigma + 1}{\sigma - \tau + 1} \frac{\lambda^\tau \left(\frac{\sigma}{\sigma+1} - \lambda\right)}{\lambda - \frac{\sigma - \tau}{\sigma - \tau + 1}}, \tag{6.23}$$

$$y(\lambda) = \frac{\tau}{\sigma - \tau + 1} \frac{\lambda^\sigma (\lambda - 1)^2}{\lambda - \frac{\sigma - \tau}{\sigma - \tau + 1}}, \tag{6.24}$$

for $\lambda \in (0, \overline{\lambda})$ or $\lambda \in (\overline{\lambda}, \infty)$ respectively. We have

$$(x(0^+), y(0^+)) = (0, 0).$$

Furthermore, if $\sigma > \tau$, then

$$(x(+\infty), y(+\infty)) = (-\infty, \infty),$$
$$(x(\overline{\lambda}^-), y(\overline{\lambda}^-)) = (-\infty, -\infty),$$
$$(x(\overline{\lambda}^+), y(\overline{\lambda}^+)) = (\infty, \infty),$$

and if $\tau > \sigma + 1$, then

$$(x(+\infty), y(+\infty)) = (\infty, -\infty),$$
$$(x(\overline{\lambda}^-), y(\overline{\lambda}^-)) = (-\infty, \infty),$$
$$(x(\overline{\lambda}^+), y(\overline{\lambda}^+)) = (\infty, -\infty).$$

Also,

$$x'(\lambda) = \frac{-\tau(\sigma + 1)}{\sigma - \tau + 1} \frac{\lambda^{\tau - 1}}{\left(\lambda - \frac{\sigma - \tau}{\sigma - \tau + 1}\right)^2} h(\lambda)$$

and

$$y'(\lambda) = \frac{\tau(\sigma + 1)}{\sigma - \tau + 1} \frac{(\lambda - 1)\lambda^{\sigma - 1}}{\left(\lambda - \frac{\sigma - \tau}{\sigma - \tau + 1}\right)^2} h(\lambda),$$

where

$$h(\lambda) = \lambda^2 - \frac{2\sigma^2 - 2\sigma\tau + 2\sigma - \tau - 1}{(\sigma + 1)(\sigma - \tau + 1)}\lambda + \frac{\sigma(\sigma - \tau)}{(\sigma + 1)(\sigma - \tau + 1)}. \tag{6.25}$$

The roots of $h(\lambda)$ are α and β given respectively by

$$\alpha = \frac{2\sigma^2 - 2\sigma\tau + 2\sigma - \tau - 1 - \sqrt{1 - 4\sigma + 2\tau + \tau^2 - 4\sigma^2 + 4\sigma\tau}}{2(\sigma + 1)(\sigma - \tau + 1)}$$

$$= \frac{\sigma}{\sigma + 1} - \frac{(\tau + 1) + \sqrt{(\tau + 1)^2 - 4\sigma(\sigma - \tau + 1)}}{2(\sigma + 1)(\sigma - \tau + 1)}$$

$$= \frac{\sigma - \tau}{\sigma - \tau + 1} + \frac{\tau - 1 - \sqrt{(\tau - 1)^2 - 4(\sigma - \tau)(\sigma + 1)}}{2(\sigma + 1)(\sigma - \tau + 1)} \tag{6.26}$$

and

$$\beta = \frac{2\sigma^2 - 2\sigma\tau + 2\sigma - \tau - 1 + \sqrt{1 - 4\sigma + 2\tau + \tau^2 - 4\sigma^2 + 4\sigma\tau}}{2(\sigma+1)(\sigma-\tau+1)}$$

$$= \frac{\sigma}{\sigma+1} - \frac{(\tau+1) - \sqrt{(\tau+1)^2 - 4\sigma(\sigma-\tau+1)}}{2(\sigma+1)(\sigma-\tau+1)}. \tag{6.27}$$

Let Φ be the set of positive roots of h. Since $\sigma - \tau + 1 \neq 0$, the set of positive roots of x' is also equal to Φ. Hence,

$$\frac{dy}{dx} = \lambda^{\sigma-\tau}(1-\lambda), \ \lambda \in (0,\infty)\backslash\Phi$$

and

$$\frac{d^2y}{dx^2} = \frac{(\sigma-\tau+1)^2}{\tau(\sigma+1)}\left(\lambda - \frac{\sigma-\tau}{\sigma-\tau+1}\right)^3 \lambda^{\sigma-2\tau}\frac{1}{h(\lambda)}, \ \lambda \in (0,\infty)\backslash\Phi.$$

When $\tau > \sigma + 1$ or $\sigma > \tau$, the roots of $h(\lambda)$ behave differently. Therefore we consider the following cases: (1) $\sigma \geq \rho(\tau)$ and $\sigma > 1$, (2) $8 \leq \tau < \sigma < \rho(\tau)$, (3) $\tau > \sigma + 1$, where

$$\rho(\tau) = \frac{\sqrt{2}\sqrt{\tau^2+1} + \tau - 1}{2}, \ \tau = 1, 2, \dots . \tag{6.28}$$

Note that $\rho(1) = 1$, $\rho(2) = 2.081...$, ..., $\rho(6) = 6.801...$, $\rho(7) = 8$, ... Furthermore, $\rho'(x) > 1$ for $x > 0$. Thus $\rho(\tau) = \tau$ for $\tau = 1$, $\rho(\tau) > \tau$ for $2 \leq \tau \leq 6$, $\rho(\tau) = \tau + 1$ for $\tau = 7$ and $\rho(\tau) > \tau + 1$ for $\tau \geq 8$. These show that the above conditions are indeed exhaustive.

In the following results, we will use Theorem 4.1 to approach our problem as follows. We need to find characteristic regions A such that f does not have any roots in $(0,\overline{\lambda})$, B such that f does not have any roots in $(\overline{\lambda},\infty)$, and the set D of (p,q) such that $f(\overline{\lambda}|p,q) \neq 0$, and then the desired dual set of order 0 of f is $A \cap B \cap D$.

Theorem 6.12. *Suppose $\sigma \geq \rho(\tau)$ and $\sigma > 1$. Let the curve G be defined by*

$$x(\lambda) = \frac{\sigma+1}{\sigma-\tau+1}\frac{\lambda^\tau\left(\frac{\sigma}{\sigma+1} - \lambda\right)}{\lambda - \frac{\sigma-\tau}{\sigma-\tau+1}}, y(\lambda) = \frac{\tau}{\sigma-\tau+1}\frac{\lambda^\sigma(\lambda-1)^2}{\lambda - \frac{\sigma-\tau}{\sigma-\tau+1}}$$

for $\lambda > (\sigma-\tau)/(\sigma-\tau+1)$. Then the $\mathbf{C}\backslash(0,\infty)$-characteristic region of (6.21) is $\vee(G)$.

Proof. The $\mathbf{C}\backslash(0,\infty)$-characteristic region of $f(\lambda|x,y)$ defined by (6.21) is the intersection of the $\mathbf{C}\backslash(0,\overline{\lambda})$-, $\mathbf{C}\backslash\{\overline{\lambda}\}$- and $\mathbf{C}\backslash(\overline{\lambda},\infty)$-characteristic regions of $f(\lambda|x,y)$. The $\mathbf{C}\backslash\{\overline{\lambda}\}$-characteristic region is just the set of points that do not lie on the straight line $L_{\overline{\lambda}}: f(\overline{\lambda}|x,y) = 0$, that is, $\mathbf{C}\backslash L_{\overline{\lambda}}$. As for the other two regions, the $\mathbf{C}\backslash(0,\overline{\lambda})$-characteristic region is just the dual set of order 0 of the envelope G_1 of the family $\Phi_{(0,\overline{\lambda})} = \{L_\lambda | \lambda \in (0,\overline{\lambda})\}$ and the $\mathbf{C}\backslash(\overline{\lambda},\infty)$-characteristic region is

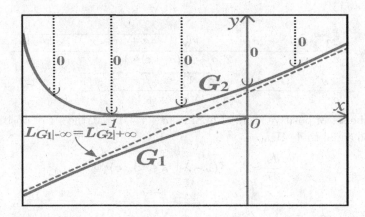

Fig. 6.7

the dual set of order 0 of the envelope G_2 of the family $\Phi_{(\overline{\lambda},\infty)} = \{L_\lambda|\ \lambda \in (\overline{\lambda},\infty)\}$, where L_λ is the straight line defined by $f(\lambda|x,y) = 0$ for $\lambda \in (0,\overline{\lambda}) \cup (\overline{\lambda},\infty)$.

When $\sigma \geq \rho(\tau)$ and $\sigma > 1$, we see that $\sigma > \tau$. Since the last term in (6.25) is positive, and since the discriminant

$$
\left\{\frac{2\sigma^2 - 2\sigma\tau + 2\sigma - \tau - 1}{(\sigma+1)(\sigma-\tau+1)}\right\}^2 - \frac{4\sigma(\sigma-\tau)}{(\sigma+1)(\sigma-\tau+1)}
$$
$$
= \frac{2(\tau^2+1) - (2\sigma-\tau+1)^2}{(\sigma+1)^2(\sigma-\tau+1)^2} \leq \frac{2(\tau^2+1) - 2(\tau^2+1)}{(\sigma+1)^2(\sigma-\tau+1)^2} = 0,
$$

(where we have used the fact that $\sigma \geq \rho(\tau)$ if, and only if, $2\sigma - \tau + 1 \geq \sqrt{2}\sqrt{\tau^2+1}$), by Theorems 6.1 and 5.2, we see that $h(\lambda)$ has at most one positive root. Thus the function h in (6.25) satisfies $h(\lambda) \geq 0$ for $\lambda > 0$.

As in Example 2.2, these and other easily obtained information show that the envelope G_1 is described by the parametric functions $x(\lambda)$ and $y(\lambda)$ over the interval $(0,\overline{\lambda})$, while the envelope G_2 by the same functions but over the interval $(\overline{\lambda},\infty)$. Furthermore, G_1 and G_2 can be depicted as shown in Figure 6.7. The curve G_1 is the graph of a strictly increasing, strictly concave and smooth function $y = G_1(x)$ over $(-\infty,0)$ such that $G_1(0^+) = 0$ and $G_1'(0^+) = 0$; and G_2 is the graph of a strictly convex and smooth function $y = G_2(x)$ defined on $\mathbf{R}$ such that $G_2'(-\infty) = -\infty$. Furthermore, by Lemma 3.2, the straight line $L_{\overline{\lambda}} : f(\lambda|x,y) = 0$ is the asymptote of the function G_1 at $-\infty$ and the function G_2 at $+\infty$.

In view of the distribution maps in Figures 3.1 and 3.19, the dual sets of order 0 of G_1 and G_2 can easily be obtained. The intersection of these dual sets with $\mathbf{C}\backslash L_{\overline{\lambda}}$ is also easily found (cf. Figure A.12) so that (p,q) is in the $\mathbf{C}\backslash(0,\infty)$-characteristic region for *(6.21)* if, and only if, (p,q) lies strictly above G_2. The proof is complete.

Next we suppose $8 \leq \tau < \sigma < \rho(\tau)$. In this case, we may show that the roots α and β in (6.26) and (6.27) satisfy

$$
0 < \frac{\sigma-\tau}{\sigma-\tau+1} < \alpha < \beta < \frac{\sigma}{\sigma+1} < 1. \tag{6.29}
$$

So $h(\lambda)$ has two distinct positive roots.

Theorem 6.13. *Suppose* $8 \leq \tau < \sigma < \rho(\tau)$. *Let* α, β *be defined by (6.27) and (6.27) and let the curve* G *be defined by*

$$x(\lambda) = \frac{\sigma+1}{\sigma-\tau+1} \frac{\lambda^{\tau}\left(\frac{\sigma}{\sigma+1} - \lambda\right)}{\lambda - \frac{\sigma-\tau}{\sigma-\tau+1}}, \quad y(\lambda) = \frac{\tau}{\sigma-\tau+1} \frac{\lambda^{\sigma}(\lambda-1)^{2}}{\lambda - \frac{\sigma-\tau}{\sigma-\tau+1}}$$

for $\lambda > (\sigma-\tau)/(\sigma-\tau+1)$. *Let* $G_2^{(1)}$ *be the restriction of* G *over* $((\sigma-\tau)/(\sigma-\tau+1), \alpha]$ *and* $G_2^{(3)}$ *be the restriction of* G *over* $[\beta, \infty)$. *Then the* $\mathbf{C}\backslash(0,\infty)$-*characteristic region of (6.21) is* $\vee\left(G_2^{(3)}\right) \oplus \vee\left(G_2^{(1)}\right)$.

Proof. The proof is similar to that of Theorem 6.12 and hence will be sketched. The parametric functions, the number $\overline{\lambda}$, the straight lines L_λ, the families $\Phi_{(0,\overline{\lambda})}$ and $\Phi_{(\overline{\lambda},\infty)}$, the regions $\mathbf{C}\backslash(0,\overline{\lambda})$-, $\mathbf{C}\backslash\{\overline{\lambda}\}$- and $\mathbf{C}\backslash(\overline{\lambda},\infty)$-characteristic regions and the envelopes G_1 and G_2 are the same as in the proof of Theorem 6.12.

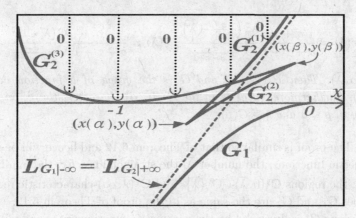

Fig. 6.8

The envelope G_1 is described by the parametric functions $x(\lambda)$ and $y(\lambda)$ defined by (6.23) and (6.24) over the interval $(0, \overline{\lambda})$, the envelope G_2 by the same functions but over the interval $(\overline{\lambda}, \infty)$. Since $h(\lambda)$ now has two distinct positive roots α and β that satisfy the relations in (6.29), we see that $h(\lambda) > 0$ for $\lambda \in (0, \alpha) \cup (\beta, \infty)$ and $h(\lambda) > 0$ for $\lambda \in (\alpha, \beta)$. Thus, $x(\lambda)$ is strictly increasing on (α, β) and strictly decreasing on $(0, \overline{\lambda}) \cup (\overline{\lambda}, \alpha) \cup (\beta, \infty)$. With these and other easily obtained information, the curves G_1 and G_2 can be depicted as shown in Figure 6.8. The curve G_1 is the graph of a function $y = G_1(x)$ over $(-\infty, 0)$ which is strictly increasing and strictly concave. The curve G_2 is made up of three pieces $G_2^{(1)}$, $G_2^{(2)}$ and $G_2^{(3)}$ which corresponds to the cases where $\lambda \in (\overline{\lambda}, \alpha]$, $\lambda \in (\alpha, \beta)$ and $\lambda \in [\beta, 1)$ respectively. The piece $G_2^{(1)}$ is the graph of a function $y = G_2^{(1)}(x)$ defined on $[x(\alpha), \infty)$ which is strictly increasing, and strictly convex. The piece $G_2^{(2)}$ is the graph of a function

$y = G_2^{(2)}(x)$ defined on $(x(\alpha), x(\beta))$ which is strictly increasing and strictly concave. The piece $G_2^{(3)}$ is the graph of a function $y = G_2^{(3)}(x)$ defined on $(-\infty, x(\beta)]$ which is strictly increasing and strictly convex and satisfies $G_2^{(3)'}(-\infty) = -\infty$. By Lemma 3.2, the straight line $L_{\overline{\lambda}}$ is the asymptote of the function $G_2^{(3)}$ at $x = -\infty$ and the function $G_2^{(1)}$ at $x = +\infty$. In view of the distribution maps in Figures A.16 and 3.16, the dual sets of order 0 of the envelopes G_1 and G_2 are easy to obtain. Furthermore, the intersection of these dual sets with $\mathbf{C}\backslash L_{\overline{\lambda}}$ (cf. Figure A.27) is given by $\vee\left(G_2^{(3)}\right) \oplus \vee\left(G_2^{(1)}\right)$. The proof is complete.

Next we suppose $\tau > \sigma + 1$. In this case, we may show that the roots α, β defined by (6.26) and (6.27) satisfy

$$0 < \alpha < \frac{\sigma}{\sigma+1} < \frac{\sigma-\tau}{\sigma-\tau+1} < \beta. \tag{6.30}$$

Theorem 6.14. *Suppose $\tau > \sigma + 1$. Let α be defined by (6.27) and the curve G defined by*

$$x(\lambda) = \frac{\sigma+1}{\sigma-\tau+1} \frac{\lambda^\tau\left(\frac{\sigma}{\sigma+1} - \lambda\right)}{\lambda - \frac{\sigma-\tau}{\sigma-\tau+1}}, \quad y(\lambda) = \frac{\tau}{\sigma-\tau+1}\frac{\lambda^\sigma(\lambda-1)^2}{\lambda - \frac{\sigma-\tau}{\sigma-\tau+1}}$$

for $\lambda \in (\alpha, \overline{\lambda})$. Then $x(\alpha) > 0$ and G is the graph of a function defined on $(-\infty, x(\alpha))$. Furthermore, (p, q) is in the $\mathbf{C}\backslash(0, \infty)$-characteristic region of (6.21) if, and only if, $p \leq 0$ and $q > G(p)$.

Proof. The proof is similar to that of Theorem 6.12 and hence will be sketched. The parametric functions, the number $\overline{\lambda}$, the straight lines L_λ, the families $\Phi_{(0,\overline{\lambda})}$ and $\Phi_{(\overline{\lambda},\infty)}$, the regions $\mathbf{C}\backslash(0, \overline{\lambda})$-, $\mathbf{C}\backslash\{\overline{\lambda}\}$- and $\mathbf{C}\backslash(\overline{\lambda}, \infty)$-characteristic regions and the envelopes G_1 and G_2 are the same as in the proof of Theorem 6.12.

The envelope G_1 is described by the parametric functions $x(\lambda)$ and $y(\lambda)$ defined by (6.23) and (6.24) over the interval $(0, \overline{\lambda})$, the envelope G_2 by the same functions but over the interval $(\overline{\lambda}, \infty)$. Since $h(\lambda)$ now has two distinct positive roots α and β that satisfy the relations in (6.30), we see that $h(\lambda) > 0$ for $\lambda \in (0, \alpha) \cup (\beta, \infty)$ and $h(\lambda) < 0$ for $\lambda \in (\alpha, \beta)$. Thus, $x(\lambda)$ is strictly increasing on $(0, \alpha) \cup (\beta, \infty)$ and strictly decreasing on $(\alpha, \overline{\lambda}) \cup (\overline{\lambda}, \beta)$. With these and other easily obtained information, the curves G_1 and G_2 can be depicted as shown in Figure 6.9. The curve G_1 is made up by two pieces $G_1^{(1)}$ and $G_1^{(2)}$, while G_2 by two pieces $G_2^{(1)}$ and $G_2^{(2)}$. The curve $G_1^{(1)}$ corresponds to where $\lambda \in (0, \alpha)$ and $G_1^{(2)}$ to where $\lambda \in [\alpha, \overline{\lambda})$. The curve $G_2^{(1)}$ corresponds to where $\lambda \in (\overline{\lambda}, \beta)$ and $G_2^{(2)}$ to where $\lambda \in [\beta, \infty)$. The curve $G_1^{(1)}$ is the graph of a function $y = G_1^{(1)}(x)$ defined on $(0, x(\alpha))$ which is strictly increasing and strictly concave and $G_1^{(1)'}(0^+) = +\infty$. The curve $G_1^{(2)}$ is the graph of a function $y = G_1^{(2)}(x)$ defined on $(-\infty, x(\alpha)]$ which is strictly convex and touches the x-axis at $x = -1$. The curve $G_2^{(1)}$ is the graph of a function

$y = G_2^{(1)}(x)$ defined on $(x(\beta), \infty)$ which is strictly decreasing and strictly concave. Finally, the curve $G_2^{(2)}$ is the graph of a function $y = G_2^{(2)}(x)$ defined on $[x(\beta), \infty)$ which is strictly decreasing and strictly convex. The curves $G_2^{(1)}$ and $G_2^{(2)}$ have a common tangent line T that passes through the point $(x(\beta), y(\beta))$. We may check that $G_2^{(2)}(-\infty) = -\infty$. Furthermore, by Lemma 3.2, the straight line $L_{\overline{\lambda}}$ is the asymptote of the function $G_1^{(2)}$ at $-\infty$ and the function $G_2^{(1)}$ at $+\infty$. Since $G_2^{(1)}$ is strictly concave and $G_2^{(2)}$ is strictly convex, it is easily seen that the tangent line T has a slope which is greater than that of the asymptote $L_{\overline{\lambda}}$.

The dual sets of order 0 of G_1 and G_2, in view of the distribution maps in Figures A.6 and A.8, are easy to obtain. The intersection of these two dual sets with $\mathbf{C} \backslash L_{\overline{\lambda}}$ (cf. Figure A.26) is then equal to the set of points that lie strictly above the function $G_1^{(2)}$ restricted to the interval $(-\infty, 0]$. The proof is complete.

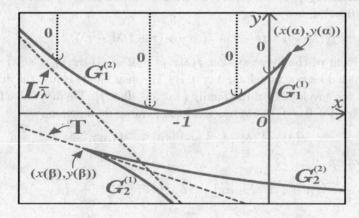

Fig. 6.9

6.2.4 $\Delta(1, 1, 0)$-*Polynomials II*

If we consider the neutral difference equation

$$\Delta(x_k + cx_{k-\tau}) + px_{k-\tau} + qx_{k+\delta} = 0, \ k \in \mathbf{N},$$

where $c, p, q \in \mathbf{R}$ and τ, δ are nonnegative integers, then we are led to Δ-polynomials of the form

$$\lambda - 1 + \lambda^{-\tau}(\lambda - 1)c + \lambda^{-\tau}p + \lambda^{\delta}q$$

or

$$\lambda - 1 + \lambda^{-\tau}(c\lambda - c + p) + \lambda^{\delta}q. \tag{6.31}$$

If $\delta = 1$, then (6.31) can be written as

$$\lambda - 1 + q + \lambda^{-\tau}(c\lambda + p - c),$$

which is a $\Delta(1,1)$-polynomial. Such a function has already been handled in Theorem 6.4.

If $\delta = -\sigma < 0$, then the resulting function is just the one discussed in the previous Section. Therefore, in this section, we consider the quasi-polynomial

$$f(\lambda|p,q) = (\lambda - 1)\lambda^\tau + c(\lambda - 1) + p + q\lambda^{\delta+\tau}, \qquad (6.32)$$

where $\tau, \delta \in \mathbf{Z}[1,\infty)$. We will be interested in determining the values of c, p, q, τ and δ for which this polynomial does not have any positive roots.

It will also be convenient to distinguish two cases: (1) $\delta > 1$ and $\tau > 1$, (2) $\delta > 1$ and $\tau = 1$.

6.2.4.1   *The Case $\delta, \tau > 1$*

Consider the family $\{L_\lambda | \lambda \in (0, \infty)\}$ of straight lines defined by $L_\lambda : f(\lambda|x,y) = 0$ where $\lambda \in (0, \infty)$. Since

$$f'_\lambda(\lambda|x,y) = (\delta+\tau)\lambda^{\delta+\tau-1}y + (\tau+1)\lambda^\tau - \tau\lambda^{\tau-1} + c,$$

the determinant of the linear system $f(\lambda|x,y) = 0 = f'_\lambda(\lambda|x,y)$ is $D(\lambda) = -(\delta + \tau)^{\delta+\tau-1}$ which does not vanish for $\lambda > 0$. By Theorem 2.6, Ω is just the dual of the envelope of the envelope G of the family $\{L_\lambda | \lambda \in (0, \infty)\}$. We may solve for x and y from $f(\lambda|x,y) = 0 = f'_\lambda(\lambda|x,y)$ to yield the parametric functions

$$x(\lambda) = \Gamma(\lambda; -\delta, \tau) \text{ and } y(\lambda) = -\Gamma(\lambda; \tau, -\delta) \qquad (6.33)$$

for $\lambda > 0$, where

$$\Gamma(x; a, b) = \frac{1}{b-a}\left[(a+1)x^{b+1} - ax^b + c(a+1-\tau)x^{b+1-\tau} - c(a-\tau)x^{b-\tau}\right], \quad \lambda > 0. \qquad (6.34)$$

Therefore,

$$x'(\lambda) = -(\delta+\tau)^{-1}F(\lambda)$$

and

$$y'(\lambda) = (\delta+\tau)^{-1}\lambda^{-(\tau+\delta)}F(\lambda)$$

where

$$F(\lambda) = (\tau+1)(\delta-1)\lambda^\tau - \tau\delta\lambda^{\tau-1} + c(\delta+\tau-1).$$

For λ such that $F(\lambda) \neq 0$, we have

$$\frac{dy}{dx}(\lambda) = -\lambda^{-(\delta+\tau)},$$

and

$$\frac{d^2y}{dx^2}(\lambda) = \frac{-(\delta+\tau)^2\lambda^{-(\delta+\tau+1)}}{(\tau+1)(\delta-1)\lambda^\tau - \tau\delta\lambda^{\tau-1} + c(\delta+\tau-1)}$$
$$= \frac{-(\delta+\tau)^2\lambda^{-(\delta+\tau+1)}}{F(\lambda)}.$$

Also

$$\lim_{\lambda \to 0^+} \frac{dy}{dx}(\lambda) = -\infty \quad \text{and} \quad \lim_{\lambda \to \infty} \frac{dy}{dx}(\lambda) = 0. \tag{6.35}$$

Then six mutually exclusive and exhaustive cases, characterized by the sizes of c, are needed:

$$c \geq \frac{\delta}{\delta + \tau - 1} \left[\frac{(\tau - 1)\delta}{(\tau + 1)(\delta - 1)} \right]^{\tau - 1}. \tag{6.36}$$

$$\left(\frac{\tau - 1}{\tau + 1} \right)^{\tau - 1} \leq c < \frac{\delta}{\delta + \tau - 1} \left[\frac{(\tau - 1)\delta}{(\tau + 1)(\delta - 1)} \right]^{\tau - 1}, \tag{6.37}$$

$$\frac{(\tau - 1)^{\tau - 1}}{\tau^\tau} < c < \left(\frac{\tau - 1}{\tau + 1} \right)^{\tau - 1} \tag{6.38}$$

$$0 < c \leq \frac{(\tau - 1)^{\tau - 1}}{\tau^\tau}, \tag{6.39}$$

$$c = 0,$$

and

$$c < 0. \tag{6.40}$$

We remark that the above classifications are valid in view of the inequalities

$$0 < \frac{(\tau - 1)^\tau}{\tau^\tau} < \left(\frac{\tau - 1}{\tau + 1} \right)^{\tau - 1} < \frac{\delta}{\delta + \tau - 1} \left[\frac{(\tau - 1)\delta}{(\tau + 1)(\delta - 1)} \right]^{\tau - 1}, \tag{6.41}$$

which can be verified by elementary means. We remark further that when $c = 0$,

$$f(\lambda | x, y) = \lambda^\tau (\lambda - 1) + x + y\lambda^{\tau + \delta} = \lambda^\tau \left\{ (\lambda - 1) + \lambda^{-\tau} x + \lambda^\delta y \right\}$$

and hence is equivalent to a $\Delta(1, 0, 0)$-polynomial with one power studied before, and hence this case will not be discussed. To facilitate discussions, we will set

$$\rho = \frac{(\tau - 1)\delta}{(\tau + 1)(\delta - 1)}, \tag{6.42}$$

which is positive in view of the assumption $\tau > 1$. Note that $F(0^+) = c(\delta + \tau - 1)$, $F(+\infty) = +\infty$, and

$$F'(\lambda) = \tau(\tau + 1)(\delta - 1)\lambda^{\tau - 2}(\lambda - \rho).$$

If (6.36) holds, then $F(0^+) > 0$, $F'(\lambda) < 0$ for $\lambda \in (0, \rho)$, $F'(\lambda) > 0$ for $\lambda > \rho$, and

$$F(\rho) = (\delta + \tau - 1) \left[c - \frac{\delta \rho^{\tau - 1}}{\delta + \tau - 1} \right] \geq 0.$$

Thus $F(\lambda) > 0$ for $\lambda > 0$ except at $\lambda = \rho$.

If (6.37) or (6.38) or (6.39) holds, then $F(0) > 0$, $F'(\lambda) < 0$ for $\lambda \in (0, \rho)$, $F'(x) > 0$ for $\lambda > \rho$,

$$F\left(\frac{\delta}{\delta - 1}\right) = \frac{\delta^\tau}{(\delta - 1)^{\tau-1}} + c(\delta + \tau - 1) > 0,$$

and $F(\rho) < 0$. Thus there exist exactly two roots λ_* and λ^* of $F(\lambda)$ in $(0, \infty)$. Furthermore, we have $0 < \lambda_* < \rho < \lambda^* < \delta/(\delta - 1)$.

If (6.40) holds, then we see that $F(0) < 0, F'(\lambda) < 0$ for $\lambda \in (0, \rho)$, $F'(\lambda) > 0$ for $\lambda > \rho$. Note that

$$F(\rho) = -\delta\rho^{\tau-1} + c(\delta + \tau - 1) < 0,$$

thus there exists a unique root $\bar{\lambda}$ of $F(\lambda)$ in (ρ, ∞) and $F(\lambda) < 0$ for $0 < \lambda < \bar{\lambda}$ as well as $F(\lambda) > 0$ for $\lambda > \bar{\lambda}$.

Theorem 6.15. *Suppose $\delta, \tau > 1$ and (6.36) holds. Let the curve G be described by the parametric functions in (6.33) for $\lambda > 0$. Then G is also the graph of a function $y = G(x)$ over the interval $(-\infty, c)$. Furthermore, (p, q) is in the $\mathbf{C}\backslash(0, \infty)$-characteristic region of (6.32) if, and only if, $p < c$ and $q < G(p)$, or $p \geq c$ and $q \geq 0$.*

Proof. By (6.33), we have

$$(x(0^+), y(0^+)) = (c, -\infty), \ (x(+\infty), y(+\infty)) = (-\infty, 0).$$

Furthermore, $dy/dx < 0$ and $d^2y/dx^2 < 0$ except possibly at $\lambda = \rho$. As in Example 2.2, these together with other easily obtained information allow us to depict the envelope G of the family $\{L_\lambda | \lambda \in (0, \infty)\}$ in Figure 6.10. The function $y = G(x)$ is a strictly decreasing, strictly concave and smooth over the interval $(-\infty, c)$ such that $G(-\infty) = 0$, $G'(-\infty) = 0$, $G(c^+) = -\infty$ and $G'(c^+) = -\infty$. Hence the asymptote $L_{G|-\infty}$ is Θ_0. In view of the distribution map in Figure 3.13, the dual set of order 0 of G is the set of points (p, q) that satisfies $p < c$ and $q < G(p)$, or $p \geq c$ and $q \geq 0$.

Theorem 6.16. *Suppose $\delta, \tau > 1$ and (6.40) holds. Let $\bar{\lambda}$ be the unique positive root of the function $F(x)$ in (ρ, ∞) where ρ is defined by (6.42). Let the curve G be defined by (6.33) for $\lambda > 0$. When restricted to the interval $(0, \bar{\lambda})$, G is also the graph of a strictly decreasing, strictly convex and smooth function $y = G_1(x)$ defined on $(c, x(\bar{\lambda}))$; and when restricted to $[\bar{\lambda}, \infty)$, G is also the graph of a strictly decreasing, strictly concave and smooth function $y = G_2(x)$ over $(-\infty, x(\bar{\lambda})]$. Furthermore, (p, q) is in the $\mathbf{C}\backslash(0, \infty)$-characteristic region of (6.32) if, and only if, $(p, q) \in \wedge(G_2\chi_{(-\infty, c]})$ or $(p, q) \in \vee(G_1) \oplus \overline{\vee}(\Theta_0\chi_{(c, +\infty)})$.*

Proof. By (6.33), we have

$$(x(0^+), y(0^+)) = (c, +\infty), \ (x(+\infty), y(+\infty)) = (-\infty, 0).$$

There exists a unique root $\bar{\lambda}$ of $F(\lambda)$ in (ρ, ∞) and $F(\lambda) < 0$ for $0 < \lambda < \bar{\lambda}$ as well as $F(\lambda) > 0$ for $\lambda > \bar{\lambda}$. As in Example 2.2, these together with other easily

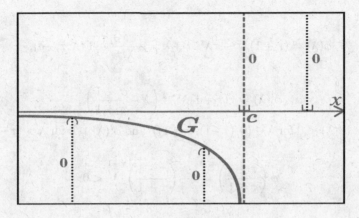

Fig. 6.10

obtained information allow us to depict the graph of the envelope G of the family $\{L_\lambda | \lambda \in (0, \infty)\}$ in Figure 6.11. The envelope G is composed of two curves G_1 and G_2 with turning point $(p_0, q_0) = (x(\bar{\lambda}), y(\bar{\lambda}))$. The first piece G_1 corresponds to the case where $\lambda \in (0, \bar{\lambda})$ and the second piece G_2 to the case where $\lambda \in [\bar{\lambda}, \infty)$. Furthermore, the function $y = G_1(x)$ is strictly decreasing, strictly convex and smooth for $x \in (c, p_0)$ such that $G_1(c^+) = +\infty$ and $G_1'(c^+) = -\infty$; and the function $y = G_2(x)$ is strictly decreasing, strictly concave and smooth for $x \in (-\infty, p_0)$ such that $G_2(-\infty) = 0$ and $G_2'(-\infty) = 0$. In view of the distribution map in Figure A.2, (p, q) is a dual point of order 0 of G if, and only, if $p \leq c$ and $q < G_2(p)$, or, $p > c$ and $(p, q) \in \vee(G_1) \oplus \triangledown(\Theta_0)$.

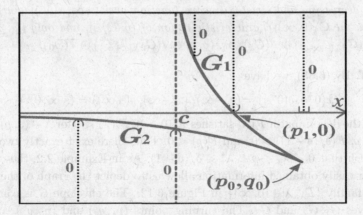

Fig. 6.11

We remark that since G_2 is strictly decreasing on $(-\infty, p_0)$ and $G_2(-\infty) = 0$, we see that $q_0 < 0$. Therefore the curve G_1 intercepts with the x-axis at some point p_1. To locate this point, note that $(p_1, 0) = (x(T), y(T))$ for some unique $T \in (0, \bar{\lambda})$.

Indeed, let

$$\phi(\lambda) = (\tau+1)\lambda^\tau - \tau\lambda^{\tau-1} + c = -\frac{\delta+\tau}{\lambda^{\delta+\tau-1}}\Gamma(\lambda;\tau,-\delta).$$

Since

$$\phi'(\lambda) = \tau(\tau+1)\lambda^{\tau-2}\left(\lambda - \frac{\tau-1}{\tau+1}\right),$$

we see that $\phi'(\lambda) < 0$ for $\lambda \in (0,(\tau-1)/(\tau+1))$, and $\phi'(\lambda) > 0$ for $\lambda > (\tau-1)/\tau+1)$. Note that

$$\phi\left(\frac{\tau-1}{\tau+1}\right) = c - \left(\frac{\tau-1}{\tau+1}\right)^{\tau-1} < 0$$

and

$$\phi\left(\overline{\lambda}\right) = \frac{\tau}{\delta-1}\left(\overline{\lambda}^{\tau-1} - c\right) > 0.$$

Thus there is a unique root T of $\phi(\lambda)$, and hence of $y(\lambda)$, in $((\tau-1)/(\tau+1),\overline{\lambda})$. We may now solve the unique root T of the equation $y(\lambda) = 0$ for $\lambda \in (0,\overline{\lambda})$ and then $p_1 = x(T)$. Therefore we may be more precise in expressing the previous conclusion, namely, $p \le c$ and $q < G_2(p)$, or, $p \in (c,p_1]$ and $q > G_1(p)$, or, $p > p_1$ and $q \ge 0$.

Theorem 6.17. *Suppose $\delta,\tau > 1$ and (6.39) holds. Let λ_* and λ^*, $0 < \lambda_* < \rho < \lambda^* < \delta/(\delta-1)$, be the two roots of $F(\lambda)$ in $(0,\infty)$. Let the curve G be defined by (6.33) for $\lambda > 0$. Let G_1, G_2 and G_3 be the restrictions of G over the intervals $(0,\lambda_*]$, (λ_*,λ^*) and $[\lambda^*,+\infty)$ respectively. Then G_1 is the graph of a strictly decreasing, strictly concave function $y = G_1(x)$ over $[x(\lambda_*),c)$, G_2 is the graph of a strictly decreasing and strictly convex function over $(x(\lambda_*),x(\lambda^*))$, and G_3 is the graph of a strictly decreasing and strictly concave function over $(-\infty,x(\lambda^*)]$. Furthermore, (p,q) is in the $\mathbf{C}\backslash(0,\infty)$-characteristic region of (6.32) if, and only if, $p < c$ and $(p,q) \in \wedge(G_3\chi_{(-\infty,c)}) \oplus \wedge(G_1)$, or, $(p,q) \in \vee(G_2\chi_{[c,+\infty)}) \oplus \overline{\vee}(\Theta_0)$.*

Proof. By (6.33), we have

$$(x(0^+),y(0^+)) = (c,-\infty), \quad (x(+\infty),y(+\infty)) = (-\infty,0).$$

However, the denominator $F(\lambda)$ satisfies $F(0) > 0$, $F'(\lambda) < 0$ for $\lambda \in (0,\rho)$, $F'(x) > 0$ for $\lambda > \rho$, $F(\delta/(\delta-1)) > 0$, and $F(\rho) < 0$. Thus there exist exactly two roots λ_* and λ^* such that $0 < \lambda_* < \rho < \lambda^* < \delta/(\delta-1)$. As in Example 2.2, these together with other easily obtained information allow us to depict the graph of the envelope G of the family $\{L_\lambda | \lambda \in (0,\infty)\}$ in Figure 6.12. The envelope G is composed of three curves G_1, G_2 and G_3. The turning points (p_*,q_*) and (p^*,q^*) correspond to λ_* and λ^* respectively, and are given by $p_* = x(\lambda_*)$, $q_* = y(\lambda_*)$, $p^* = x(\lambda^*)$ and $q^* = y(\lambda^*)$. The first piece G_1 corresponds to the case where $\lambda \in (0,\lambda_*]$, the second piece G_2 to the case where $\lambda \in (\lambda_*,\lambda^*)$ and the third piece G_3 to the case $[\lambda^*,\infty)$. Furthermore, the function $y = G_1(x)$ is strictly decreasing, strictly concave and smooth for $x \in [p_*,c)$ such that $G_1(c^-) = -\infty$ and $G_1'(c^-) = -\infty$; the function $y = G_2(x)$ is strictly decreasing, strictly convex and smooth $x \in (p_*, p^*)$, while the function $y = G_3(x)$ is strictly decreasing, strictly concave and smooth for $x \in (-\infty, p^*]$ such that $G_3(-\infty) = 0$ and $G'_3(-\infty) = 0$. In view of the distribution map in Figure A.17, (p, q) is a dual point of order 0 of G if, and only if, $p < c$ and $(p, q) \in \wedge(G_3) \oplus \wedge(G_1)$, or, $(p, q) \in \vee(G_2\chi_{[c, +\infty)}) \oplus \triangledown(\Theta_0)$. The proof is complete.

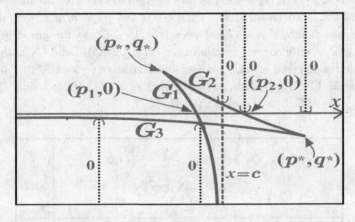

Fig. 6.12

We remark that the curve G_1 intersects the x-axis at some unique point $(p_1, 0)$, while G_2 at some unique point $(p_2, 0)$. To locate them, we first show that $y(\lambda) = 0$ has a unique root T_1 in $(0, (\tau - 1)/(\tau + 1))$ and a unique root T_2 in $((\tau - 1)/\tau, \infty)$ so that $p_1 = x(T_1)$ and $p_2 = x(T_2)$. Indeed, let

$$\phi(\lambda) \equiv (\tau + 1)\lambda^\tau - \tau\lambda^{\tau-1} + c = -\frac{\delta + \tau}{\lambda^{\delta + \tau - 1}}y(\lambda). \tag{6.43}$$

Since $\phi(0) = c > 0$, $\phi(+\infty) = \infty$,

$$\phi'(\lambda) \equiv \tau(\tau + 1)\lambda^{\tau-2}\left(\lambda - \frac{\tau - 1}{\tau + 1}\right),$$

we see that $\phi'(\lambda) < 0$ for $\lambda \in (0, (\tau-1)/(\tau+1))$, $\phi'(\lambda) > 0$ for $\lambda > (\tau-1)/(\tau+1)$,

$$\phi\left(\frac{\tau - 1}{\tau + 1}\right) = c - \left(\frac{\tau - 1}{\tau + 1}\right)^{\tau-1} < 0, \tag{6.44}$$

and

$$\phi\left(\frac{\tau - 1}{\tau}\right) = c - \frac{(\tau - 1)^{\tau-1}}{\tau^\tau} \leq 0. \tag{6.45}$$

Thus there exist exactly two roots T_1, T_2 of $\phi(\lambda)$, and hence of $y(\lambda)$, in $(0, (\tau - 1)/(\tau + 1))$ and $[(\tau - 1)/\tau, \infty)$ respectively. Finally, note that

$$x(T_1) = \Gamma(T_1; -\delta, \tau) < \lim_{\lambda \to 0+} \Gamma(\lambda; -\delta, \tau) = c$$

$$\leq c + \frac{\tau}{\tau + 1}T_2\left(\frac{1}{\tau}T_2^{\tau-1} - c\right) = \Gamma(T_2; -\delta, \tau) = x(T_2),$$

we see further that $p_1 < c \le p_2$. Therefore, if we desire, we may be more precise in expressing the conclusion of the above result.

Theorem 6.18. *Suppose $\delta, \tau > 1$ and (6.38) holds. Let λ_* and λ^*, $0 < \lambda_* < \rho < \lambda^* < \delta/(\delta - 1)$, be the two roots of $F(\lambda)$ in $(0, \infty)$. Let the curve G be defined by (6.33) for $\lambda > 0$. Let G_1, G_2 and G_3 be the restrictions of G over the intervals $(0, \lambda_*]$, (λ_*, λ^*) and $[\lambda^*, +\infty)$ respectively. Then G_1 is the graph of a strictly decreasing, strictly concave function $y = G_1(x)$ over $[x(\lambda_*), c)$, G_2 is the graph of a strictly decreasing and strictly convex function over $(x(\lambda_*), x(\lambda^*))$, and G_3 is the graph of a strictly decreasing and strictly concave function over $(-\infty, x(\lambda^*)]$. Furthermore, (p, q) is in the $\mathbf{C}\backslash(0, \infty)$-characteristic region of (6.32) if, and only if, $(\bar{p}, q) \in \wedge(G_3\chi_{(-\infty, c)}) \oplus \wedge(G_1)$, or, $(p, q) \in \nabla(\Theta_0\chi_{[c, +\infty)})$.*

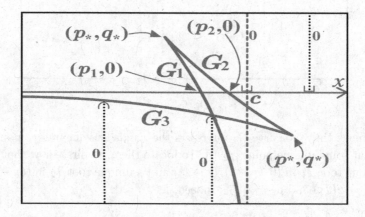

Fig. 6.13

The first two paragraphs of the proof of this result are exactly the same as that of Theorem 6.17, except that the corresponding envelope now appears as that in Figure 6.13. Indeed, we assert that the curve G_1 intersects the x-axis at some unique point $(p_1, 0)$, while G_2 at some unique point $(p_2, 0)$, and $p_2 < c$. To see this, we note that the function ϕ defined by (6.43) satisfies the same properties as in the proof of Theorem 6.17, except that (6.45) is now replaced with

$$\phi\left(\frac{\tau - 1}{\tau}\right) = c - \frac{(\tau - 1)^{\tau - 1}}{\tau^\tau} > 0. \tag{6.46}$$

Hence there exist exactly two roots T_1, T_2 of $y(\lambda)$ in $(0, (\tau - 1)/(\tau + 1))$ and $((\tau - 1)/(\tau + 1), (\tau - 1)/\tau)$ respectively. Since $T_2 < (\tau - 1)/\tau$, we see further that

$$p_2 = x(T_2) = \Gamma(T_2; -\delta, \tau) = c + \frac{\tau}{\tau + 1}T_2\left(\frac{1}{\tau}T_2^{\tau - 1} - c\right) < c.$$

In view of Figure A.17, the dual set of order 0 of G is the set of points (p, q) that satisfies $p \le x(\lambda_*)$ and $q < G_3(p)$, or, $x(\lambda_*) < p < c$ and $q < \min\{G_3(p), G_1(p)\}$, or, $p \ge c$ and $q \ge 0$.

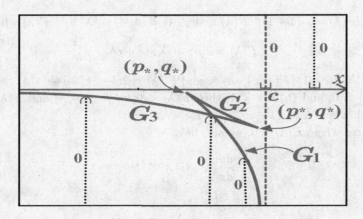

Fig. 6.14

Theorem 6.19. *Suppose* $\delta, \tau > 1$ *and (6.37) holds. Let* λ_* *and* λ^*, $0 < \lambda_* < \rho < \lambda^* < \delta/(\delta - 1)$, *be the two roots of* $F(\lambda)$ *in* $(0, \infty)$. *Let the curve* G *be defined by (6.33) for* $\lambda > 0$. *Let* G_1, G_2 *and* G_3 *be the restrictions of* G *over the intervals* $(0, \lambda_*]$, (λ_*, λ^*) *and* $[\lambda^*, +\infty)$ *respectively. Then* G_1 *is the graph of a strictly decreasing, strictly concave function* $y = G_1(x)$ *over* $[x(\lambda_*), c)$, G_2 *is the graph of a strictly decreasing and strictly convex function over* $(x(\lambda_*), x(\lambda^*))$, *and* G_3 *is the graph of a strictly decreasing and strictly concave function over* $(-\infty, x(\lambda^*)]$. *Furthermore,* (p, q) *is in the* C\(0, ∞)-*characteristic region of (6.32) if, and only if,* $(p, q) \in \wedge(G_3\chi_{(-\infty,c)}) \oplus \wedge(G_1)$, *or,* $(p, q) \in \nabla(\Theta_0\chi_{[c,+\infty)})$.

Again, the first two paragraphs of the proof of this result are exactly the same as that of Theorem 6.17, except that the corresponding envelope now appears as that in Figure 6.14. Indeed, we assert that the curve G_1 lies completely in the interior of the lower half plane unless $c = ((\tau - 1)/(\tau + 1))^{\tau - 1}$. To see this, we note that the function ϕ defined by (6.43) satisfies the same properties as in the proof of Theorem 6.17, except that (6.44) is now replaced with

$$\phi\left(\frac{\tau - 1}{\tau + 1}\right) = c - \left(\frac{\tau - 1}{\tau + 1}\right)^{\tau - 1} \geq 0,$$

and (6.45) with (6.46).

6.2.4.2 *The Case* $\delta > 1$ *and* $\tau = 1$

In this section, we assume that $\delta > 1$ and $\tau = 1$. Then only two distinct cases arise: $c \geq 1$ and $c < 1$. As in the last section, the envelope G is given by

$$x(\lambda) = \Gamma(\lambda; -\delta, 1) \text{ and } y(\lambda) = \Gamma(\lambda; 1, -\delta), \tag{6.47}$$

where

$$\Gamma(x; a, b) = \frac{1}{b - a} \left[(a + 1)x^{b+1} - ax^b + cax^b - c(a - 1)x^{b-1}\right] \text{ for } \lambda > 0.$$

Note that $x'(\lambda) = -(\delta+1)^{-1}F(\lambda)$ and $y'(\lambda) = (\delta+1)^{-1}\lambda^{-(\tau+\delta)}F(\lambda)$ where

$$F(\lambda) = 2(\delta-1)\lambda - \delta + c\delta.$$

Since $F(0) = \delta(c-1)$, $F(+\infty) = +\infty$ and $F'(\lambda) = 2(\delta-1)$, we see that (a) if $c \geq 1$, then $F(\lambda) > 0$, and (b) if $c < 1$, then $F(\lambda) < 0$ for $\lambda \in (0, \overline{\lambda})$ and $F(\lambda) > 0$ for $\lambda \in (\overline{\lambda}, \infty)$, where $\overline{\lambda} = \delta(1-c)/2(\delta-1)$.

At places where $F(\lambda) \neq 0$, we also have

$$\frac{dy}{dx}(\lambda) = -\lambda^{-(\delta+1)}$$

and

$$\frac{d^2y}{dx^2}(\lambda) = \frac{-(\delta+1)^2\lambda^{-(\delta+2)}}{F(\lambda)}.$$

Thus

$$\lim_{\lambda\to 0^+} \frac{dy}{dx} = -\infty \text{ and } \lim_{\lambda\to\infty} \frac{dy}{dx} = 0.$$

Theorem 6.20. *Suppose $\tau = 1, \delta > 1$ and $c \geq 1$. Let the curve G be defined by (6.47) for $\lambda > 0$. Then G is also the graph of a strictly decreasing, strictly concave and smooth function $y = G(x)$ over the interval $(-\infty, c)$ such that $G(-\infty) = 0$, $G'(-\infty) = 0$, $G(c^-) = -\infty$ and $G'(c^-) = -\infty$. Furthermore, (p, q) is in the $\mathbf{C}\backslash(0, \infty)$-characteristic region of (6.32) if, and only if, $p \geq c$ and $q \geq 0$, or, $p < c$ and $q < G(p)$.*

Indeed, since $F(\lambda) > 0$ for $\lambda > 0$, the proof is exactly the same as the proof of Theorem 6.15, except that now $F(\lambda) > 0$ for $\lambda > 0$ instead of $F(\lambda) > 0$ for $\lambda > \rho$. The corresponding graph of G is similar to that in Figure 6.10.

Theorem 6.21. *Suppose $\tau = 1, \delta > 1$ and $c < 1$. Let $\overline{\lambda}$ be the unique positive root of the function $F(\lambda)$ in $(0, \infty)$. Let the curve G defined by (6.47) for $\lambda > 0$. When restricted to the interval $(0, \overline{\lambda})$, G is also the graph of a strictly decreasing, strictly convex and smooth function $y = G_1(x)$ defined on $(c, x(\overline{\lambda}))$; and when restricted to $[\overline{\lambda}, \infty)$, G is also the graph of a strictly decreasing, strictly concave and smooth function $y = G_2(x)$ over $(-\infty, x(\overline{\lambda})]$. Furthermore, (p, q) is in the $\mathbf{C}\backslash(0, \infty)$-characteristic region of (6.32) if, and only if, $(p, q) \in \wedge(G_2\chi_{(-\infty, c]})$, or, $(p, q) \in \vee(G_1\chi_{(c, \infty)}) \oplus \nabla(\Theta_0)$.*

Indeed, since $F(\lambda) < 0$ for $\lambda \in (0, \overline{\lambda})$ and $F(\lambda) > 0$ for $\lambda \in (\overline{\lambda}, \infty)$, the proof is similar to the proof of Theorem 6.16. The corresponding graph is similar to that in Figure 6.11.

6.2.5 $\Delta(n, n, 0)$-*Polynomials*

Recall that a $\Delta(n, n, 0)$-polynomial takes the form

$$f_0(\lambda) + \lambda^{\tau_1} f_1(\lambda) + \lambda^{\tau_2} c,$$

where c is a nonzero real number, τ_1 and τ_2 are mutually distinct nonzero integers and f_0, f_1 are polynomials of degree n. Such a function is difficult to handle. However, if we consider the neutral difference equation of the form

$$\Delta^n (x_k + px_{k-\tau}) + qx_{k-\sigma} = 0, \quad k \in \mathbf{N}, \tag{6.48}$$

where $n, \tau \in \mathbf{Z}[1, \infty)$, $\sigma \in \mathbf{N}$, and p, q are real numbers, then seeking geometric sequence solutions of the form $\{\lambda^k\}$ will lead us to the special $\Delta(n, n, 0)$-polynomial

$$(\lambda - 1)^n + p\lambda^{-\tau}(\lambda - 1)^n + q\lambda^{-\sigma}. \tag{6.49}$$

In this section, we will consider the $\mathbf{C}\backslash(0,\infty)$-characteristic region of (6.49).

First, note that if $n = 1$, then the above function is the same $\Delta(1, 1, 0)$-polynomial discussed in Section 6.2.3. If $\sigma = 0$, then the above function is a $\Delta(n, n)$-polynomial. However, this function has not been handled in our previous discussions. Therefore, in this section, we will assume throughout that $n > 1$, $\sigma \geq 0$ and $\tau \geq 1$.

Note that the $\mathbf{C}\backslash(0,\infty)$-characteristic region of (6.49) is the same as that of the following function

$$\lambda^{\tau+\sigma}(\lambda - 1)^n + p\lambda^{\sigma}(\lambda - 1)^n + q\lambda^{\tau} \tag{6.50}$$

obtained by multiplying (6.49) by $\lambda^{\sigma+\tau}$.

The parity of the integer n is crucial and hence two groups of different cases are needed.

6.2.5.1   *The case where n is odd*

In this section, we assume that n is an odd integer in $\mathbf{Z}[3, \infty)$. To facilitate discussions, we will also classify equation into five different cases by means of the following exhaustive and mutually exclusive conditions according to the sizes of the positive integers τ and σ: (i) $\sigma = 0$, (ii) $\sigma = \tau = 1$, (iii) $\sigma = \tau > 1$, (iv) $0 < \sigma < \tau$, and (v) $\sigma \geq \tau + 1$.

Theorem 6.22. *Suppose $\sigma = 0$ and n is an odd integer in $\mathbf{Z}[3, \infty)$. Let the curve G be defined by the parametric functions*

$$x(\lambda) = \frac{-n\lambda^{\tau+1}}{(n-\tau)\lambda + \tau}, \quad y(\lambda) = \frac{\tau(\lambda-1)^{n+1}}{(n-\tau)\lambda + \tau} \tag{6.51}$$

over the interval $(0, \infty)$. If $\tau \leq n$, then G is also the graph of a function $y = G(x)$ defined over $(-\infty, 0)$, furthermore, (p, q) is a point in the $\mathbf{C}\backslash(0,\infty)$-characteristic region of the function

$$h(\lambda|p, q) = (\lambda - 1)^n + p\lambda^{-\tau}(\lambda - 1)^n + q, \tag{6.52}$$

if, and only if, $p < 0$ and $q > G(p)$, or $p = 0$ and $q \geq 1$. If $\tau > n$, then the restriction of the curve G over the interval $(0, \tau/(\tau-n))$ is also the graph of a function $y = S(x)$ defined over $(-\infty, 0)$, furthermore, (p, q) is a point in the $\mathbf{C} \backslash (0, \infty)$-characteristic region of the function (6.52) if, and only if, $p < 0$ and $q > S(p)$, or $p = 0$ and $q \geq 1$.

Proof. For each $\lambda \in (0, \infty)$, let L_λ be the straight line defined by

$$p\lambda^{-\tau}(\lambda - 1)^n x + y = -(\lambda - 1)^n,$$

i.e., $L_\lambda : h(\lambda|x, y) = 0$. When $\sigma = 0$,

$$h'_\lambda(\lambda|x, y) = -\tau\lambda^{-\tau-1}(\lambda - 1)^n x + n\lambda^{-\tau}(\lambda - 1)^{n-1}x + n(\lambda - 1)^{n-1}.$$

The determinant of the linear system $h(\lambda|x, y) = 0 = h'_\lambda(\lambda|x, y)$ is $D(\lambda) = \lambda^{-\tau-1}(\lambda - 1)^{n-1}((n - \tau)\lambda + \tau)$, which has the positive root 1 and the additional positive root $\tau/(\tau - n)$ if $\tau > n$.

We need to consider two sub-cases: (1) $\tau \leq n$, and (2) $n < \tau$.

Suppose $\tau \leq n$. Then $D(\lambda)$ has the unique root 1 in $(0, \infty)$. By Theorem 4.1, the $\mathbf{C} \backslash (0, \infty)$-characteristic region of $h(\lambda|p, q)$ is the intersection of the $\mathbf{C} \backslash \{1\}$-, $\mathbf{C} \backslash (0, 1)$- and $\mathbf{C} \backslash (1, \infty)$-characteristic regions of $h(\lambda|p, q)$. The $\mathbf{C} \backslash \{1\}$-characteristic region, since $h(1|p, q) = q$, is just the set

$$\{(p, q) \in \mathbf{R}^2 | q \neq 0\}, \tag{6.53}$$

that is, $\mathbf{C} \backslash \mathbf{R}$. As for the other two characteristic regions, the $\mathbf{C} \backslash (0, 1)$-characteristic region is the dual set of order 0 of the envelope G_1 of the family $\Phi_{(0,1)} = \{L_\lambda | \lambda \in (0, 1)\}$, and the $\mathbf{C} \backslash (1, \infty)$-characteristic region is the dual set of order 0 of the envelope G_2 of the family $\Phi_{(1,\infty)} = \{L_\lambda | \lambda \in (1, \infty)\}$.

By Theorem 2.3, we may show that the envelope G_1 of the family $\Phi_{(0,1)}$ is described by the parametric functions in (6.51) for $\lambda \in (0, 1)$, while the envelope G_2 of the family $\Phi_{(1,\infty)}$ by the same parametric functions but for $\lambda \in (1, \infty)$. Indeed, the properties of the functions $x(t)$ and $y(t)$ over the whole interval $(0, \infty)$ are easy to obtain. We have

$$(x(0^+), y(0^+)) = (0, 1), \ (x(1), y(1)) = (-1, 0), \ (x(+\infty), y(+\infty)) = (-\infty, +\infty),$$

$$x'(\lambda) = -\frac{n\tau\lambda^\tau\left[(n - \tau)\lambda + \tau + 1\right]}{\left[(n - \tau)\lambda + \tau\right]^2},$$

$$y'(\lambda) = \frac{\tau(\lambda - 1)^n\left[n(n - \tau)\lambda + n(\tau + 1)\right]}{\left[(n - \tau)\lambda + \tau\right]^2},$$

$$\frac{dy}{dx} = -(\lambda - 1)^n\lambda^{-\tau},$$

and

$$\frac{d^2y}{dx^2} = \frac{(\lambda - 1)^{n-1}\left[(n - \tau)\lambda + \tau\right]^3}{n\tau\lambda^{2\tau+1}\left[(n - \tau)\lambda + \tau + 1\right]},$$

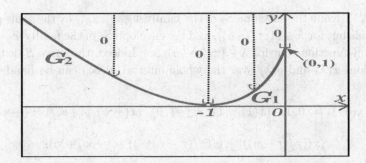

Fig. 6.15

for $\lambda \in (0,\infty)$. We may thus infer from these properties that the curve described by $(x(\lambda), y(\lambda))$ over the interval $(0,\infty)$ is also the graph of a function $y = G(x)$ defined over $(-\infty, 0)$ such that $G(0^-) = 1$, $G'(0^-) = +\infty$, $G(-\infty) = +\infty$ and $G'(-\infty) = -\infty$. Furthermore, the envelope G_1 (as part of G) is the graph of a function $y = G_1(x)$ which is strictly increasing, strictly convex, and smooth over the interval $(x(1^-), 0)$ (which is equal to $(-1, 0)$) and G_2 (as part of G) is the graph of a function $y = G_2(x)$ which is strictly decreasing, strictly convex, and smooth over $(-\infty, x(1^+))$ (which is equal to $(-\infty, -1)$). See Figure 6.15. Moreover, $\lim_{\lambda \to 0} dy/dx = \infty$ and $\lim_{\lambda \to \infty} dy/dx = -\infty$, $(x(1^-), y(1^-)) = (x(1^+), y(1^+)) = (-1, 0)$, $G_1'((-1)^+) = G_2'((-1)^-) = 0$, and $G_1'(0^-) = +\infty$, as well as $G_2'(-\infty) = +\infty$. By the distribution maps in Figures 3.5 and 3.11, the dual sets of order 0 of G_1 and G_2 can easily be found. The intersection of these dual sets and $\mathbf{C}\backslash\mathbf{R}$, in view of Theorem 3.22 and the distribution map in Figure 3.14, is the set of points (p, q) that satisfies $p < 0$ and $q > G(p)$, or $p = 0$ and $q \geq 1$.

Next, we assume that $\tau > n$. Then $D(\lambda)$ has the positive roots 1 and $\tau/(\tau - n)$. As in the previous case, the $\mathbf{C}\backslash(0,\infty)$-characteristic region is the intersection of the $\mathbf{C}\backslash(0,1)$-, $\mathbf{C}\backslash(1, \tau/(\tau - n))$-, $\mathbf{C}\backslash(\tau/(\tau - n), \infty)$-, $\mathbf{C}\backslash\{1\}$- and $\mathbf{C}\backslash\{\tau/(\tau - n)\}$-characteristic regions of h. The $\mathbf{C}\backslash\{1\}$-characteristic region of $h(\lambda|p, q)$, since $h(1|p, q) = q$, is just the set $\mathbf{C}\backslash\mathbf{R}$, while the $\mathbf{C}\backslash\{\tau/(\tau - n)\}$-characteristic region is the set of points (p, q) that does not lie on the straight line $L_{\tau/(\tau-n)}$, that is, $\mathbf{C}\backslash L_{\tau/(\tau-n)}$.

The $\mathbf{C}\backslash(0, 1)$-characteristic region is the dual set of order 0 of the envelope of the family $\Phi_{(0,1)} = \{L_\lambda| \lambda \in (0, 1)\}$, the $\mathbf{C}\backslash(1, \tau/(\tau - n))$-characteristic region is the dual set of order 0 of the envelope of the family $\Phi_{(1,\tau/(\tau-n))} = \{L_\lambda| \lambda \in (1, \tau/(\tau - n))\}$, and the $\mathbf{C}\backslash(\tau/(\tau-n), \infty)$-characteristic region is the dual set of order 0 of the envelope of the family $\Phi_{(\tau/(\tau-n),\infty)} = \{L_\lambda| \lambda \in (\tau/(\tau - n), \infty)\}$. By Theorem 2.3, the envelope S_1 of the family $\Phi_{(0,1)}$ is described by the parametric functions (6.51), or,

$$x(\lambda) = \frac{n}{\tau - n} \frac{\lambda^{\tau+1}}{\lambda - \frac{\tau}{\tau-n}} \text{ and } y(\lambda) = -\frac{\tau}{\tau - n} \frac{(\lambda - 1)^{n+1}}{\lambda - \frac{\tau}{\tau-n}}$$

for $\lambda \in (0,1)$, while the envelope S_2 of the family $\Phi_{(1,\tau/(\tau-n))}$ by the same parametric functions but for $\lambda \in (1, \tau/(-n))$, and the envelope S_3 of the family $\Phi_{(\tau/(\tau-n),\infty)}$ by same functions but for $\lambda \in (\tau/(\tau-n),\infty)$. Indeed, the curve S described by the functions $x(\lambda)$ and $y(\lambda)$ over the whole interval $(0,\infty)$ can be handled easily. We have

$$(x(0^+), y(0^+)) = (0,1), \ (x(1), y(1)) = (-1, 0), \ (x(+\infty), y(+\infty)) = (+\infty, -\infty),$$

$$(x((\tau/(\tau-n))^-), y((\tau/(\tau-n))^-)) = (-\infty, +\infty),$$

$$(x((\tau/(\tau-n))^+), x((\tau/(\tau-n))^+)) = (+\infty, -\infty),$$

$$x'(\lambda) = -\frac{n\tau\lambda^\tau\,[(n-\tau)\lambda + \tau + 1]}{[(n-\tau)\lambda + \tau]^2}$$

and

$$y'(\lambda) = \frac{n\tau(\lambda-1)^n\,[(n-\tau)\lambda + (\tau+1)]}{[(n-\tau)\lambda + \tau]^2}$$

for $\lambda \in (0,\infty)\backslash\{1, \tau/(\tau-n)\}$. Furthermore, $x'(\lambda)$ has a unique positive root $(\tau + 1)/(\tau - n)$. Thus

$$\frac{dy}{dx} = -(\lambda - 1)^n \lambda^{-\tau},$$

$$\frac{d^2y}{dx^2} = \frac{(\lambda-1)^{n-1}\,[(n-\tau)\lambda + \tau]^3}{n\tau\lambda^{2\tau+1}\,[(n-\tau)\lambda + \tau + 1]},$$

for $\lambda \in (0,\infty)\backslash\{1, \tau/(\tau-n), (\tau+1)/(\tau-n)\}$. Let $\alpha = x((\tau+1)/(\tau-n))$ and $\beta = y((\tau+1)/(\tau-n))$. The envelope S_1, which corresponds to the restriction of S over the interval $\lambda \in (0,1)$, is also the graph of a strictly increasing, strictly convex and smooth function $y = S_1(x)$ defined on $(-1,0)$ such that $S_1(0^+) = 1$, $S_1'(0^+) = +\infty$, $S_1((-1)^+) = 0$ and $S_1'((-1)^+) = 0$; the envelope S_2, which corresponds to the restriction of S over the interval $\lambda \in (1, \tau/(\tau-n))$, is also the graph of a strictly decreasing, strictly convex and smooth function $y = S_2(x)$ defined on $(-\infty, -1)$ such that $S_2((-1)^-) = 0$, $S_2'((-1)^-) = 0$ and $S_2(-\infty) = +\infty$; the envelope S_3 is made up of two pieces $S_3^{(1)}$ and $S_3^{(2)}$ which correspond to the case where $\lambda \in (\tau/(\tau-n), (\tau+1)/(\tau-n))$ and where $\lambda \in [(\tau+1)/(\tau-n), \infty)$ respectively. $S_3^{(1)}$ is the graph of a function $y = S_3^{(1)}(x)$ which is strictly decreasing, strictly concave and smooth over (α, ∞) and $S_3^{(2)}$ is the graph of a function $y = S_3^{(2)}(x)$ which is strictly decreasing, strictly convex and smooth over $[\alpha, \infty)$. The turning point (α, β) is given by

$$(\alpha, \beta) = \left(x\left(\frac{\tau+1}{\tau-n}\right), y\left(\frac{\tau+1}{\tau-n}\right)\right) = \left(\frac{n(\tau+1)^{\tau+1}}{(\tau-n)^{+1}}, \frac{-\tau(n+1)^{n+1}}{(\tau-n)^{n+1}}\right),$$

and in view of our assumption that $\tau > n$, is located in the fourth open quadrant. See Figure 6.16. Furthermore, by Lemma 3.2, $L_{\tau/(\tau-n)}$ is the asymptote of S_2 at

$-\infty$ and $S_3^{(1)}$ at $+\infty$. This asymptote intercepts the x-axis at the point $-\tau^\tau/(\tau-n)^n$ which is strictly less than -1. It also separates the curves S_1 and S_2 from $S_3^{(1)}$. Finally, the common tangent of $S_3^{(1)}$ and $S_3^{(2)}$ is given by

$$T : y = -\frac{(n+1)^n(\tau-n)^{\tau-n}}{(\tau+1)^n}x - \left(\frac{n+1}{n-\tau}\right)^n$$

which passes through (α,β). Note that the slope of T is greater than the slope of $L_{\tau/(\tau-n)}$ in view of $\tau > n$. Note further that since the turning point (α,β) is below the straight line $L_{\tau/(\tau-n)}$ and since (α,β) is in the interior of the fourth quadrant, the negatively sloped lines $L_{\tau/(\tau-n)}$ and T cross at some point in the interior of the fourth open quadrant also. See Figure 6.16.

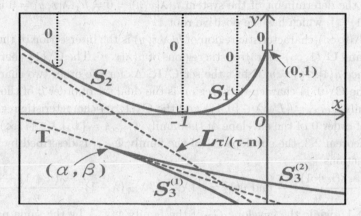

Fig. 6.16

In view of the distribution maps in Figures 3.5, 3.11 and A.8c, the dual sets of order 0 of the envelopes S_1, S_2 and S_3 are easily obtained. The intersection of these dual sets and $\mathbf{C}\backslash\mathbf{R}$ and $\mathbf{C}\backslash L_{\tau/(\tau-n)}$, in view of Theorem 3.22 and the distribution map in Figure A.23, is also easily found so that (p,q) lies in our desired characteristic region if, and only if, $p < 0$ and $q > S(p)$, or $p = 0$ and $q \geq 1$, where S is the restriction of the function $(x(\lambda),y(\lambda))$ over the interval $(0,\tau/(\tau-n))$. The proof is complete.

Corollary 6.1. *When n is an odd integer in $\mathbf{Z}[3,+\infty)$ and $\sigma = p = 0$, $h(\lambda|p,q)$ in (6.52) does not have any positive roots if, and only if, $q \geq 1$.*

Corollary 6.2. *Suppose n is odd in $\mathbf{Z}[3,+\infty)$, $\sigma = 0$ and $\tau = n$. Then $h(\lambda|p,q)$ in (6.52) does not have any positive roots if, and only if, $p = 0$ and $q \geq 1$, or, $p < 0$ and $q > \left[1 - (-p)^{1/(n+1)}\right]^{n+1}$.*

Theorem 6.23. *Suppose n is an odd integer in $\mathbf{Z}[3,\infty)$ and $\sigma = \tau = 1$. Then (p,q) is in the $\mathbf{C}\backslash(0,\infty)$-characteristic region of (6.49) if, and only if, either $q \geq p \geq 1/n$,* or, $p < 1/n$ and

$$q > \frac{n^n}{(n+1)^{n+1}}(1+p)^{n+1}.$$

Proof. It suffices to consider the polynomial

$$h(\lambda|x,y) = (\lambda-1)^n\lambda + (\lambda+1)^n x + y$$

obtained from (6.49) by substituting $\tau = \sigma = 1$ and $p = x$ and $q = y$ into it and then multiplying the subsequent function by λ. Let L_λ be the straight line defined by $h(\lambda|x,y) = 0$. Note that

$$h'_\lambda(\lambda|x,y) = (\lambda-1)^n + n\lambda(\lambda-1)^{n-1} + n(\lambda-1)^{n-1}x.$$

Therefore the determinant of the system $h(\lambda|x,y) = 0 = h'_\lambda(\lambda|x,y) = 0$ is $D(\lambda) = -n\lambda(\lambda-1)^{n-1}$, which has the positive root 1.

The $\mathbf{C}\backslash(0,\infty)$-characteristic region of $h(\lambda|x,y)$ is the intersection of the $\mathbf{C}\backslash\{1\}$-, $\mathbf{C}\backslash(0,1)$- and $\mathbf{C}\backslash(1,\infty)$-characteristic regions of $h(\lambda|x,y)$. The $\mathbf{C}\backslash\{1\}$-characteristic region, since $h(1|x,y) = y$, is just the set $\mathbf{C}\backslash\mathbf{R}$. As for the other two characteristic regions, the $\mathbf{C}\backslash(0,1)$-characteristic region is the dual set of order 0 of the envelope of the family $\Phi_{(0,1)} = \{L_\lambda|\ \lambda \in (0,1)\}$ and the $\mathbf{C}\backslash(1,\infty)$-characteristic region is the dual set of order 0 of the envelope of the family $\Phi_{(1,\infty)} = \{L_\lambda|\ \lambda \in (1,\infty)\}$.

By Theorem 2.3, the envelope G_1 of the family $\Phi_{(0,1)}$ is described by the parametric functions

$$x(\lambda) = \frac{-\lambda(n+1)+1}{n} \text{ and } y(\lambda) = -(\lambda-1)^n\lambda - (\lambda-1)^n\frac{-\lambda(n+1)+1}{n} \quad (6.54)$$

for $\lambda \in (0,1)$, while the envelope G_2 of the family $\Phi_{(1,\infty)}$ by the same parametric functions but for $\lambda \in (1,\infty)$. Indeed, the properties of the functions $x(\lambda)$ and $y(\lambda)$ over the whole interval $(0,\infty)$ are easy to obtain. The linear function $x(\lambda)$ maps $(0,\infty)$ onto $(-\infty,1/n)$ and $x(1) = -1$. By solving $x = (1 - \lambda(n+1))/n$ for λ, and then substituting $\lambda = (1-nx)/(n+1)$ into $y(\lambda)$, we see that the curve G described by $(x(\lambda),y(\lambda))$ over $(0,\infty)$ is also the graph of the function $y = G(x)$ defined by

$$G(x) = (-1)^{n+1}\frac{n^n}{(n+1)^{n+1}}(1+x)^{n+1}, \ x \in (-\infty,1/n). \quad (6.55)$$

Note further that

$$G((1/n)^-) = 1/n, \ G(-1) = 0, \ G(-\infty) = \infty,$$

$$\frac{dG}{dx} = \frac{n^n}{(n+1)^n}(1+x)^n,$$

and

$$\frac{d^2G}{dx^2} = \frac{n^{n+1}}{(n+1)^n}(1+x)^{n-1}.$$

Since $G'(-\infty) = -\infty$, by Lemma 3.4, the function $G(x)$ satisfy $G \sim H_{-\infty}$. Since $G'((1/n)^-) = 1$, Υ is the tangent line of G at $x = (1/n)^-$. By means of the

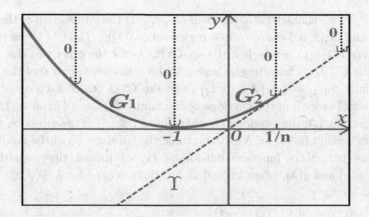

Fig. 6.17

above information, we see that the function G is strictly convex on $(-\infty, 1/n)$, is strictly above the x-axis except at $x = -1$ at which $G(-1) = 0$ and $G'((-1)^-) = G'((-1)^+) = 0$ and the tangent at $x = 1/n$ is given by Υ. Furthermore, the envelope G_2 is the restriction of the function $y = G(x)$ over the interval $(-\infty, -1)$, and the envelope G_1 is the restriction of the function $y = G(x)$ over the interval $(-1, 1/n)$. See Figure 6.17.

In view of the distribution map in Figure 3.1 and 3.11, the dual sets of order 0 of G_1 and G_2 can easily be found. Furthermore, the intersections of these dual sets and $\mathbf{C}\backslash\mathbf{R}$, in view of Theorem 3.22 and Figure 3.11, is also easily found so that (p, q) belongs to it if, and only if, $p < 1/n$ and $q > G(p)$, or, $p \geq 1/n$ and $q \geq p$.

Theorem 6.24. *Suppose n is an odd integer in $\mathbf{Z}[3,\infty)$ and $\sigma = \tau > 1$. Let the curve S be described by the parametric functions*

$$x(\lambda) = \frac{n+\tau}{n}\lambda^{\tau-1}\left(\frac{\tau}{n+\tau} - \lambda\right), \quad y(\lambda) = \frac{\tau}{n}(\lambda-1)^{n+1}\lambda^{\tau-1} \quad (6.56)$$

over $((\tau - 1)/(n + \tau), \infty)$. Then S is also the graph of a function $y = S(x)$ defined on $\left(-\infty, \frac{(\tau-1)^{\tau-1}}{n(n+\tau)^{\tau-1}}\right)$. Furthermore, the $\mathbf{C}\backslash(0,\infty)$-characteristic region of (6.49) is $\vee(S) \oplus \vee(\Upsilon)$.

Proof. It suffices to consider the polynomial

$$h(\lambda|x,y) = (\lambda - 1)^n\lambda^\tau + (\lambda - 1)^n x + y$$

obtained from (6.49) by substituting $\tau = \sigma$ and $p = x$ and $q = y$ into it and multiplying the subsequent equation by λ^τ. Let L_λ be the straight line defined by $h(\lambda|x,y) = 0$. Note that

$$h'_\lambda(\lambda; x, y) = n(\lambda - 1)^{n-1}x + n(\lambda - 1)^{n-1}\lambda^\tau + \tau\lambda^{\tau-1}(\lambda - 1)^n.$$

Therefore the determinant of the system $h(\lambda|x,y) = 0 = h'_\lambda(\lambda|x,y)$ is $D(\lambda) = n\lambda(\lambda - 1)^{n-1}$, which has the positive root 1.

The $\mathbf{C}\backslash(0,\infty)$-characteristic region of $h(\lambda|x,y)$ is the intersection of the $\mathbf{C}\backslash\{1\}$-, $\mathbf{C}\backslash(0,1)$- and $\mathbf{C}\backslash(1,\infty)$-characteristic regions of $h(\lambda|x,y)$. The $\mathbf{C}\backslash\{1\}$-characteristic region, since $h(1|x,y) = y$, is just the set $\mathbf{C}\backslash\mathbf{R}$. As for the other two characteristic regions, the $\mathbf{C}\backslash(0,1)$-characteristic region is the dual set of order 0 of the envelope of the family $\Phi_{(0,1)} = \{L_\lambda|\ \lambda \in (0,1)\}$, and the $\mathbf{C}\backslash(1,\infty)$-characteristic region is the dual set of order 0 of the envelope of the family $\Phi_{(1,\infty)} = \{L_\lambda|\ \lambda \in (1,\infty)\}$.

By Theorem 2.3, the envelope G_1 of the family $\Phi_{(0,1)}$ is described by the parametric functions in (6.56) for $\lambda \in (0,1)$, while the envelope G_2 of the family $\Phi_{(1,\infty)}$ by the same parametric functions but for $\lambda \in (1,\infty)$. Indeed, the properties of the functions $x(\lambda)$ and $y(\lambda)$, when treated as functions defined for $\lambda \in (0,\infty)$, are easy to obtain. We have

$$(x(0^+),y(0^+)) = (0,0),\ (x(1),y(1)) = (-1,0),\ (x(+\infty),y(+\infty)) = (-\infty,+\infty),$$

and

$$x'(\lambda) = -\frac{(n+\tau)\tau}{n}\lambda^{\tau-2}\left(\lambda - \frac{\tau-1}{n+\tau}\right).$$

Note that the unique positive root of $x'(\lambda)$ is $\lambda^* = (\tau-1)/(n+\tau)$ which is less than 1. If λ is not equal to the root λ^*, then

$$\frac{dy}{dx} = -(\lambda-1)^n,$$

and

$$\frac{d^2y}{dx^2} = \frac{n^2}{(n+\tau)\tau}\lambda^{2-\tau}\frac{(\lambda-1)^{n-1}}{\lambda - \frac{\tau-1}{n+\tau}}.$$

Let $p_1 = x(\lambda^*)$ and $q_1 = y(\lambda^*)$. Then the curve G described by $(x(\lambda),y(\lambda))$ defined for $\lambda \in (0,\infty)$, as shown in Figure 6.18, is made up of three pieces $G^{(1)}$, $G^{(2)}$ and $G^{(3)}$.

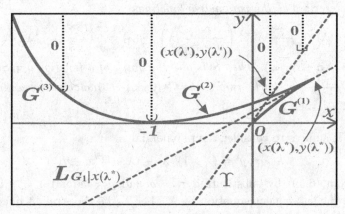

Fig. 6.18

The first piece $G^{(1)}$ is the restriction of G over the interval $(0,\lambda^*)$, the second $G^{(2)}$ over the interval $[\lambda^*,1)$ and the third $G^{(3)}$ over $[1,\infty)$. The turning point $(x(\lambda^*), y(\lambda^*))$ corresponds to the unique maximum of $x(\lambda)$ over $(0, \infty)$. The curve $G^{(1)}$ is also the graph of a strictly increasing, strictly concave and smooth function $y = G^{(1)}(x)$ defined over $(0, x(\lambda^*))$ such that $G^{(1)}(0^+) = 0$ and $G^{(1)'}(0^+) = (-1)^{n+1} = 1$; $G^{(2)}$ is the graph of a strictly increasing, strictly convex and smooth function $y = G^{(2)}(x)$ defined over $(-1, x(\lambda^*)]$ such that $G^{(2)}((-1)^+) = 0$ and $G^{(2)}((-1)^+) = 0$; and $G^{(3)}$ is the graph of a strictly decreasing, strictly convex and smooth function $y = G^{(3)}(x)$ defined over $(-\infty, -1]$ such that $G^{(3)}((-1)^-) = 0$, $G^{(3)'}((-1)^-) = 0$, $G^{(3)}(-\infty) = +\infty$ and $G^{(3)'}(-\infty) = -\infty$. Hence $L_{G^{(1)}|0} = \Upsilon$ and $G^{(3)} \sim H_{-\infty}$. Furthermore, the quasi-tangent line $L_{G^{(1)}|0}$ intersects with $G^{(2)}$ at an unique point $(x(\lambda'), y(\lambda'))$, where λ' is obtained by solving the unique root in $(\lambda^*, 1)$ of the equation $x(\lambda) = y(\lambda)$, i.e.,

$$\frac{\tau}{n}(\lambda - 1)^{n+1} = \frac{\tau}{n} - \frac{n+\tau}{n}\lambda.$$

In view of the above discussions, the envelope G_1 is made up of $G^{(1)}$ and $G^{(2)}$, while G_2 is $G^{(3)}$ with the point $(-1, 0)$ removed. The dual sets of $G^{(1)}, G^{(2)}$ and G_2 are, in view of the distribution maps in Figure 3.1 and 3.11, are easy to find. Furthermore, their intersection with $\mathbf{C}\backslash\mathbf{R}$, in view of Theorem 3.22, is also easily found so that a point (p, q) belongs to it if, and only if, $p \le x(\lambda')$ and $q > G_1(p)$, or, $p > x(\lambda')$ and $q > p$. In other words, $(p, q) \in \vee(G_1) \oplus \vee(\Upsilon)$.

We remark that in the above Theorem, it is not difficult to see from the concavity of the curve G_1 and the convexity of the curve G_2 that the first coordinate p' of the point $(x(\lambda'), y(\lambda'))$ of intersection is strictly between 0 and $x(\lambda^*)$. In particular, when $\sigma = \tau = 2$, $p' = x(\lambda') \in (0, 1/(n+2)n)$. Note that in this case, $x(\lambda') = p'$ becomes

$$(\lambda')^2 - \frac{2}{n+2}\lambda' + \frac{np'}{n+2} = 0,$$

and for $p' \le x(\lambda^*)$, it has the unique solution

$$\lambda' = \frac{1}{n+2}\left(1 + \sqrt{1 - n(n+2)p'}\right)$$

in (λ^*, ∞).

Theorem 6.25. *Suppose n is an odd integer in $\mathbf{Z}[3, \infty)$ and $0 < \sigma < \tau$. Let the curve G be defined by the parametric functions*

$$x(\lambda) = -(n+\sigma)\lambda^\tau \frac{\lambda - \frac{\sigma}{n+\sigma}}{(n+\sigma-\tau)\lambda + \tau - \sigma}, \; y(\lambda) = \frac{\tau(\lambda-1)^{n+1}\lambda^\sigma}{(n+\sigma-\tau)\lambda + \tau - \sigma} \quad (6.57)$$

over $[\sigma/(\sigma+n), \infty)$. Then S is the graph of a function $y = S(x)$ defined over $(-\infty, 0]$. Furthermore, (p, q) is a point of the $\mathbf{C}\backslash(0, \infty)$-characteristic region of (6.49) if, and only if, $p \le 0$ and $q > S(p)$.

Proof. It suffices to consider the polynomial

$$h(\lambda|x,y) = (\lambda - 1)^n \lambda^\tau + (\lambda - 1)^n x + y\lambda^{\tau - \sigma}$$

obtained from (6.49) by substituting $\sigma = \tau$ and $p = x$ and $q = y$ into it and then multiplying the resulting function by λ^τ. Let L_λ be the straight line defined by $h(\lambda|x,y) = 0$. Note that

$$h'_\lambda(\lambda|x,y) = \lambda^{\tau - 1}(\lambda - 1)^{n-1}((\tau + n)\lambda - \tau) + n(\lambda - 1)^{n-1}x + (\tau - \sigma)\lambda^{\tau - \sigma - 1}y.$$

Hence the determinant of the system $h(\lambda|x,y) = 0 = h'_\lambda(\lambda|x,y)$ is $D(\lambda) = \lambda^{\tau - \sigma - 1}(\lambda - 1)^{n-1}((\tau - \sigma - n)\lambda - (\tau - \sigma))$, which has the positive root 1 and $(\tau - \sigma)/(\tau - \sigma - n)$ if $\tau - \sigma > n$.

We need to consider three sub-cases: (1) $\tau - \sigma = n$, (2) $\tau - \sigma < n$, and (3) $\tau - \sigma > n$.

Suppose $\tau - \sigma = n$. Then $D(\lambda)$ has the unique positive root 1. The $\mathbf{C}\backslash(0,\infty)$-characteristic region of $h(\lambda|x,y)$ is the intersection of the $\mathbf{C}\backslash\{1\}$-, $\mathbf{C}\backslash(0,1)$- and the $\mathbf{C}\backslash(1,\infty)$-characteristic region of $h(\lambda|x,y)$. The $\mathbf{C}\backslash\{1\}$-characteristic region, since $h(1|x,y) = y$, is just the set $\mathbf{C}\backslash\mathbf{R}$. As for the other two regions, the $\mathbf{C}\backslash(0,1)$-characteristic region is the dual set of order 0 of the envelope G_1 of the family $\Phi_{(0,1)} = \{L_\lambda|\ \lambda \in (0,1)\}$, and the $\mathbf{C}\backslash(1,\infty)$-characteristic region is the dual set of order 0 of the envelope G_2 of the family $\Phi_{(1,\infty)} = \{L_\lambda|\ \lambda \in (1,\infty)\}$.

By Theorem 2.3, the envelope G_1 is described by the parametric functions (6.57) for $\lambda \in (0,1)$ and G_2 by the same functions but for $\lambda \in (1,\infty)$. Indeed, the properties of the functions $x(\lambda)$ and $y(\lambda)$, when treated as functions defined over $(0,\infty)$, are easy to obtain. We have

$$(x(0^+), y(0^+)) = (0,0),\ (x(1), y(1)) = (-1,0),\ (x(+\infty), y(+\infty)) = (-\infty, +\infty),$$

and

$$\frac{dx}{d\lambda} = -\frac{\tau(\tau + 1)}{n}\lambda^{\tau - 1}\left(\lambda - \frac{\sigma}{\tau + 1}\right).$$

The function $x'(\lambda)$ has the unique positive root $\lambda^* = \sigma/(\tau + 1)$ which is less than 1. If $\lambda \in (0,\infty)\backslash\{\lambda^*\}$, then

$$\frac{dy}{dx} = -(\lambda - 1)^n \lambda^{\sigma - \tau},$$

and

$$\frac{d^2y}{dx^2} = \frac{n^2}{\tau(\tau + 1)}(\lambda - 1)^{n-1}\lambda^{\sigma - 2\tau}\frac{1}{\lambda - \frac{\sigma}{\tau + 1}}.$$

We may now infer from these properties that the curve G described by the parametric functions in (6.57) over the interval $(0,\infty)$ is composed of three pieces $G^{(1)}, G^{(2)}$ and $G^{(3)}$ as shown in Figure 6.19. The turning point $(x(\lambda^*), y(\lambda^*))$ corresponds to the unique maximum of $x(\lambda)$ over $(0,\infty)$. The piece $G^{(1)}$ corresponds to the case where $\lambda \in (0,\lambda^*)$, $G^{(2)}$ to the case where $\lambda \in [\lambda^*, 1)$ and $G^{(3)}$ to the case where $\lambda \in [1,\infty)$. The curve $G^{(1)}$ is the graph of a function

$y = G^{(1)}(x)$ which is strictly increasing, strictly concave and smooth on $(0, x(\lambda^*))$ such that $G^{(1)}(0^+) = 0$ and $G^{(1)'}(0^+) = +\infty$. The curve $G^{(2)}$ has a horizontal slope at $x = -1^+$ and intersects with the y-axis at $\lambda = \sigma/(n + \sigma)$ and is the graph of a function $y = G^{(2)}(x)$ which is strictly increasing, strictly convex and smooth on $(-1, x(\lambda^*)]$ such that $G^{(2)}((-1)^+) = 0$ and $G^{(2)'}((-1)^+) = 0$. The curve $G^{(3)}$ is the graph of a function $y = G^{(3)}(x)$ which is strictly decreasing, strictly convex and smooth on $(-\infty, -1]$ such that $G^{(3)}((-1)^-) = 0$, $G^{(3)'}((-1)^-) = 0$, $G^{(3)}(-\infty) = +\infty$ and $G^{(3)'}(-\infty) = -\infty$. Hence $G^{(3)} \sim H_{-\infty}$. In view of the above discussions, the envelope G_1 is made up of $G^{(1)}$ and $G^{(2)}$, while G_2 is $G^{(3)}$ with the point $(-1, 0)$ removed. The dual sets of order 0 of $G^{(1)}, G^{(2)}$ and G_2 are, in view of the distribution maps in Figures 3.5, 3.1 and 3.11, are easy to find. Furthermore, the intersection of these dual sets with $\mathbf{C}\backslash\mathbf{R}$, in view of Theorem 3.22, is also easily obtained so that a point (p, q) belongs to it if, and only if, $p \le 0$ and $q > G(p)$.

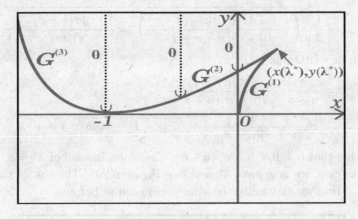

Fig. 6.19

Next suppose $\tau - \sigma < n$. The first part of the proof is similar to that of the previous case. However, we now have

$$(x(0^+), y(0^+)) = (0, 0), \quad (x(1), y(1)) = (-1, 0), \quad (x(+\infty), y(+\infty)) = (-\infty, +\infty),$$

and

$$\frac{dx}{d\lambda} = \frac{-\tau(n + \sigma)\lambda^{\tau-1}}{(n + \sigma - \tau)\left(\lambda + \frac{\tau-\sigma}{n+\sigma-\tau}\right)^2} Q(\lambda), \tag{6.58}$$

$$\frac{dy}{dx} = -(\lambda - 1)^n \lambda^{\sigma-\tau}, \tag{6.59}$$

and

$$\frac{d^2y}{dx^2} = \frac{(n + \sigma - \tau)^2}{\tau(n + \sigma)} \frac{(\lambda - 1)^{n-1}\lambda^{\sigma-2\tau}\left(\lambda + \frac{\tau-\sigma}{n+\sigma-\tau}\right)^3}{Q(\lambda)}, \tag{6.60}$$

where

$$Q(\lambda) = \lambda^2 + \frac{n(\tau+1) - 2\sigma(n+\sigma-\tau)}{(n+\sigma)(n+\sigma-\tau)}\lambda - \frac{\sigma(\tau-\sigma)}{n+\sigma-\tau}. \qquad (6.61)$$

Note that since $(\tau-\sigma)/(n+\sigma-\tau) > 0$, the quadratic equation $Q(\lambda) = 0$ has two real roots

$$\alpha = -\frac{n(\tau+1) - 2\sigma(n+\sigma-\tau)}{2(n+\sigma)(n+\sigma-\tau)}$$
$$-\frac{1}{2}\sqrt{\left(\frac{n(\tau+1) - 2\sigma(n+\sigma-\tau)}{(n+\sigma)(n+\sigma-\tau)}\right)^2 + \frac{4\sigma(\tau-\sigma)}{(n+\sigma)(n+\sigma-\tau)}}, \qquad (6.62)$$

and

$$\beta = -\frac{n(\tau+1) - 2\sigma(n+\sigma-\tau)}{2(n+\sigma)(n+\sigma-\tau)}$$
$$+\frac{1}{2}\sqrt{\left(\frac{n(\tau+1) - 2\sigma(n+\sigma-\tau)}{(n+\sigma)(n+\sigma-\tau)}\right)^2 + \frac{4\sigma(\tau-\sigma)}{(n+\sigma)(n+\sigma-\tau)}}. \qquad (6.63)$$

Since β can be written as

$$\beta = \frac{\sigma}{n+\sigma} - \frac{n(\tau+1)}{2(n+\sigma)(n+\sigma-\tau)}$$
$$+\frac{1}{2(n+\sigma)(n+\sigma-\tau)}\sqrt{n^2(\tau+1)^2 - 4n\sigma(n+\sigma-\tau)},$$

we see further that $\alpha < 0 < \beta < \sigma/(n+\sigma)$. Therefore, instead of $\lambda^* = \sigma/(\tau+1)$ in the previous case, we now have $\lambda^* = \beta$ (see Figure 6.20). The rest of the proof is the same, so that we may make the same conclusion as before.

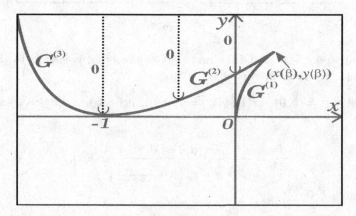

Fig. 6.20

Next suppose $\tau - \sigma > n$. The determinant $D(\lambda)$ has the positive roots 1 and $(\tau-\sigma)/(\tau-\sigma-n)$. The $\mathbf{C}\backslash(0,\infty)$-characteristic region of $h(\lambda|x,y)$ is the intersection of the $\mathbf{C}\backslash\{1\}$-, $\mathbf{C}\backslash\{(\tau-\sigma)/(\tau-\sigma-n)\}$-, $\mathbf{C}\backslash(0,1)$-, $\mathbf{C}\backslash(1,(\tau-\sigma)/(\tau-\sigma-n))$-

and $\mathbf{C} \backslash ((\tau - \sigma)/(\tau - \sigma - n), \infty)$-characteristic regions of $h(\lambda | x, y)$. The $\mathbf{C} \backslash \{1\}$-characteristic region is given by $\mathbf{C} \backslash \mathbf{R}$ as before. The $\mathbf{C} \backslash (0, 1)$-characteristic region is the dual set of order 0 of the envelope G_1 of the family $\Phi_{(0,1)} = \{L_\lambda | \lambda \in (0, 1)\}$. The $\mathbf{C} \backslash (1, (\tau - \sigma)/(\tau - \sigma - n))$-characteristic region is the dual set of order 0 of the envelope G_2 of the family $\Phi_{(0,1)} = \{L_\lambda | \lambda \in (1, (\tau - \sigma)/(\tau - \sigma - n))\}$. The $\mathbf{C} \backslash ((\tau - \sigma)/(\tau - \sigma - n), \infty)$-characteristic region is the dual set of order 0 of the envelope G_3 of the family $\Phi_{(0,1)} = \{L_\lambda | \lambda \in ((\tau - \sigma)/(\tau - \sigma - n), \infty)\}$.

By Theorem 2.3, the envelope G_1 is defined by the parametric functions (6.57) for $\lambda \in (0, 1)$, G_2 by the same functions for $\lambda \in (1, (\tau - \sigma)/(\tau - \sigma - n))$, and G_3 by the same functions for $\lambda \in ((\tau - \sigma)/(\tau - \sigma - n), \infty)$. The properties of the functions $x(\lambda)$ and $y(\lambda)$, when treated as functions over the whole interval $(0, \infty)$, are easy to obtain. Indeed, the derivatives $dx/d\lambda$, dy/dx and d^2y/dx^2 are of the same form as in the previous case, but the roots α and β of the equation $Q(\lambda) = 0$ now satisfy

$$0 < \alpha < \frac{\sigma}{n + \sigma} < 1 < \frac{\tau - \sigma}{\tau - \sigma - n} < \beta < \infty.$$

We have

$$(x(0^+), y(0^+)) = (0, 0), \quad (x(1), y(1)) = (-1, 0), \quad (x(+\infty), y(+\infty)) = (+\infty, -\infty).$$

Furthermore,

$$\lim_{\lambda \to ((\tau - \sigma)/(\tau - \sigma - n))^+} x(\lambda) = -\infty \qquad (6.64)$$

and

$$\lim_{\lambda \to ((\tau - \sigma)/(\tau - \sigma - n))^-} x(\lambda) = +\infty. \qquad (6.65)$$

We may now infer from these properties to conclude that the parametric curve corresponding to the parametric functions $x(\lambda)$ and $y(\lambda)$ over the interval $(0, \infty)$ is composed of five pieces $G^{(1)}, G^{(2)}, G^{(3)}, G^{(4)}$ and $G^{(5)}$ as shown in Figure 6.21.

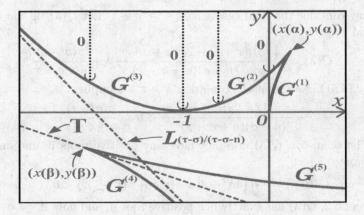

Fig. 6.21

The piece $G^{(1)}$ corresponds to the case where $\lambda \in (0, \alpha)$, the piece $G^{(2)}$ corresponds to where $\lambda \in [\alpha, 1)$, the piece $G^{(3)}$ to where $\lambda \in [1, (\tau - \sigma)/(\tau - \sigma - n))$, the piece $G^{(4)}$ to where $\lambda \in ((\tau - \sigma)/(\tau - \sigma - n), \beta)$, and the piece $G^{(5)}$ to where $\lambda \in [\beta, \infty)$.

The piece $G^{(1)}$ is the graph of a strictly increasing, strictly concave and smooth function $y = G^{(1)}(x)$ over $(0, x(\alpha))$ such that $G^{(1)}(0^+) = 0$ and $G^{(1)'}(0^+) = +\infty$. The piece $G^{(2)}$ intercepts with the y-axis at $\lambda = \sigma/(n+\sigma)$ and has a horizontal slope at $x = (-1)^+$ and is the graph of a strictly increasing, strictly convex and smooth function over $(-1, x(\alpha)]$ such that $G^{(2)}((-1)^+) = 0$ and $G^{(2)'}((-1)^+) = 0$. The piece $G^{(3)}$ is the graph of a strictly decreasing, strictly convex and smooth function $y = G^{(3)}(x)$ over $(-\infty, -1]$ such that $G^{(3)}((-1)^-) = 0$ and $G^{(3)'}((-1)^-) = 0$. The piece $G^{(4)}$ is the graph of a strictly decreasing, strictly concave and smooth function $y = G^{(4)}(x)$ over $(x(\beta), +\infty)$ such that $G^{(4)}(+\infty) = -\infty$; and $G^{(5)}$ is the graph of a strictly decreasing, strictly convex and smooth function $y = G^{(5)}(x)$ over $(x(\beta), +\infty)$ such that $G^{(5)}(+\infty) = -\infty$ and $G^{(5)'}(+\infty) = 0$ (and hence $G^{(5)} \sim H_{+\infty}$). The pieces $G^{(4)}$ and $G^{(5)}$ have a common tangent T which passes through the turning point $(x(\beta), y(\beta))$. In view of Lemma 3.2 and (6.64) as well as (6.65), we see further that $L_{(\tau-\sigma)/(\tau-\sigma-n)}$ is the asymptote of $G^{(3)}$ at $-\infty$ and the asymptote of $G^{(4)}$ at $+\infty$. Since $G^{(4)}$ is concave and $G^{(5)}$ is convex, it is not difficult to see that the tangent T has a slope which is greater than that of the asymptote.

In view of the above discussions, the envelope G_1 is made up of $G^{(1)}$ and $G^{(2)}$, the envelope G_2 is $G^{(3)}$ with the point $(-1, 0)$ removed, and G_3 is made up of $G^{(4)}$ and $G^{(5)}$. The corresponding dual sets of order 0 of G_1, G_2 and G_3, in view of the distribution maps in Figures 3.5, 3.1, 3.10 and A.8c are easily obtained. Since the $\mathbf{C}\backslash\{(\tau - \sigma)/(\tau - \sigma - n)\}$-characteristic region is $\mathbf{C}\backslash L_{(\tau-\sigma)/(\tau-\sigma-n)}$, the intersection of the $\mathbf{C}\backslash\{1\}$-, $\mathbf{C}\backslash\{(\tau - \sigma)/(\tau - \sigma - n)\}$-, $\mathbf{C}\backslash(0, 1)$-, $\mathbf{C}\backslash(1, (\tau - \sigma)/(\tau - \sigma - n))$- and $\mathbf{C}\backslash((\tau - \sigma)/(\tau - \sigma - n), \infty)$-characteristic regions of $h(\lambda|x, y)$, in view of Theorem 3.22, is easily obtained so that (p, q) is in it if, and only if, $p \le 0$ and $q > G(p)$. The proof is complete.

We now consider the final case where $\sigma \ge \tau + 1$. Let $Q(\lambda)$ be the quadratic polynomial

$$Q(\lambda) = \lambda^2 + \frac{n(\tau + 1) - 2\sigma(n + \sigma - \tau)}{(n + \sigma)(n + \sigma - \tau)}\lambda + \frac{\sigma(\sigma - \tau)}{n + \sigma - \tau}$$

defined by (6.61). Note that the condition $\sigma \ge \tau + 1$ implies

$$\frac{n(\tau + 1) - 2\sigma(n + \sigma - \tau)}{(n + \sigma)(n + \sigma - \tau)} < 0 < \frac{\sigma(\sigma - \tau)}{n + \sigma - \tau}.$$

Thus by Theorem 5.1, $Q(\lambda)$ does not have any positive roots if, and only if, the discriminant

$$H(\sigma) = -n \left(4\sigma^2 + 4(n - \tau)\sigma - n(\tau + 1)^2\right) < 0.$$

By Theorem 5.2, $Q(\lambda)$ has exactly one positive root if, and only if,

$$H(\sigma) = 0,$$

and it has two distinct positive roots if, and only if,

$$0 < H(\sigma).$$

The function $H(t)$ satisfies $H(0) > 0$ and $H(+\infty) = -\infty$. Furthermore, $H'(t) = -n(8t + 4(n - \tau))$, which does not have any positive root. Thus $H(t)$ is a strictly decreasing function on $(0, \infty)$. The equation $H(t) = 0$ has the unique root

$$r = \frac{1}{2} \left(\tau - n + \sqrt{n+1} \sqrt{\tau^2 + n} \right).$$

Hence $H(t) > 0$ for $t \in (0, r)$ and $H(t) < 0$ for $\sigma \in (r, \infty)$.

We may now conclude that $Q(\lambda)$ has two distinct positive roots if, and only if,

$$\tau + 1 \leq \sigma < \frac{1}{2} \left(\tau - n + \sqrt{n+1} \sqrt{\tau^2 + n} \right), \tag{6.66}$$

it has at most one positive root if, and only if,

$$\sigma \geq \max \left\{ \tau + 1, \frac{1}{2} \left(\tau - n + \sqrt{n+1} \sqrt{\tau^2 + n} \right) \right\}. \tag{6.67}$$

Note that the condition (6.66) is not vacuous since

$$\frac{1}{2} \left(\tau - n + \sqrt{n+1} \sqrt{\tau^2 + n} \right) - \tau = \frac{n(\tau - 1)^2}{2(\tau + n + \sqrt{n+1} \sqrt{\tau^2 + n})} \geq 0.$$

Let

$$\lambda^* \equiv \frac{\sigma - \tau}{\sigma - \tau + n}. \tag{6.68}$$

If $Q(\lambda)$ has two mutually distinct positive roots, then they can be formally written as α and β defined formally by (6.62) and (6.63) respectively. It is easy to verify that

$$0 < \lambda^* < \alpha < \beta < \frac{\sigma}{n + \sigma} < 1. \tag{6.69}$$

Theorem 6.26. *Suppose n is an odd integer in $\mathbf{Z}[3, \infty)$ and $\sigma \geq \tau + 1$. Suppose further that condition (6.67) holds. Let the curve G be defined by (6.57) for $\lambda > (\sigma - \tau)/(\sigma - \tau + n)$. Then G is also the graph of a strictly convex function defined on $\mathbf{R}$. Furthermore, (p, q) is in the $\mathbf{C}\backslash(0, \infty)$-characteristic region of (6.49) if, and only if, $q > G(p)$*

Proof. It suffices to consider the polynomial

$$h(\lambda|x, y) \equiv (\lambda - 1)^n \lambda^\sigma + x\lambda^{\sigma - \tau}(\lambda - 1)^n + y,$$

obtained from (6.49) by replacing $p = x$ and $q = y$ into it and then multiplying the resulting equation by λ^σ. Let L_λ be the straight line defined by $h(\lambda|x, y) = 0$. Note that

$$h'_\lambda(\lambda|x, y) = \lambda^{\sigma - \tau - 1}(\lambda - 1)^{n-1}((n + \sigma - \tau)\lambda + \tau - \sigma) x$$
$$+ \lambda^{\sigma - 1}(\lambda - 1)^{n-1}((n + \sigma)\lambda - \sigma).$$

Thus the determinant of the system $h(\lambda|x,y) = 0 = h'_\lambda(\lambda|x,y)$ is

$$D(\lambda) = \lambda^{\sigma-\tau-1}(\lambda-1)^{n-1}\left((n+\sigma-\tau)\lambda+\tau-\sigma\right),$$

which has the positive roots 1 and $(\sigma-\tau)/(n+\sigma-\tau)$. For the sake of convenience, let us denote

$$\lambda^* = \frac{\sigma-\tau}{n+\sigma-\tau}.$$

The $\mathbf{C}\backslash(0,\infty)$-characteristic region of $h(\lambda|x,y)$ is the intersection of the $\mathbf{C}\backslash(0,\lambda^*)$-, $\mathbf{C}\backslash\{\lambda^*\}$-, $\mathbf{C}\backslash(\lambda^*,1)$-, $\mathbf{C}\backslash\{1\}$- and $\mathbf{C}\backslash(1,\infty)$-characteristic regions of $h(\lambda|x,y)$.

The $\mathbf{C}\backslash\{1\}$-characteristic region, since $h(1|x,y) = y$, is $\mathbf{C}\backslash\mathbf{R}$. The $\mathbf{C}\backslash\{\lambda^*\}$-characteristic region is the set of points that do not lie on the straight line L_{λ^*}, that is, $\mathbf{C}\backslash L_{\lambda^*}$. As for the other three regions, the $\mathbf{C}\backslash(0,\lambda^*)$-characteristic region is the dual set of order 0 of the envelope of the family $\Phi_{(0,\lambda^*)} = \{L_\lambda|\ \lambda \in (0,\lambda^*)\}$, the $\mathbf{C}\backslash(\lambda^*,1)$-characteristic region is the dual set of order 0 of the envelope of the family $\Phi_{(\lambda^*,1)} = \{L_\lambda|\ \lambda \in (\lambda^*,1)\}$, and the $\mathbf{C}\backslash(1,\infty)$-characteristic region is the dual set of order 0 of the envelope of the family $\Phi_{(1,\infty)} = \{L_\lambda|\ \lambda \in (1,\infty)\}$.

By Theorem 2.3, the envelope G_1 of the family $\Phi_{(0,\lambda^*)}$ is described by the functions in (6.57) for $\lambda \in (0,\lambda^*)$, the envelope G_2 by the same functions but for $\lambda \in (\lambda^*,1)$, and the envelope G_3 by the same functions but for $\lambda \in (1,\infty)$. Indeed, the properties of the functions $x(\lambda)$ and $y(\lambda)$, when treated as functions on $(0,\infty)$, can easily be obtained. The rational functions $x(\lambda)$ and $y(\lambda)$ have a common singularity at $\lambda = \lambda^*$. The derivatives $dx/d\lambda$, dy/dx and d^2y/dx^2 are formally the same as those given by (6.58), (6.59) and (6.60) respectively.

Under the condition (6.67), $Q(\lambda) > 0$ for $\lambda \in (0,\infty)\backslash\{\lambda^*\}$ except perhaps at one point. Thus $x'(\lambda)$, as can be seen from (6.58), is less than 0 for $\lambda \in (0,\infty)\backslash\{\lambda^*\}$ except perhaps at one point. We may now infer from these properties that the curve described by $(x(\lambda),y(\lambda))$ over $(0,\infty)$ is made up of three pieces $G^{(1)}, G^{(2)}$ and $G^{(3)}$ which correspond to the cases where $\lambda \in (0,\lambda^*)$, $\lambda \in (\lambda^*,1)$ and $\lambda \in [1,\infty)$ respectively. See Figure 6.22.

The piece $G^{(1)}$ is the graph of a strictly increasing, strictly concave and smooth function $y = G^{(1)}(x)$ defined over $(-\infty,0)$ such that $G^{(1)}(0^-) = 0$ and $G^{(1)'}(0^-) = 0$. The piece $G^{(2)}$ is the graph of a strictly increasing, strictly convex and smooth function $y = G^{(2)}(x)$ defined over $(-1,\infty)$ such that $G^{(2)}((-1)^+) = 0$, $G^{(2)'}((-1)^+) = 0$ and $G^{(2)}(+\infty) = +\infty$. The piece $G^{(3)}$ is the graph of a strictly decreasing and strictly convex function $y = G^{(3)}(x)$ defined over $(-\infty,-1]$ such that $G^{(3)}((-1)^-) = 0$, $G^{(3)'}((-1)^-) = 0$, $G^{(3)}(-\infty) = +\infty$ and $G^{(3)'}(-\infty) = -\infty$. Since $\lim_{\lambda\to(\lambda^*)-} x(\lambda) = -\infty$ and $\lim_{\lambda\to(\lambda^*)+} x(\lambda) = +\infty$, in view of Lemma 3.2, we also see that L_{λ^*} is the asymptote of $G^{(1)}$ at $-\infty$ and of $G^{(2)}$ at $+\infty$.

The envelope G_1 is $G^{(1)}$, the envelope G_2 is $G^{(2)}$ while the envelope G_3 is $G^{(3)}$ with the point $(-1,0)$ removed. The dual sets of order 0 of G_1, G_2 and G_3 are, in view of the distribution map in Figure A.12, easy to obtain. Their intersection with

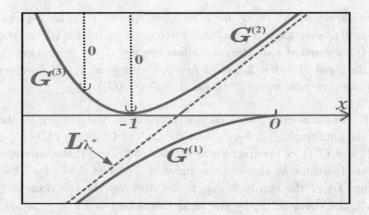

Fig. 6.22

the $\mathbf{C}\backslash\{\lambda^*\}$-, and $\mathbf{C}\backslash\{1\}$-characteristic sets mentioned above are also easy to find so that (p, q) is in it if, and only if, $q > G(p)$. The proof is complete.

In case $\tau = 1$ in the above Theorem, the equation $x(\lambda) = p$ can be written as

$$\lambda^2 + \frac{(n + \sigma - 1)p - \sigma}{n + \sigma}\lambda - \frac{\sigma - 1}{n + \sigma}p = 0.$$

It is easily verified that this equation has the root

$$\lambda = \frac{\sigma - (n + \sigma - 1)p}{2(n + \sigma)} + \frac{\sqrt{[(n + \sigma - 1)p + \sigma]^2 - 4np}}{2(n + \sigma)}$$

in the interval $((\sigma - 1)/(n + \sigma - 1), \infty)$. Thus $y > G(p)$ can be written as

$$q > \frac{\left\{2n + \sigma + (n + \sigma - 1)p - \sqrt{[(n + \sigma - 1)p + \sigma]^2 - 4np}\right\}^n}{2^{n + \sigma}(n + \sigma)^{n + \sigma}}$$

$$\times \left\{\sigma - (n + \sigma - 1)p + \sqrt{[(n + \sigma - 1)p + \sigma]^2 - 4np}\right\}^{\sigma - 1}$$

$$\times \left\{(n + \sigma + 1)p + \sigma + \sqrt{[(n + \sigma - 1)p + \sigma]^2 - 4np}\right\} \tag{6.70}$$

The following is now clear.

Corollary 6.3. *Suppose n is an odd integer in $\mathbf{Z}[3, \infty)$ and $\sigma \geq \tau + 1$. Then (6.49) does not have any positive roots if, and only if, the inequality (6.70) holds.*

The proof of the following result is similar to the previous one, and hence will be sketched.

Theorem 6.27. *Suppose n is an odd integer in $\mathbf{Z}[3, \infty)$ and $\sigma \geq \tau + 1$. Suppose further that condition (6.66) holds. Let α and β be defined by (6.62) and (6.63) and* *the curve G be defined by (6.57) for $\lambda > (\sigma - \tau)/(\sigma - \tau + n)$. Let G_2 and S be the restrictions of G over the intervals $((\sigma - \tau)/(\sigma - \tau + n), \alpha)$ and (β, ∞) respectively. Then G_2 is the graph of a strictly convex function $y = G_2(x)$ defined over $(x(\alpha), \infty)$ and S is the graph of a strictly convex function over $(-\infty, x(\beta))$. Furthermore, the $\mathbf{C}\backslash(0, \infty)$-characteristic region of (6.49) is $\vee(S) \oplus \vee(G_2)$.*

Proof. The parametric functions, the determinant, the number λ^*, the straight lines L_λ, the families $\Phi_{(0,\lambda^*)}, \Phi_{(\lambda^*,1)}$ and $\Phi_{(1,\infty)}$, the $\mathbf{C}\backslash(0, \lambda^*)$-, $\mathbf{C}\backslash\{\lambda^*\}$-, $\mathbf{C}\backslash\{1\}$-, $\mathbf{C}\backslash(\lambda^*, 1)$- and $\mathbf{C}\backslash(1, \infty)$-characteristic regions of $h(\lambda|x, y)$, the envelopes G_1, G_2 and G_3, are the same as those in the proof of Theorem 6.26. By Theorem 2.3, the envelope G_1 of the family $\Phi_{(0,\lambda^*)}$ is described by the functions in (6.57) for $\lambda \in (0, \lambda^*)$, the envelope G_2 by the same functions but for $\lambda \in (\lambda^*, 1)$, and the envelope G_3 by the same functions but for $\lambda \in (1, \infty)$. The properties of the functions $x(\lambda)$ and $y(\lambda)$, when treated as functions on $(0, \infty)$, can easily be obtained. Indeed, the curve described by $(x(\lambda), y(\lambda))$ over $(0, \infty)$ is made up of five pieces $G^{(1)}, G^{(2)}, G^{(3)}, G^{(4)}$ and $G^{(5)}$ as shown in Figure 6.23. They correspond to the cases where $\lambda \in (0, \lambda^*), (\lambda^*, \alpha], (\alpha, \beta), [\beta, 1)$ and $(1, \infty)$ respectively, where α and β are defined by (6.62) and (6.63) and under the condition (6.66), they satisfy (6.69).

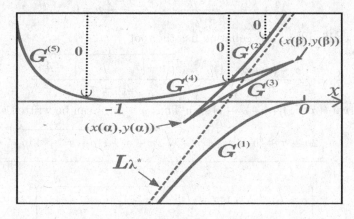

Fig. 6.23

The piece $G^{(1)}$ is the graph of a strictly increasing, strictly concave and smooth function $y = G^{(1)}(x)$ defined on $(-\infty, 0)$ such that $G^{(1)}(0^-) = 0$ and $G^{(1)'}(0^-) = 0$. The piece $G^{(2)}$ is the graph of a strictly increasing, strictly convex and smooth function $y = G^{(2)}(x)$ defined on $[x(\alpha), +\infty)$ such that $G^{(2)}(+\infty) = +\infty$. The piece $G^{(3)}$ is the graph of a strictly increasing, strictly convex and smooth function $y = G^{(3)}(x)$ defined on $(x(\alpha), x(\beta))$. The piece $G^{(4)}$ is the graph of a strictly increasing, strictly convex and smooth function $y = G^{(4)}(x)$ defined on $(-1, x(\beta)]$ such that $G^{(4)}((-1)^+) = 0$ and $G^{(4)'}((-1)^+) = 0$. The piece $G^{(5)}$ is the graph of a strictly decreasing, strictly convex function $y = G^{(5)}(x)$ defined on $(-\infty, -1]$ such that $G^{(5)}(-\infty) = +\infty$ and $G^{(5)'}(-\infty) = -\infty$. The straight line L_{λ^*} is the asymptote of the function $G^{(1)}$ at $-\infty$ and the function $G^{(2)}$ at $+\infty$.

The envelope G_1 is $G^{(1)}$, the envelope G_2 is made up by $G^{(2)}$, $G^{(3)}$ and $G^{(4)}$, while the envelope G_3 is $G^{(5)}$ with the point $(-1, 0)$ removed. The dual sets of order 0 of G_1, G_2 and G_3 are, in view of the distribution map in Figure A.27, easy to obtain. Their intersection with the $\mathbf{C}\backslash\{\lambda^*\}$-, and $\mathbf{C}\backslash\{1\}$-characteristic sets, in view of Theorem 3.22, are also easy to find so that (p, q) is in it if, and only if, $\vee(G^{(5)}) \oplus \vee(G^{(4)}) \oplus \vee(G^{(2)}) \oplus \vee(\Theta_0)$. The proof is complete.

6.2.5.2 *The case where n is even*

In this section, we assume throughout that the positive integer n is even. To facilitate discussions, we will also classify (6.49) into four cases: (i) $0 \le \sigma < \tau$, (ii) $\sigma = \tau = 1$, (iii) $\sigma = \tau > 1$, and (iv) $\sigma \ge \tau + 1$.

Theorem 6.28. *Suppose n is even and $0 \le \sigma < \tau$. Then equation (6.49) does not have any positive solutions if, and only if, $p \ge 0$ and $q > 0$.*

Proof. It suffices to consider the polynomial

$$h(\lambda|p, q) \equiv (\lambda - 1)^n \lambda^\tau + (\lambda - 1)^n p + q\lambda^{\tau-\sigma}.$$

Since $\lim_{\lambda\to\infty} h(\lambda|p, q) = \infty$, $h(1|p, q) = q$ and $h(0|p, q) = p$, thus when $p < 0$ or $q \le 0$, (6.49) will have at least one positive root. When $p \ge 0$ and $q > 0$, it is easy to see that $h(\lambda|p, q) > 0$ for all $\lambda > 0$. The proof is complete.

Theorem 6.29. *Suppose n is even and $\sigma = \tau = 1$. Let $G(x)$ be the function defined by*

$$G(x) = \frac{n^n}{(n+1)^{n+1}}(1+x)^{n+1}, \ x \in (-\infty, 1/n).$$

Then the $\mathbf{C}\backslash(0, \infty)$-characteristic region of (6.49) is $\vee(G) \oplus \triangledown(-\Upsilon) \oplus \vee(\Theta_0)$.

Proof. When $\sigma = \tau = 1$, it suffices to consider the polynomial

$$h(\lambda|x, y) = (\lambda - 1)^n \lambda + (\lambda - 1)^n x + y$$

obtained from (6.49) by substituting $\sigma = \tau = 1$ and $p = x$ and $q = y$ into it. Let L_λ be the straight line defined by $h(\lambda|x, y) = 0$. We have

$$h_\lambda'(\lambda|x, y) = n(\lambda - 1)^{n-1} x + (\lambda - 1)^{n-1}((n+1)\lambda - 1).$$

Since the determinant of the system $h(\lambda|x, y) = 0 = h_\lambda'(\lambda|x, y)$ is $D(\lambda) = n(\lambda - 1)^{n-1}$ which has the unique positive root 1.

The $\mathbf{C}\backslash(0, \infty)$-characteristic region of $h(\lambda|x, y)$ is the intersection of the $\mathbf{C}\backslash(0, 1)$-, $\mathbf{C}\backslash\{1\}$- and the $\mathbf{C}\backslash(1, \infty)$-characteristic region of $h(\lambda|x, y)$. The $\mathbf{C}\backslash\{1\}$-characteristic region, in view of the fact that $h(1|x, y) = y$, is just $\mathbf{C}\backslash\mathbf{R}$. As for the other two regions, the $\mathbf{C}\backslash(0, 1)$ is just the dual set of order 0 of the envelope of the family of straight lines $\Phi_{(0,1)} = \{L_\lambda | \lambda \in (0, 1)\}$, and the $\mathbf{C}\backslash(1, \infty)$-characteristic region is the dual set of order 0 of the envelope of the family of straight lines $\Phi_{(1,\infty)} = \{L_\lambda | \lambda \in (1,\infty)\}$.

By Theorem 2.3, the envelope G_1 of the family $\Phi_{(0,1)}$ is described by the parametric functions

$$x(\lambda) = \frac{-\lambda(n+1)+1}{n} \text{ and } y(\lambda) = -(\lambda-1)^n\lambda - (\lambda-1)^n\frac{-\lambda(n+1)+1}{n} \quad (6.71)$$

for $\lambda \in (0,1)$, while the envelope G_2 of the family $\Phi_{(1,\infty)}$ by the same parametric functions but $\lambda \in (1,\infty)$. Indeed, the properties of the functions $x(\lambda)$ and $y(\lambda)$, when treated as functions defined for $\lambda \in (0,\infty)$ are easy to obtain. The linear function $x(\lambda)$ maps $(0,\infty)$ onto $(-\infty, 1/n)$ and $x(1) = -1$. By solving $x = (1 - \lambda(n+1))/n$ for λ, and then substituting $\lambda = (1 - nx)/(n + 1)$ into $y(\lambda)$, we see that the curve G described by $(x(\lambda), y(\lambda))$ over $(0,\infty)$ is also the graph of the function $y = G(x)$ defined by

$$G(x) = \frac{(-1)^{n+1}n^n}{(n+1)^{n+1}}(1+x)^{n+1}, \ x \in (-\infty, 1/n). \quad (6.72)$$

Note further that

$$G((1/n)^-) = 1/n, \ G(-1) = 0 \text{ and } G(-\infty) = \infty,$$

$$\frac{dG}{dx} = -\frac{n^n}{(n+1)^n}(1+x)^n,$$

and

$$\frac{d^2G}{dx^2} = -\frac{n^{n+1}}{(n+1)^n}(1+x)^{n-1}.$$

Since $G'(-\infty) = -\infty$, by Lemma 3.4, the function $G(x)$ satisfies $G \sim H_{-\infty}$. Since $G'((1/n)^-) = -1$, $-\Upsilon$ is the tangent line of G at $x = (1/n)^-$. By means of the above information, we may plot the curve G as shown in Figure 6.24.

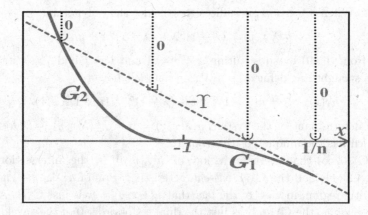

Fig. 6.24

The envelope G_2 is the restriction of the function $y = G(x)$ over the interval $(-\infty, -1)$ which is also strictly decreasing and strictly convex, and the envelope G_1 is the restriction of the function $y = G(x)$ over the interval $(-1, 1/n)$ which is strictly decreasing and strictly concave.

In view of the distribution maps in Figures 3.1 and 3.11, the dual sets of order 0 of G_1 and G_2 can easily be found. Furthermore, the intersections of these dual sets and $\mathbf{C}\backslash\mathbf{R}$, in view of Theorem 3.22, is also easily found to be $\vee(G_2) \oplus \triangledown(T) \oplus \vee(\Theta_0)$.

We remark that the tangent line $-\Upsilon$ intercepts with G_2 at a unique point $(p^*, -p^*)$ which can be found by solving for the unique root p^* in $(-\infty, -1)$ of the equation

$$p = \frac{n^n}{(n+1)^{n+1}}(1+p)^{n+1}.$$

In terms of p^*, the conclusion of Theorem 6.29 asserts that (p, q) is a point in the $\mathbf{C}\backslash(0,\infty)$-characteristic region of $h(\lambda|x, y)$ if, and only if, $p \leq p^*$ and $q > G(p)$, or, $p \in (p^*, 0)$ and $q \geq -p$, or, $p \geq 0$ and $q > 0$.

As another remark, note that the proof of Theorem 6.29 is basically the same as that of Theorem 6.23. The main difference is that n is odd in Theorem 6.23, while n is even in Theorem 6.29. Therefore, the restriction of the function in (6.55) over the interval $(-\infty, -1)$ is the same as the function in (6.72) over the interval $(-\infty, -1)$. But the restriction of the function in (6.72) over the interval is the negative of the restriction of (6.55) over the interval $(-1, 1/n)$ (cf. Figures 6.17 and 6.24). This fact can also be explained by observing the behavior of the parametric functions $y(\lambda)$ in (6.54) and (6.71) for $\lambda \in (0, 1)$.

Therefore it is not surprising that similar proofs are valid for the following results that corresponds to Theorems 6.24, 6.25, 6.26 and 6.27 respectively. We will skip these proofs. Instead, we will draw the corresponding figures that help to understand the statement of the Theorems.

Theorem 6.30. *Suppose n is even and $\sigma = \tau > 1$. Let the curve S be defined by*

$$x(\lambda) = \frac{n+\tau}{n}\lambda^{\tau-1}\left(\frac{\tau}{n+\tau} - \lambda\right) \text{ and } y(\lambda) = \frac{\tau}{n}(\lambda - 1)^{n+1}\lambda^{\tau-1} \qquad (6.73)$$

for $\lambda > (\tau - 1)/(n + \tau)$. Then S is also the graph of a function $y = S(x)$ defined on $\left(-\infty, \frac{(\tau-1)^{\tau-1}}{n(n+\tau)^{\tau-1}}\right)$. Furthermore, the $\mathbf{C}\backslash(0,\infty)$-characteristic region of (6.49) is $\vee(S) \oplus \triangledown(-\Upsilon) \oplus \vee(\Theta_0)$ (see Figure 6.25).

Theorem 6.31. *Suppose n is even, $\sigma \geq \tau + 1$ and condition (6.67) holds. Let the curve G be defined by (6.57) for $\lambda \in (0,\infty)\backslash\{(\sigma - \tau)/(\sigma - \tau + n)\}$. Then the restriction $G^{(1)}$ of G over $(0, (\sigma - \tau)/(\sigma - \tau + n))$ is the graph of a strictly convex function $y = G^{(1)}(x)$ defined over $(-\infty, 0)$ and the restriction $G^{(3)}$ of G over $(1, \infty)$ is the graph of a strictly convex function $y = G^{(3)}(x)$ defined over $(-\infty, -1)$. Furthermore, the $\mathbf{C}\backslash(0,\infty)$-characteristic region of the function (6.49) is $\vee(G^{(1)}) \oplus \vee(G^{(3)}) \oplus \vee(\Theta_0)$ (see Figure 6.26).*

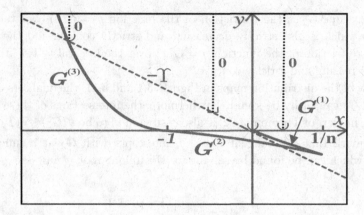

Fig. 6.25

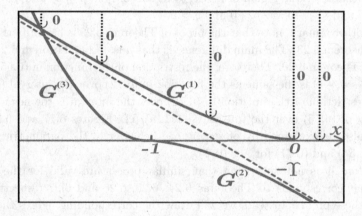

Fig. 6.26

Theorem 6.32. *Suppose n is even, $\sigma \geq \tau + 1$ and condition (6.66) holds. Let the curve G be defined by (6.57) for $\lambda \in (0, \infty) \backslash \{(\sigma - \tau)/(\sigma - \tau + n)\}$. Then the restriction $G^{(1)}$ of G over $(0, (\sigma - \tau)/(\sigma - \tau + n))$ is the graph of a strictly convex function $y = G^{(1)}(x)$ defined over $(-\infty, 0)$ and the restriction $G^{(5)}$ of G over $(1, \infty)$ is the graph of a strictly convex function $y = G^{(3)}(x)$ defined over $(-\infty, -1)$. Furthermore, the $\mathbf{C} \backslash (0, \infty)$-characteristic region of the function (6.49) is $\vee(G^{(1)}) \oplus \vee(G^{(3)}) \oplus \vee(\Theta_0)$ (see Figure 6.27).*

6.3   Δ-Polynomials Involving Three Powers

If we consider the difference equation

$$x_{k+1} - x_k + p x_{k-\tau} + q x_{k-\sigma} + r x_{k-\delta} = 0, \ n \in \mathbf{N}, \tag{6.74}$$

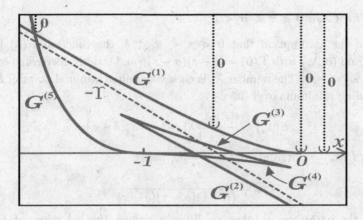

Fig. 6.27

where p, q, r are real numbers and τ, σ, δ are integers, then we are led to Δ-polynomials of the form

$$\lambda - 1 + \lambda^{-\tau} p + \lambda^{-\sigma} q + \lambda^{-\delta} r.$$

We will find the $\mathbf{C} \backslash (0, \infty)$-characteristic regions of the above function in cases where $0 < \tau < \sigma < \delta$ and where $\tau < \sigma < \delta < -1$. For this purpose, let

$$f(\lambda | x, y) = \lambda - 1 + \lambda^{-\tau} p + \lambda^{-\sigma} x + \lambda^{-\delta} y. \tag{6.75}$$

We will determine the $\mathbf{C} \backslash (0, \infty)$-characteristic region for $f(\lambda | x, y)$ for different values of p.

To facilitate discussions, in this section, we will let

$$\Gamma(t; x, y) = \frac{t^{y - \tau}}{\delta - \sigma} \left[(x + 1) t^{\tau + 1} - x t^\tau + (x - \tau) p \right],$$

and let the functions $x(\lambda)$ and $y(\lambda)$ be defined respectively by

$$x(\lambda) = -\Gamma(\lambda; \delta, \sigma) = \frac{\delta \lambda^\sigma - (\delta + 1) \lambda^{\sigma+1} - (\delta - \tau) p \lambda^{\sigma - \tau}}{\delta - \sigma} \tag{6.76}$$

and

$$y(\lambda) = \Gamma(\lambda; \sigma, \delta) = \frac{(\sigma + 1) \lambda^{\delta+1} - \sigma \lambda^\delta + (\sigma - \tau) p \lambda^{\delta - \tau}}{\delta - \sigma}. \tag{6.77}$$

Since

$$x'(\lambda) = \frac{\delta \sigma \lambda^{\sigma-1} - (\delta + 1)(\sigma + 1) \lambda^\sigma - (\delta - \tau)(\sigma - \tau) p \lambda^{\sigma - \tau - 1}}{\delta - \sigma},$$

if we let

$$F(\lambda) = -\frac{(\delta - \sigma) x'(\lambda)}{\lambda^{\sigma - \tau - 1}} = (\delta + 1)(\sigma + 1) \lambda^{\tau+1} - \sigma \delta \lambda^\tau + (\delta - \tau)(\sigma - \tau) p, \tag{6.78}$$

then

$$F'(\lambda) = (\tau + 1)(\sigma + 1)(\delta + 1) \lambda^{\tau-1} (\lambda - \lambda_*),$$

where

$$\lambda_* = \frac{\tau \sigma \delta}{(\tau + 1)(\sigma + 1)(\delta + 1)}. \tag{6.79}$$

6.3.1   *The Case $0 < \tau < \sigma < \delta$*

In view of the assumption that $0 < \tau < \sigma < \delta$, the function $F(\lambda)$ is strictly decreasing on $(0, \lambda_*)$ with $F(0) = (\delta - \tau)(\sigma - \tau)p$ and strictly increasing on (λ_*, ∞) with $F(+\infty) = +\infty$. The number λ_* is then the unique minimal point of F and the corresponding minimum over $[0, \infty)$ is

$$F(\lambda_*) = -\frac{\sigma\delta}{\tau + 1}\lambda_*^\tau + (\sigma - \tau)(\delta - \tau)p.$$

Note that $F(\lambda_*) = 0$ if, and only if, $p = p_*$, where

$$p_* = \frac{\sigma\delta}{(\tau + 1)(\sigma - \tau)(\delta - \tau)}\lambda_*^\tau.$$

When the parameter p takes on different values, the behaviors of F are also different. We will let

$$\lambda^* = \frac{\tau\sigma}{(\tau + 1)(\sigma + 1)},$$

and

$$\tilde{\lambda} = \frac{\tau}{\tau + 1},$$

and consider five different and exhaustive cases:

$$p \le 0, \tag{6.80}$$

$$0 < p \le \frac{(\tilde{\lambda})^\tau}{\tau + 1}, \tag{6.81}$$

$$\frac{(\tilde{\lambda})^\tau}{\tau + 1} < p \le \frac{\sigma(\lambda^*)^\tau}{(\sigma - \tau)(\tau + 1)} \tag{6.82}$$

$$\frac{\sigma(\lambda^*)^\tau}{(\sigma - \tau)(\tau + 1)} < p < p_*, \tag{6.83}$$

and

$$p \ge p_*. \tag{6.84}$$

Under the condition that $p \le 0$, we have $F(0^+) = (\sigma - \tau)(\delta - \tau)p \le 0$ and

$$F(\lambda_*) = -\frac{\sigma\delta}{\tau + 1}\lambda_*^\tau + (\sigma - \tau)(\delta - \tau)p < 0.$$

Thus the function $F(\lambda)$ has a unique root $\lambda_1^{(1)}$ in (λ_*, ∞) so that $F(\lambda) < 0$ on $(0, \lambda_1^{(1)})$ and $F(\lambda) > 0$ on $(\lambda_1^{(1)}, \infty)$. We remark that in view of (6.78), $x'(\lambda)$ has the same unique root $\lambda_1^{(1)}$ in (λ_*, ∞), and $x'(\lambda) > 0$ on $(0, \lambda_1^{(1)})$ as well as $x'(\lambda) < 0$ on $(\lambda_1^{(1)}, \infty)$. Similar remarks will apply for the other three cases to be discussed below.

Under the condition (6.81), that is, $0 < p \leq \tau^\tau/(\tau+1)^{\tau+1}$, we have

$$F(0^+) = (\sigma - \tau)(\delta - \tau)p > 0 \qquad .$$

and

$$F(\lambda_*) \leq (\delta+1)(\sigma+1)\lambda_*^{\tau+1} - \sigma\delta\lambda_*^\tau + (\delta-\tau)(\sigma-\tau)\frac{\tau^\tau}{(\tau+1)^{\tau+1}}$$

$$\leq \frac{-\sigma\delta\tau^\tau}{(\tau+1)^{\tau+1}}\left[\frac{\sigma^\tau\delta^\tau}{(\sigma+1)^\tau(\delta+1)^\tau} - \frac{(\sigma-\tau)(\delta-\tau)}{\sigma\delta}\right].$$

We assert that the last term is negative. Indeed, consider the function

$$\psi(t) = \frac{t-\tau}{t}\left(\frac{t+1}{t}\right)^\tau, \ t > \tau.$$

Since $\psi(\tau+) = 0$, $\psi(+\infty) = 1$ and

$$\psi'(t) = \tau(\tau+1)\frac{(t+1)^{\tau-1}}{t^{\tau+2}} \geq 0,$$

thus $\psi(t) < 1$ for $t > \tau$. This implies

$$\frac{t-\tau}{t} < \frac{t^\tau}{(t+1)^\tau},$$

and consequently

$$0 < \frac{\sigma-\tau}{\sigma} < \frac{\sigma^\tau}{(\sigma+1)^\tau}, \ 0 < \frac{\delta-\tau}{\delta} < \frac{\delta^\tau}{(\delta+1)^\tau},$$

and $F(\lambda_*) < 0$. Thus the function $F(\lambda)$ has two unique roots $\lambda_1^{(2)}, \lambda_2^{(2)}$ satisfying $0 < \lambda_1^{(2)} < \lambda_* < \lambda_2^{(2)} < \infty$ so that $F(\lambda) > 0$ on $(0, \lambda_1^{(2)})$, $F(\lambda) < 0$ on $(\lambda_1^{(2)}, \lambda_2^{(2)})$ and $F(\lambda) > 0$ on $(\lambda_2^{(2)}, \infty)$.

Under the conditions (6.82) or (6.83), we have $F(0^+) = (\sigma - \tau)(\delta - \tau)p > 0$ and

$$F(\lambda_*) \leq (\delta+1)(\sigma+1)\lambda_*^{\tau+1} - \sigma\delta\lambda_*^\tau + \frac{\sigma\delta}{\tau+1}\lambda_*^\tau < 0.$$

As in the previous case, the function $F(\lambda)$ has two unique roots $\lambda_1^{(3)}, \lambda_2^{(3)}$ satisfying $0 < \lambda_1^{(3)} < \lambda_* < \lambda_2^{(3)} < \infty$ so that $F(\lambda) > 0$ on $(0, \lambda_1^{(3)})$, $F(\lambda) < 0$ on $(\lambda_1^{(3)}, \lambda_2^{(3)})$ and $F(\lambda) > 0$ on $(\lambda_2^{(3)}, \infty)$. In addition, if the condition (6.83) holds, we have

$$F(\lambda^*) = (\delta+1)(\sigma+1)\left[\frac{\tau\sigma}{(\tau+1)(\sigma+1)}\right]^{\tau+1}$$

$$-\sigma\delta\left[\frac{\tau\sigma}{(\tau+1)(\sigma+1)}\right]^\tau + (\delta-\tau)(\sigma-\tau)p$$

$$= -\frac{(\delta-\tau)\sigma}{\tau+1}\left[\frac{\tau\sigma}{(\tau+1)(\sigma+1)}\right]^\tau + (\delta-\tau)(\sigma-\tau)p$$

$$> -\frac{(\delta-\tau)\sigma}{\tau+1}\left[\frac{\tau\sigma}{(\tau+1)(\sigma+1)}\right]^\tau + (\delta-\tau)\frac{\sigma}{\tau+1}\left[\frac{\tau\sigma}{(\tau+1)(\sigma+1)}\right]^\tau$$

$$= 0,$$

so we also know that $\lambda_* < \lambda_2^{(3)} < \lambda^*$.

Under the condition (6.84), we have $F(0^+) = (\sigma - \tau)(\delta - \tau)p > 0$ and

$$F(\lambda_*) = -\frac{\sigma\delta}{\tau+1}\lambda_*^\tau + (\sigma - \tau)(\delta - \tau)p \geq -\frac{\sigma\delta}{\tau+1}\lambda_*^\tau + \frac{\sigma\delta}{\tau+1}\lambda_*^\tau = 0.$$

Thus the function $F(\lambda)$ is positive on $(0, \infty)$ except possibly at λ_*.

Next note that the function $y(\lambda)$ defined by (6.77) is similar in form to the function $x(\lambda)$ defined by (6.76). Therefore, similar properties are expected. Indeed,

$$\frac{(\delta - \sigma)y'(\lambda)}{\lambda^{\delta-\tau-1}} = (\sigma + 1)(\delta + 1)\lambda^{\tau+1} - \delta\sigma\lambda^\tau + (\delta - \tau)(\sigma - \tau)p = F(\lambda).$$

Furthermore, it is easily verified that

$$\frac{dy}{dx} = -\lambda^{\delta-\sigma},$$

and

$$\frac{d^2y}{dx^2} = \frac{(\delta - \sigma)^2\lambda^{\delta-2\sigma+\tau}}{F(\lambda)}$$

for positive λ that are different from the roots of $F(\lambda)$.

Lemma 6.1. *The following properties of the functions $x(\lambda)$ and $y(\lambda)$ defined by (6.76) and (6.77) hold:*

(i) *under condition (6.80), $x'(\lambda)$ and $y'(\lambda)$ have the unique root $\lambda_1^{(1)}$ in (λ_*, ∞), and $x'(\lambda) > 0, y'(\lambda) < 0$ on $(0, \lambda_1^{(1)})$, as well as $x'(\lambda) < 0, y'(\lambda) > 0$ on $(\lambda_1^{(1)}, \infty)$;*

(ii) *under the condition (6.81), $x'(\lambda)$ and $y'(\lambda)$ have the two unique roots $\lambda_1^{(2)}, \lambda_2^{(2)}$ satisfying $0 < \lambda_1^{(2)} < \lambda_* < \lambda_2^{(2)} < \infty$, and $x'(\lambda) < 0, y'(\lambda) > 0$ on $(0, \lambda_1^{(2)})$, and $x'(\lambda) > 0, y'(\lambda) < 0$ on $(\lambda_1^{(2)}, \lambda_2^{(2)})$, as well as $x'(\lambda) < 0, y'(\lambda) > 0$ on $(\lambda_2^{(2)}, \infty)$;*

(iii) *under the condition (6.82), $x'(\lambda)$ and $y'(\lambda)$ has the two unique roots $\lambda_1^{(3)}, \lambda_2^{(3)}$ satisfying $0 < \lambda_1^{(3)} < \lambda_* < \lambda_2^{(3)} < \infty$, and $x'(\lambda) < 0, y'(\lambda) > 0$ on $(0, \lambda_1^{(3)})$, and $x'(\lambda) > 0, y'(\lambda) < 0$ on $(\lambda_1^{(3)}, \lambda_2^{(3)})$, as well as $x'(\lambda) < 0, y'(\lambda) > 0$ on $(\lambda_2^{(3)}, \infty)$;*

(iv) *under the condition (6.83), $x'(\lambda)$ and $y'(\lambda)$ has the two unique roots $\lambda_1^{(3)}, \lambda_2^{(3)}$ satisfying $0 < \lambda_1^{(3)} < \lambda_* < \lambda_2^{(3)} < \lambda^* < \infty$, and $x'(\lambda) < 0, y'(\lambda) > 0$ on $(0, \lambda_1^{(3)})$, and $x'(\lambda) > 0, y'(\lambda) < 0$ on $(\lambda_1^{(3)}, \lambda_2^{(3)})$, as well as $x'(\lambda) < 0, y'(\lambda) > 0$ on $(\lambda_2^{(3)}, \infty)$; and finally*

(v) *under the condition (6.84), $x'(\lambda)$ is negative while $y'(\lambda)$ is negative on $(0, \infty)$, except possibly at λ_**

Theorem 6.33. *Suppose $p \leq 0$. Let G_2 be the curve described by the parametric functions $x(\lambda) = -\Gamma(\lambda; \delta, \sigma)$ and $y(\lambda) = \Gamma(\lambda; \sigma, \delta)$ for $\lambda \in [\lambda_1^{(1)}, \infty)$. Then G_2 is also the graph of a strictly convex function $y = G_2(x)$ defined over $(-\infty, -\Gamma(\lambda_1^{(1)}; \delta, \sigma)]$ and the $\mathbf{C}\backslash(0, \infty)$-characteristic region of (6.75) is $\vee(G_2) \oplus \triangledown(\Theta_0)$.*

Proof. It suffices to consider the polynomial

$$h(\lambda|x, y) = \lambda^{\delta+1} - \lambda^{\delta} + p\lambda^{\delta-\tau} + x\lambda^{\delta-\sigma} + y. \tag{6.85}$$

Since

$$h'_\lambda(\lambda|x, y) = (\delta + 1)\lambda^{\delta} - \delta\lambda^{\delta-1} + (\delta - \tau)p\lambda^{\delta-\tau-1} + (\delta - \sigma)x\lambda^{\delta-\sigma-1},$$

the determinant of the linear system $h(\lambda|x, y) = 0 = h'_\lambda(\lambda|x, y)$ is $D(\lambda) = -(\delta - \sigma)\lambda^{\delta-\sigma-1}$ which does not vanish for $\lambda > 0$. By Theorem 2.6, the $\mathbf{C}\backslash(0, \infty)$-characteristic region for $h(\lambda|x, y)$ is the dual set of order 0 of the envelope G of the family of straight lines $\{L_\lambda| \lambda \in (0, \infty)\}$ defined by $L_\lambda : h(\lambda|x, y) = 0$. To find the envelope G, we solve from $h(\lambda|x, y) = 0 = h'_\lambda(\lambda|x, y)$ for x and y to yield the parametric functions

$$x(\lambda) = \frac{\lambda^{\sigma-\tau}}{\delta - \sigma}\left[\delta\lambda^\tau - (\delta + 1)\lambda^{\tau+1} - (\delta - \tau)p\right] = -\Gamma(\lambda; \delta, \sigma), \tag{6.86}$$

$$y(\lambda) = \frac{\lambda^{\delta-\tau}}{\delta - \sigma}\left[(\sigma + 1)\lambda^{\tau+1} - \sigma\lambda^\tau + (\sigma - \tau)p\right] = \Gamma(\lambda; \sigma, \delta), \tag{6.87}$$

where $\lambda > 0$. The polynomials $x(\lambda)$ and $y(\lambda)$ have been defined previously by (6.76) and (6.77) respectively, and their properties have also been discussed. Therefore, as in Example 2.2, we may infer from Lemma 6.1 and the derivatives dy/dx and d^2y/dx^2 as well as other easily obtained information that the envelope G is composed of two curves G_1 and G_2 as shown in Figure 6.28. The turning point (q_1, r_1) corresponds to the point which is the unique maximal point of the function $x(\lambda)$ in the interval $(0, \infty)$ and is therefore the unique root $\lambda_1^{(1)}$ in the interval (λ_*, ∞) of the equation $x'(\lambda) = 0$. The first piece G_1 corresponds to the case where $\lambda \in (0, \lambda_1^{(1)})$ and the second piece G_2 to the case where $\lambda \in [\lambda_1^{(1)}, \infty)$. Furthermore, G_1 is the graph of a strictly decreasing, strictly concave and smooth function $y = G_1(x)$ over the interval $(0, q_1)$ such that $G_1(0^+) = 0$ and $G_1'(0^+) = 0$, while G_2 is the graph of a strictly decreasing, strictly convex and smooth function $y = G_2(x)$ the interval $(-\infty, q_1)$ such that $G_2(-\infty) = +\infty$ and $G_2'(-\infty) = -\infty$. In view of the distribution map in Figure A.3, the dual set of order 0 of G is $\vee(G_2) \oplus \triangledown(\Theta_0)$. The proof is complete.

We remark, in view of Figure 6.28, that the domain of the function G_2 may be restricted to a smaller interval. Indeed, note that the curve G_2 intersects the x-axis at some unique point $(q_2, 0)$. To locate it, we first show that $y(\lambda) = 0$ has a unique root in (λ^*, ∞). To see this, we may let

$$\phi(\lambda) = \lambda^{\tau-\delta}(\delta - \sigma)y(\lambda) = \lambda^{\tau-\delta}(\delta - \sigma)\Gamma(\lambda; \sigma, \delta). \tag{6.88}$$

Since

$$\phi'(\lambda) = (\tau + 1)(\sigma + 1)\lambda^{\tau-1}(\lambda - \lambda^*), \quad \lambda^* = \frac{\tau\sigma}{(\tau + 1)(\sigma + 1)},$$

we see that the function $\phi(\lambda)$ satisfies $\phi(+\infty) = \infty$, $\phi'(\lambda^*) = 0$, $\phi'(\lambda) < 0$ for $\lambda \in (0, \lambda^*)$, $\phi(0) = (\sigma - \tau)p \leq 0$, and

$$\phi(\lambda^*) = -\sigma(\lambda^*)^\tau + (\sigma - \tau)p < 0.$$

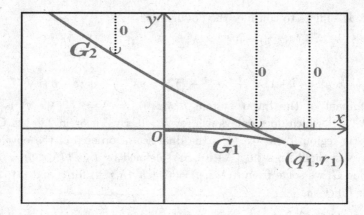

Fig. 6.28

Thus $\phi(\lambda) < 0$ for $\lambda \in (0, \lambda^*)$, $\phi'(\lambda) > 0$ for $\lambda \in (\lambda^*, \infty)$. These imply that there is a unique root t_* of $y(\lambda)$, or $\Gamma(\lambda; \sigma, \delta)$, in (λ^*, ∞) as required. Once we have determined t_* by solving for the unique root in (λ^*, ∞) of the equation $\Gamma(\lambda; \sigma, \delta) = 0$, we can then calculate $q_2 = x(t_*) = -\Gamma(t_*; \delta, \sigma)$. Then (q, r) is a point in the $\mathbf{C} \backslash (0, \infty)$-characteristic region of $h(\lambda | x, y)$ if, and only if, $q \leq q_2$ and $r > G_2(q)$, or, $q > q_2$ and $r \geq 0$.

Theorem 6.34. *Suppose $0 < p \leq \tau^\tau/(\tau + 1)^{\tau+1}$. Let G_3 be the curve described by the parametric functions $x(\lambda) = -\Gamma(\lambda; \delta, \sigma)$ and $y(\lambda) = \Gamma(\lambda; \sigma, \delta)$ for $\lambda \in [\lambda_2^{(2)}, \infty)$. Then G_3 is also the graph of a strictly convex function $y = G_3(x)$ defined over $(-\infty, -\Gamma(\lambda_2^{(2)}; \delta, \sigma)]$ and the $\mathbf{C} \backslash (0, \infty)$-characteristic region of (6.75) is $\vee(G_3) \oplus \triangledown(\Theta_0)$.*

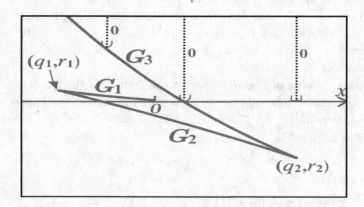

Fig. 6.29

Proof. The proof is similar to that of Theorem 6.33 and hence will be sketched. The parametric functions of the envelope G are given by (6.86) and (6.87) as before.

However, the envelope G is now composed of three curves G_1, G_2 and G_3 as depicted in Figure 6.29.

The turning points (q_1, r_1) and (q_2, r_2) correspond to the unique minimal and maximal points $\lambda_1^{(2)}$ and $\lambda_2^{(2)}$ of the function $x(\lambda)$ in $(0, \infty)$. As explained earlier, we have $0 < \lambda_1^{(2)} < \lambda_* < \lambda_2^{(2)} < \infty$. The first curve G_1 corresponds to the case where $\lambda \in (0, \lambda_1^{(2)})$, the second curve G_2 the case where $\lambda \in [\lambda_1^{(2)}, \lambda_2^{(2)})$ and the third curve G_3 the case where $\lambda \in [\lambda_2^{(2)}, \infty)$. Furthermore, G_1 is the graph of a strictly decreasing, strictly convex and smooth function $y = G_1(x)$ over $(q_1, 0)$ such that $G_1(0^-) = 0$ and $G_1'(0^-) = 0$, G_2 is the graph of a strictly decreasing, strictly concave and smooth function $y = G_2(x)$ over $[q_1, q_2)$, and G_3 is the graph of a strictly decreasing, strictly convex and smooth function $y = G_3(x)$ over $(-\infty, q_2]$ such that $G_3(-\infty) = +\infty$ and $G_3'(-\infty) = -\infty$. The curve G_3 intersects the x-axis at a unique point $(q_3, 0)$ where $q_3 > 0$. Indeed, let us consider the same function $\phi(\lambda)$ defined by (6.88) again. Since $\phi'(\lambda) > 0$ for $\lambda \in (\lambda^*, \infty)$, $\phi(+\infty) = \infty$, and

$$\phi(\tilde\lambda) = (\sigma+1)\tilde\lambda^{\tau+1} - \sigma\tilde\lambda^\tau + (\sigma-\tau)p = (\sigma-\tau)\left[p - \frac{\tau^\tau}{(\tau+1)^{\tau+1}}\right] \leq 0,$$

we see that $y(\lambda)$ has a unique root t_* in $[\tilde\lambda, \infty)$. Furthermore,

$$q_3 = x(t_*) = -\frac{t_*^{\sigma-\tau}}{\delta-\sigma}\cdot[(\delta+1)t_*^{\tau+1} - \delta t_*^\tau + (\delta-\tau)p]$$

$$= -\frac{t_*^{\sigma-\tau}}{(\delta-\sigma)(\sigma+1)}\Big[(\sigma+1)(\delta+1)t_*^{\tau+1} - (\sigma+1)\delta t_*^\tau + (\sigma+1)(\delta-\tau)p$$
$$\qquad -(\sigma+1)(\delta+1)t_*^{\tau+1} + \sigma(\delta+1)t_*^\tau - (\sigma-\tau)(\delta+1)p\Big]$$

$$= -\frac{t_*^{\sigma-\tau}}{(\delta-\sigma)(\sigma+1)}[-(\delta-\sigma)t_*^\tau + (\delta-\sigma)(\tau+1)p]$$

$$= \frac{t_*^{\sigma+1}}{\sigma+1}[t_*^\tau - (\tau+1)p]. \qquad (6.89)$$

Under condition (6.81), we see further that

$$q_3 = \frac{t_*^{\sigma+1}}{\sigma+1}[t_*^\tau - (\tau+1)p] \geq 0$$

as desired. These show that the curves G_1, G_2 and G_3 are nonintersecting. Again, in view of Figure A.13, we see that the $\mathbf{C}\backslash(0,\infty)$-characteristic region of $h(\lambda|x, y)$ is $\nabla(G_3) \oplus \nabla(\Theta_0)$. The proof is complete.

We remark that, the domain of the function G_3 in the above result can be restricted to the smaller and precise interval $(-\infty, q_3]$.

Theorem 6.35. *Suppose condition (6.82) is satisfied. Let G be the curve described by the parametric functions $x(\lambda) = -\Gamma(\lambda; \delta, \sigma)$ and $y(\lambda) = \Gamma(\lambda; \sigma, \delta)$ for $\lambda \in (0, \infty)$. Let G_1 be the restriction of G over the interval $(0, \lambda_1^{(3)})$ and G_3 the restriction of G over the interval $(\lambda_2^{(3)}, \infty)$. Then G_1 is the graph of a strictly convex function $y = G_1(x)$ defined over $\left(-\Gamma(\lambda_1^{(3)}; \delta, \sigma), 0\right)$ and G_3 is the graph of a strictly convex* *function* $y = G_3(x)$ *defined over* $\left(-\infty, -\Gamma(\lambda_2^{(3)}; \delta, \sigma)\right)$. *Furthermore, the* $\mathbf{C} \backslash (0, \infty)$-*characteristic region of (6.75) is* $\vee(G_3) \oplus \vee G_1) \oplus \nabla(\Theta_0)$.

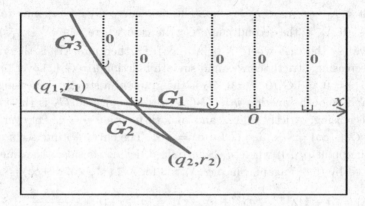

Fig. 6.30

Proof. The proof is similar to that of Theorem 6.33 and hence will be sketched. The parametric functions of the envelope G are given by (6.86) and (6.87) as before. However, the envelope G is now composed of three curves G_1, G_2 and G_3 as depicted in Figure 6.30.

The turning points (q_1, r_1) and (q_2, r_2) correspond to the unique minimal and maximal points $\lambda_1^{(3)}$ and $\lambda_2^{(3)}$ of the function $x(\lambda)$ in $(0, \infty)$. As explained earlier, we have $0 < \lambda_1^{(3)} < \lambda_* < \lambda_2^{(3)} < \infty$. The first curve G_1 corresponds to the case where $\lambda \in (0, \lambda_1^{(3)})$, the second curve G_2 the case where $\lambda \in [\lambda_1^{(3)}, \lambda_2^{(3)}]$ and the third curve G_3 the case where $\lambda \in (\lambda_2^{(3)}, \infty)$. Furthermore, G_1 is the graph of a strictly decreasing, strictly convex and smooth function $y = G_1(x)$ over $(q_1, 0)$ such that $G_1(0^-) = 0$ and $G_1'(0^-) = 0$; G_2 is the graph of a strictly decreasing, strictly concave and smooth function $y = G_2(x)$ over $[q_1, q_2]$; and G_3 is the graph of a strictly decreasing, strictly convex and smooth function $y = G_3(x)$ over $(-\infty, q_2)$ such that $G_3(-\infty) = +\infty$ and $G_3'(-\infty) = -\infty$. The curve G_3 intersects the x-axis at a unique point $(q_3, 0)$ where $q_3 < 0$. Indeed, let us consider the same function $\phi(\lambda)$ defined by (6.88) again. Recall that (see the proof of Theorem 6.33) $\phi'(\lambda) > 0$ for $\lambda \in (\lambda^*, \infty)$, $\phi(+\infty) = \infty$,

$$\phi(\lambda^*) = (\sigma + 1)(\lambda^*)^{\tau+1} - \sigma(\lambda^*)^\tau + (\sigma - \tau)p$$

$$= -\frac{\sigma}{\tau+1} \left(\frac{\tau}{\tau+1}\right)^\tau \left(\frac{\sigma}{\sigma+1}\right)^\tau + (\sigma - \tau)p$$

$$\leq -\frac{\sigma}{\tau+1} \left(\frac{\tau}{\sigma+1}\right)^\tau \left(\frac{\sigma}{\sigma+1}\right)^\tau + (\sigma - \tau)\frac{\sigma}{(\sigma-\tau)(\tau+1)} \left[\frac{\tau\sigma}{(\tau+1)(\sigma+1)}\right]^\tau$$

$$< 0,$$

and

$$\phi(\tilde{\lambda}) = (\sigma+1)\tilde{\lambda}^{\tau+1} - \sigma\tilde{\lambda}^{\tau} + (\sigma-\tau)p = (\sigma-\tau)\left[p - \frac{\tau^{\tau}}{(\tau+1)^{\tau+1}}\right] > 0,$$

we see that $y(\lambda)$ has a unique root t_* in $(\lambda^*, \tilde{\lambda})$. Furthermore, in view of (6.89) and condition (6.82), we see that

$$q_3 = \frac{t_*^{\sigma+1}}{\sigma+1}[t_*^{\tau} - (\tau+1)p] < 0$$

as required. These show that the curves G_1 and G_3 intersects at a unique point in the interior of the second quadrant and that the first coordinate of the point of intersection is strictly between q_1 and $x(t_*)$. In view of the distribution map in Figure A.13, the $\mathbf{C}\backslash(0,\infty)$-characteristic region of $h(\lambda|x,y)$ is $\vee(G_3)\oplus\vee G_1)\oplus\nabla(\Theta_0)$.

We remark that the domains of G_1 and G_3 in the above result can be restricted to the smaller but more precise intervals $(q_3, 0)$ and $(-\infty, q_3]$ respectively.

Theorem 6.36. *Suppose the condition (6.83) is satisfied. Let G be the curve described by the parametric functions $x(\lambda) = -\Gamma(\lambda; \delta, \sigma)$ and $y(\lambda) = \Gamma(\lambda; \sigma, \delta)$ for $\lambda \in (0, \infty)$. Let G_1 be the restriction of G over the interval $(0, \lambda_1^{(3)})$ and G_3 the restriction of G over the interval $(\lambda_2^{(3)}, \infty)$. Then G_1 is the graph of a strictly convex function $y = G_1(x)$ defined over $\left(-\Gamma(\lambda_1^{(3)}; \delta, \sigma), 0\right)$ and G_3 is the graph of a strictly convex function $y = G_3(x)$ defined over $\left(-\infty, -\Gamma(\lambda_2^{(3)}; \delta, \sigma)\right)$. Furthermore, the $\mathbf{C}\backslash(0,\infty)$-characteristic region of (6.75) is $\vee(G_3) \oplus \vee G_1) \oplus \nabla(\Theta_0)$.*

Proof. The proof is similar to that of Theorem 6.33 and hence will be sketched. The parametric functions of the envelope G are given by (6.86) and (6.87) as before. However, the envelope G is now composed of three curves G_1, G_2 and G_3 as depicted in Figure 6.31.

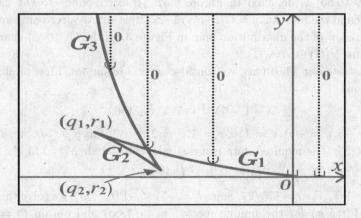

Fig. 6.31

The turning points (q_1, r_1) and (q_2, r_2) correspond to the unique minimal and maximal points $\lambda_1^{(3)}$ and $\lambda_2^{(3)}$ of the function $x(\lambda)$ in $(0, \infty)$. As explained earlier, we have $0 < \lambda_1^{(3)} < \lambda_* < \lambda_2^{(3)} < \lambda^* < \infty$. The first curve G_1 corresponds to the case where $\lambda \in (0, \lambda_1^{(3)})$, the second curve G_2 the case where $\lambda \in [\lambda_1^{(3)}, \lambda_2^{(3)})$ and the third curve G_3 the case where $\lambda \in [\lambda_2^{(3)}, \infty)$. Furthermore, the curve G_1 is the graph of a strictly decreasing, strictly convex and smooth function $y = G_1(x)$ over $(q_1, 0)$ such that $G_1(0^-) = 0$ and $G_1'(0^-) = 0$; G_2 is the graph of a strictly decreasing, strictly concave and smooth function $y = G_2(x)$ over $[q_1, q_2]$; and G_3 is the graph of a strictly decreasing and strictly convex function $y = G_3(x)$ over $(-\infty, q_2]$ such that $G_3(-\infty) = +\infty$ and $G_3'(-\infty) = -\infty$. The curve G_2 (and hence G_1 as well as G_3) is situated above the horizontal axis. Indeed, this follows from the fact that the function $\phi(\lambda)$ defined by (6.88) satisfies $\phi(\lambda) < 0$ for $\lambda \in (0, \lambda^*]$ and that $\lambda_2^{(3)} < \lambda^*$. In view of the distribution map in Figure A.13, we see that the $\mathbf{C}\backslash(0, \infty)$-characteristic region is $\vee(G_3) \oplus \vee(G_1) \oplus \triangledown(\Theta_0)$.

We remark that the conclusions in Theorems 6.35 and 6.36 are the same, and therefore the conditions (6.82) and (6.83) can be combined into one:

$$\frac{\tau^\tau}{(\tau+1)^{\tau+1}} < p \le \frac{\sigma\delta}{(\sigma-\tau)(\delta-\tau)(\tau+1)} \left[\frac{\tau\sigma\delta}{(\tau+1)(\sigma+1)(\delta+1)}\right]^\tau,$$

and a corresponding result can be written. However, the proofs of Theorems 6.35 and 6.36 provide additional information.

Theorem 6.37. *Suppose the condition (6.84) holds. Let G be the curve described by the parametric functions $x(\lambda) = -\Gamma(\lambda; \delta, \sigma)$ and $y(\lambda) = \Gamma(\lambda; \sigma, \delta)$ for $\lambda \in (0, \infty)$. Then G is also the graph of a strictly convex function $y = G(x)$ defined on $(-\infty, 0)$. Furthermore, (q, r) is a point in the nonpositive characteristic region of (6.75) if, and only if, $q < 0$ and $r > G(q)$, or, $q \ge 0$ and $r \ge 0$.*

The proof of this case is relatively easy. The envelope G of the family of curves defined by (6.85) is depicted in Figure 6.32. It corresponds to the case where $\lambda \in (0, \infty)$ and the function $y = G(x)$ over $(-\infty, 0)$ is strictly decreasing and strictly convex. In view of the distribution map in Figure 3.11, the $\mathbf{C}\backslash(0, \infty)$-characteristic region is just $\vee(G) \oplus \triangledown(\Theta_0)$.

To illustrate our Theorems, we consider several examples. First of all, consider the function

$$\lambda - 1 - 5\lambda^{-1} - 51\lambda^{-2} + 150\lambda^{-3}. \tag{6.90}$$

Here, $p = -5, q = -51, r = 150, \tau = 1, \sigma = 2, \delta = 3$. Since $p \le 0$, we may apply Theorem 6.33. The required data can easily be calculated: $\lambda^* = 1/3$,

$$\Gamma(\lambda; \sigma, \delta) = \lambda^2 \left\{3\lambda^2 - 2\lambda - 5\right\},$$

$t_* = 5/3, -\Gamma(t_*; \delta, \sigma) = 175/27$. Since $q = -51 \le -\Gamma(t_*; \delta, \sigma)$, we solve the equation $-51 = -\Gamma(t; \delta, \sigma)$ for the unique root t^* in $[5/3, \infty)$ and obtain $t^* = 3$. Since $r = 150 > \Gamma(t^*; \sigma, \delta) = 144$, we see that (6.90) does not have any positive roots.

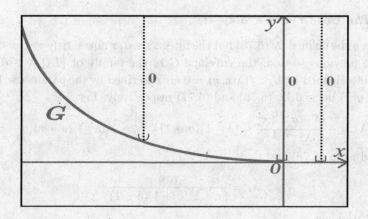

Fig. 6.32

Next consider the equation

$$\lambda - 1 + \frac{41}{1323}\lambda^{-1} + \frac{102355}{1354752}\lambda^{-2} + \frac{1}{50}\lambda^{-3}. \tag{6.91}$$

Since $0 < p \leq \tau^\tau/(\tau+1)^{\tau+1} = 1/4$, we may apply Theorem 6.34. The required data can easily be calculated: $\tilde{\lambda} = 1/2$,

$$\Gamma(\lambda; \sigma, \delta) = \lambda^2 \left\{ 3\lambda^2 - 2\lambda - \frac{41}{1323} \right\},$$

$t_* = 41/63$, $-\Gamma(t_*; \delta, \sigma) = 31939/320047$. Since $q = 102355/1354752 < 31939/320047$, we solve the equation $q = -\Gamma(; \delta, \sigma)$ for the unique root t^* in $[t_*, \infty)$ and obtain $t^* = 11/16$. Since $r = 1/50 < \Gamma(t^*; \sigma, \delta)$, we see that (6.91) has a positive root.

Next consider the equation

$$\lambda - 1 + \frac{4}{25}\lambda^{-2} + 5\lambda^{-5} + 27\lambda^{-7}. \tag{6.92}$$

Since

$$\frac{4}{27} = \frac{\tau^\tau}{(\tau+1)^{\tau+1}} < p < \frac{\sigma}{(\sigma-\tau)(\tau+1)}\left[\frac{\tau\sigma}{(\tau+1)(\sigma+1)}\right]^\tau = \frac{125}{729},$$

we may apply Theorem 6.35 to conclude that 6.92)) does not have any positive roots.

Finally, we consider the equation

$$\lambda - 1 + 2\lambda^{-1} - \frac{96}{25}\lambda^{-3} + 2\lambda^{-4}. \tag{6.93}$$

Since

$$p \geq \frac{\sigma\delta}{(\sigma-\tau)(\delta-\tau)(\tau+1)}\left[\frac{\tau\sigma\delta}{(\tau+1)(\sigma+1)(\delta+1)}\right]^\tau = \frac{3}{10},$$

we may apply Theorem 6.36. Since $q \leq 0$, we solve the equation $q = -\Gamma(\lambda; 4, 3)$ for the unique root t^* in $(0, \infty)$ and obtain $t^* = 4/5$. Since $r = 2 \leq \Gamma(4/5; 3, 4)$, we see that equation (6.93) has at least one positive root.

6.3.2 *The Case $\tau < \sigma < \delta < -1$*

Let $f(\lambda|x, y)$ be defined by (6.75) but the integers τ, σ, τ now satisfy $\tau < \sigma < \delta < -1$. As in the previous section, the envelope G of the family of $\{L_\lambda | \ \lambda \in (0, \infty)\}$ of straight lines defined by $L_\lambda : f(\lambda|x, y) = 0$ are described by the parametric functions $x(\lambda)$ and $y(\lambda)$ defined by (6.76) and (6.77) respectively. Let

$$F(\lambda) = -\frac{(\delta - \sigma)x'(\lambda)}{\lambda^{\sigma - \tau - 1}} = (\delta + 1)(\sigma + 1)\lambda^{\tau + 1} - \sigma\delta\lambda^\tau + (\delta - \tau)(\sigma - \tau)p$$

be defined as in (6.78) and

$$\lambda_* = \frac{\tau\sigma\delta}{(\tau + 1)(\sigma + 1)(\delta + 1)}$$

as in (6.79). Then

$$F(0^+) = -\infty \quad \text{and} \quad F(+\infty) = (\delta - \tau)(\sigma - \tau)p. \tag{6.94}$$

Since

$$F'(\lambda) = (\tau + 1)(\sigma + 1)(\delta + 1)\lambda^{\tau - 1}(\lambda - \lambda_*), \tag{6.95}$$

the function $F(\lambda)$ is increasing on $(0^+, \lambda^*)$ with $F(0^+) = -\infty$ and then decreasing on (λ^*, ∞) with $F(+\infty) = (\delta - \tau)(\sigma - \tau)p$. Therefore λ_* is the unique maximal point of F and the corresponding maximum over $(0, \infty)$ is

$$F(\lambda_*) = -\frac{\sigma\delta}{\tau + 1}\lambda_*^\tau + (\sigma - \tau)(\delta - \tau)p. \tag{6.96}$$

Note that $F(\lambda_*) = 0$ if, and only if, $p = \rho$, where

$$\rho = \frac{\sigma\delta}{(\sigma - \tau)(\delta - \tau)(\tau + 1)}\lambda_*^\tau. \tag{6.97}$$

Recall further that

$$\frac{dy}{dx} = -\lambda^{\delta - \sigma} < 0 \tag{6.98}$$

and

$$\frac{d^2y}{dx^2}(\lambda) = \frac{(\delta - \sigma)^2\lambda^{\delta - 2\sigma + \tau}}{F(\lambda)}$$

for $x'(\lambda) \neq 0$.

We consider four cases: Case 1: $p = 0$; Case 2: $p > 0$; Case 3: $\rho < p < 0$; and Case 4: $p \leq \rho$.

Case 1. When $p = 0$, (6.75) is then a $\Delta(1, 0, 0)$-polynomial, and this case has already been discussed in detail in Section 6.2.2.

Case 2. Suppose $p > 0$. Then $\lim_{\lambda \to \infty} F(\lambda) = (\delta - \tau)(\sigma - \tau)p > 0$. Hence there is $\lambda_1^{(1)} > 0$ such that $F(\lambda_1^{(1)}) = 0$, $F(\lambda) < 0$ on $(0, \lambda_1^{(1)})$ and $F(\lambda) > 0$ on $(\lambda_1^{(1)}, \infty)$. Since

$$(x(0^+), y(0^+)) = (-\infty, +\infty), \ (x(+\infty), y(+\infty)) = (-\infty, +\infty),$$

we may now see that the envelope G is made up of two pieces G_1 and G_2 corresponding to $\lambda \in (0, \lambda_1^{(1)}]$ and $\lambda \in (\lambda_1^{(1)}, \infty)$ respectively and the turning point corresponds to $(\alpha, \beta) = (x(\lambda_1^{(1)}), y(\lambda_1^{(1)}))$ as shown in Figure 6.33. The piece G_1 is also the graph of a strictly decreasing, strictly concave and smooth function on $(-\infty, \alpha]$ such that $G_1(-\infty) = +\infty$ and $G_1'(-\infty) = 0$; and G_2 is also the graph of a strictly decreasing, strictly convex and smooth function on $(-\infty, \alpha)$ such that $G_2(-\infty) = +\infty$ and $G_2'(-\infty) = -\infty$. Thus $-G_1 \sim H_{-\infty}$ and $G_2 \sim H_{-\infty}$. In view of the distribution maps in Figures 3.11 and 3.17, the dual sets of order 0 of G_1 and G_2 are disjoint. Hence their intersection is empty.

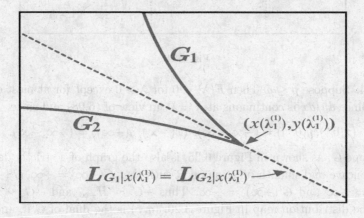

Fig. 6.33

Case 3. Suppose $\rho < p < 0$. Then $\lim_{\lambda \to \infty} F(\lambda) = (\delta - \tau)(\sigma - \tau)p < 0$ and

$$F(\lambda^*) = -\frac{\sigma\delta}{\tau+1}\lambda_*^\tau + (\sigma-\tau)(\delta-\tau)p > -\frac{\sigma\delta}{\tau+1}\lambda_*^\tau + (\sigma-\tau)(\delta-\tau)\rho = 0.$$

Hence there are $\lambda_1^{(2)} > 0$ and $\lambda_2^{(2)} > 0$ such that $F(\lambda_1^{(2)}) = F(\lambda_2^{(2)}) = 0$, $F(\lambda) < 0$ on $(0, \lambda_1^{(2)}) \cup (\lambda_2^{(2)}, \infty)$ and $F(\lambda) > 0$ on $(\lambda_1^{(2)}, \lambda_2^{(2)})$. Since

$$(x(0^+), y(0^+)) = (-\infty, +\infty), \quad (x(+\infty), y(+\infty)) = (+\infty, -\infty),$$

the envelope G is now made up of three pieces G_1, G_2 and G_3 corresponding to $\lambda \in (0, \lambda_1^{(2)}]$, $\lambda \in (\lambda_1^{(2)}, \lambda_2^{(2)})$ and $\lambda \in [\lambda_2^{(2)}, \infty)$ respectively and the turning points correspond to $(\alpha_1, \beta_1) = (x(\lambda_1^{(2)}), y(\lambda_1^{(2)}))$ and $(\alpha_2, \beta_2) = (x(\lambda_2^{(2)}), y(\lambda_2^{(2)}))$, as shown in Figure 6.34. The piece G_1 is the graph of a strictly decreasing, strictly concave and smooth function on $(-\infty, \alpha_1]$ such that $G_1(-\infty) = +\infty$ and $G_1'(-\infty) = 0$; the piece G_2 is the graph of a strictly decreasing, strictly convex and smooth function on (α_2, α_1) and the piece G_3 is the graph of a strictly decreasing, strictly concave and smooth function on $[\alpha_2, \infty)$ such that $G_3(+\infty) = -\infty$ and $G_3'(+\infty) = -\infty$. Thus $-G_1 \sim H_{-\infty}$ and $-G_3 \sim H_{+\infty}$. In view of the distribution map in Figure A.14, the dual set of order 0 of G is $\wedge(G_1) \oplus \wedge(G_3)$.

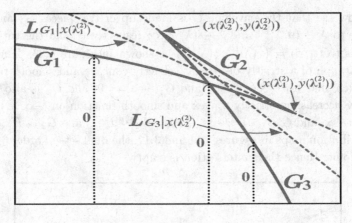

Fig. 6.34

Case 4. Suppose $p \leq \rho$. Then $F(\lambda) < 0$ for $\lambda > 0$ except for at most one point $\lambda = \lambda^*$. Since dy/dx is continuous at $\lambda = \lambda^*$ in view of (6.98) and since

$$(x(0^+), y(0^+)) = (-\infty, +\infty), \quad (x(+\infty), y(+\infty)) = (+\infty, -\infty),$$

the envelope G, as shown in Figure 6.35, is also the graph of a strictly decreasing, strictly concave and smooth function on $\mathbf{R}$ such that $G(-\infty) = +\infty$, $G'(-\infty) = 0$, $G(+\infty) = -\infty$ and $G'(+\infty) = -\infty$. Thus $-G \sim H_{-\infty}$ and $-G \sim H_{+\infty}$. In view of the distribution map in Figure 3.20, (q, r) is the dual of G if, and only if, $(q, r) \in \wedge(G)$.

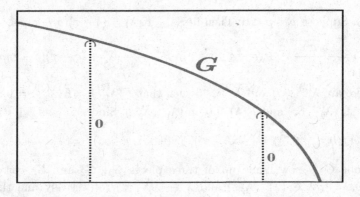

Fig. 6.35

Theorem 6.38. *Suppose $\tau < \sigma < \delta < -1$ and $p > 0$. Then the polynomial (6.75) does not have any positive roots.*

Theorem 6.39. *Suppose $\tau < \sigma < \delta < -1$ and $\rho < p < 0$, where ρ is defined by (6.97). Let λ_1 and λ_2 be the unique positive roots of the equation $F(\lambda) = 0$ defined* *by (6.88). Let the curve* G_1 *be defined by the parametric functions* $x(\lambda)$ *and* $y(\lambda)$ *in (6.76) and (6.77) over* $\lambda \in (0, \lambda_1)$ *and the curve* G_3 *be defined over* $\lambda \in (\lambda_2, \infty)$. *Then* G_1 *is the graph of a strictly decreasing, strictly concave and smooth function* $y = G_1(x)$ *defined over* $(-\infty, x(\lambda_1))$ *and* G_3 *the graph of a strictly decreasing, strictly concave and smooth function defined over* $(x(\lambda_2)), \infty)$. *Furthermore, the* $\mathbf{C}\backslash(0,\infty)$-*characteristic region of (6.75) is* $\wedge(G_1) \oplus \wedge(G_3)$.

Theorem 6.40. *Suppose* $p \leq \rho$ *and* $\tau < \sigma < \delta < -1$ *where* ρ *is defined by (6.97). Let the curve* G *be defined by the parametric functions* $x(\lambda)$ *and* $y(\lambda)$ *in (6.76) and (6.77) over* $\lambda \in (0, \infty)$. *Then* G *is the graph of a strictly decreasing, strictly concave and smooth function* $y = G(x)$ *defined over* $\mathbf{R}$. *Furthermore, the* $\mathbf{C}\backslash(0,\infty)$-*characteristic region of (6.75) is* $\wedge(G)$.

6.4   Notes

Most of the results in this Chapter are based on results published recently (since 1996), and hence they will be useful as references for readers who are concerned with characteristic functions. In particular, Theorems 6.8-6.10 are originated from [24]; Theorems 6.11-6.13 and 6.14 from [7]; Theorems 6.15-6.21 from [11]; Theorems 6.22-6.32 from [9]; and Theorems 6.33-37 from [8]. Theorems 6.38-6.40 are new, and they replace the results stated in [3] which seem to contain errors.

Chapter 7

C\R-Characteristic Regions of ∇-Polynomials

7.1 ∇-Polynomials Involving One Power

There are good reasons to study ∇-polynomials of the form

$$f_0(\lambda) + e^{-\lambda\sigma} f_1(\lambda) \tag{7.1}$$

involving one power. For instance, if we consider exponential solutions $\{e^{\lambda t}\}$ of the differential equation

$$x'(t) + px(t - \tau) = 0, \ t \in \mathbf{R},$$

then we are led to the ∇-polynomial

$$\lambda + e^{-\tau\lambda} p.$$

The **C\R**-characteristic region of the general ∇-polynomial (7.1) is difficult to find. For this reason, we will restrict ourselves to some special cases. In the first special case, we assume the degrees of f_0 and f_1 are less than or equal to 1:

$$F(\lambda) = a\lambda + b + e^{-\lambda\sigma}(c\lambda + d).$$

where $a, b, c, d, \sigma \in \mathbf{R}$. By definition, the power σ cannot be 0 in F. However, if $\sigma = 0$, the corresponding **C\R**-characteristic region is easy to find and is just $\{(a, b) \in \mathbf{R}^2 | a = 0, b \neq 0\}$.

7.1.1 ∇(0,0)-Polynomials

Consider

$$F(\lambda) = a + e^{-\lambda\sigma} b,$$

where $a, b, \sigma \in \mathbf{R}$ and $\sigma \neq 0$. If $b = 0$, then $F(\lambda)$ has a real root if, and only if, $a = 0$. If $b \neq 0$, then $F(\lambda)$ has the unique real root

$$\lambda = -\frac{1}{\sigma} \ln\left(-\frac{a}{b}\right)$$

if, and only if, $a/b < 0$. Thus, $F(\lambda)$ does not have any real roots if, and only, if $(a, b) \in \{(x, y) \in \mathbf{R}^2 : xy \geq 0\} \setminus \{(0, 0)\}$.

7.1.2 $\nabla(1,0)$-*Polynomials*

Consider

$$F(\lambda) = a\lambda + b + e^{-\lambda\sigma}c,$$

where $a, b, c \in \mathbf{R}$ and $\sigma \neq 0$. If $a = 0$, F reduces to a $\nabla(0,0)$-polynomial. We therefore assume that $a \neq 0$. Then we may divide F by a to obtain

$$\lambda + \frac{b}{a} + e^{-\lambda\sigma}\frac{c}{a}.$$

Hence it suffices to consider the ∇-polynomial

$$f(\lambda|x, y, \sigma) = \lambda - x + e^{-\lambda\sigma}y. \tag{7.2}$$

where $\sigma, x, y \in \mathbf{R}$ and $\sigma \neq 0$.

For fixed $\sigma \in R\backslash\{0\}$, let $\Omega(\sigma)$ be the $\mathbf{C}\backslash\mathbf{R}$-characteristic region of the ∇-polynomial $f(\lambda|x, y, \sigma)$. That is,

$$\Omega(\sigma) = \{(x, y) \in \mathbf{R}^2 : f(\lambda|x, y, \sigma) \text{ has no real roots}\}. \tag{7.3}$$

Theorem 7.1. *Let $f(\lambda|x, y, \sigma)$ and $\Omega(\sigma)$ be defined by (7.2) and (7.3) respectively.*

(1) Suppose $\sigma > 0$. Then $(x, y) \in \Omega(\sigma)$ if, and only if, (x, y) satisfies

$$y > \frac{1}{\sigma}\exp\left\{\sigma\left(x - \frac{1}{\sigma}\right)\right\}, \ x \in \mathbf{R}.$$

(2) Suppose $\sigma < 0$. Then $(x, y) \in \Omega(\sigma)$ if, and only if, $(-x, -y) \in \Omega(-\sigma)$.

Proof. We first assume that $\sigma > 0$. Consider the family $\{L_\lambda | \ \lambda \in \mathbf{R}\}$ of straight lines defined by $L_\lambda : f(\lambda|x, y, \sigma) = 0$ where $\lambda \in \mathbf{R}$. Since

$$f'_\lambda(\lambda|x, y, \sigma) = 1 - \sigma e^{-\sigma\lambda}y,$$

the determinant of the linear system $f(\lambda|x, y, \sigma) = f'_\lambda(\lambda|x, y, \sigma) = 0$ is $\sigma e^{-\sigma\lambda}$ which does not vanish for $\lambda \in \mathbf{R}$. By Theorem 2.6, $\Omega(\sigma)$ is just the dual set of order 0 of the envelope G of the family $\{L_\lambda | \ \lambda \in \mathbf{R}\}$. We solve from $f(\lambda|x, y, \sigma) = f'_\lambda(\lambda|x, y, \sigma) = 0$ for x and y to yield

$$x(\lambda) = \lambda + \frac{1}{\sigma}, \ y(\lambda) = \frac{1}{\sigma}e^{\sigma\lambda}, \ \lambda \in \mathbf{R}.$$

Then by Theorem 2.3, it is easy to see that the envelope G is described by the parametric functions $x(\lambda)$ and $y(\lambda)$ over $\mathbf{R}$. Furthermore, it can also be described by the graph of the function $y = G(x)$ where

$$G(x) = \frac{1}{\sigma}\exp\left\{\sigma\left(x - \frac{1}{\sigma}\right)\right\}, \ x \in \mathbf{R}.$$

Since $G(-\infty) = 0$, $G(+\infty) = +\infty$,

$$G'(x) = \exp\left\{\sigma\left(x - \frac{1}{\sigma}\right)\right\} > 0, \ x \in \mathbf{R},$$

and

$$G''(x) = \sigma \exp \left\{ \sigma \left(x - \frac{1}{\sigma} \right) \right\} > 0, \, x \in \mathbf{R},$$

we see that $G'(-\infty) = 1$, $G'(+\infty) = +\infty$ and $G(x)$ is strictly increasing and strictly convex over $\mathbf{R}$. See Figure 7.1. By Lemma 3.4, $G \sim H_{+\infty}$. Then, in view of the distribution map in Figure 3.19, (x, y) is a dual point of G if, and only if, $y > G(x)$.

Next, we assume that $\sigma < 0$. Let $g(\lambda|x, y, \tau) = \lambda - x + e^{\tau \lambda} y$ where $x, y \in \mathbf{R}$ and $\tau > 0$. Note that $f(\lambda|x, y, \sigma) = -g(\lambda| - x, -y, -\sigma)$. If λ is a real root of $f(\lambda|x, y, \sigma)$, then λ is also a real root of $g(\lambda| - x, -y, -\sigma)$. The converse is also true. Hence, $(x, y) \in \Omega(\sigma)$ if, and only if, $(-x, -y) \in \Omega(-\sigma)$.

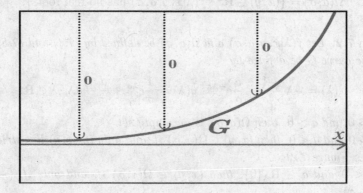

Fig. 7.1

7.1.3   ∇(0, 1)-*Polynomials*

Let us consider $\nabla(0, 1)$-polynomials of the form

$$g(\lambda|a, b, c, \sigma) = a + e^{-\lambda \sigma}(b\lambda + c),$$

where $\sigma \neq 0$ and $a, b, c \in \mathbf{R}$. If $b = 0$, then we obtain a $\nabla(0, 0)$-polynomial. We may therefore assume that $b \neq 0$. Since

$$g(\lambda|a, b, c, \sigma) = e^{-\lambda \sigma}(ae^{\lambda \sigma} + b\lambda + c),$$

we see that the **C\R**-characteristic region of $g(\lambda|a, b, c, \sigma)$ is the same as the **C\R**-characteristic region of the $\nabla(1, 0)$-polynomial $b\lambda + c + ae^{\lambda \sigma}$.

7.1.4   ∇(1, 1)-*Polynomials*

Let us consider $\nabla(1, 1)$-polynomials of the form

$$g(\lambda|a, x, c, y, \sigma) = a\lambda + x + e^{-\sigma \lambda}(c\lambda + y),$$

where $\sigma \in \mathbf{R}\backslash\{0\}$ and $a, c, x, y \in \mathbf{R}$. Again we may assume that $ac \neq 0$. Then we may divide g by a to obtain

$$\lambda + \frac{x}{a} + e^{-\sigma\lambda}\left(\frac{c}{a}\lambda + \frac{y}{a}\right).$$

Hence it suffices to consider the ∇-polynomial

$$f(\lambda|x, y, a, \sigma) = \lambda + x + e^{-\lambda\sigma}(a\lambda + y) \qquad (7.4)$$

where $a\sigma \neq 0$ and $x, y \in \mathbf{R}$.

For fixed $a, \sigma \in \mathbf{R}\backslash\{0\}$, let $\Omega(a, \sigma)$ be the $\mathbf{C}\backslash\mathbf{R}$-characteristic region of this ∇-polynomial $f(\lambda|x, y, a, \sigma)$. That is,

$$\Omega(a, \sigma) = \{(x, y) \in \mathbf{R}^2 : f(\lambda|x, y, a, \sigma) \text{ has no real roots}\}. \qquad (7.5)$$

Theorem 7.2. *Let $f(\lambda|x, y, a, \sigma)$ and $\Omega(a, \sigma)$ be defined by (7.4) and (7.5). Let the parametric curve G be defined by*

$$x(\lambda) = -\lambda - \frac{1}{\sigma} - \frac{a}{\sigma}e^{-\lambda\sigma}, \ y(\lambda) = \frac{1}{\sigma}e^{\lambda\sigma} + \frac{a}{\sigma} - a\lambda, \ \lambda \in \mathbf{R}.$$

(1) If $\sigma > 0$ and $a > 0$, then $\Omega(a, \sigma)$ is an empty set.

(2) If $\sigma > 0$ and $a < 0$, then $(x, y) \in \Omega(a, \sigma)$ if, and only if, (x, y) lies strictly above G (See Figure 7.2).

(3) If $\sigma < 0$ and $a \in \mathbf{R}\backslash\{0\}$, then $(x, y) \in \Omega(a, \sigma)$ if, and only if, $(-x, -y) \in \Omega(a, -\sigma)$.

Proof. We first note that $f(\lambda|x, y, a, \sigma) = -f(\lambda|-x, -y, a, -\sigma)$. For any $a \in \mathbf{R}\backslash\{0\}$, $(x, y) \in \Omega(a, \sigma)$ if, and only if, $(-x, -y) \in \Omega(a, -\sigma)$. So we only need to consider the two cases: $a > 0$ and $\sigma > 0$, or, $a < 0$ and $\sigma > 0$.

Consider the family $\{L_\lambda| \ \lambda \in \mathbf{R}\}$ of straight lines defined by $L_\lambda : f(\lambda|x, y, a, \sigma) = 0$ where $\lambda \in \mathbf{R}$. Since

$$f'_\lambda(\lambda|x, y, a, \sigma) = 1 - \sigma a\lambda e^{-\lambda\sigma} + ae^{-\lambda\sigma} - \sigma e^{-\lambda\sigma}y,$$

the determinant of the linear system $f(\lambda|x, y, a, \sigma) = f'_\lambda(\lambda|x, y, a, \sigma) = 0$ is $-\sigma e^{-\lambda\sigma}$ which does not vanish for $\lambda \in \mathbf{R}$. By Theorem 2.6, $\Omega(a, \sigma)$ is just the dual set of order 0 of the envelope G of the family $\{L_\lambda| \ \lambda \in \mathbf{R}\}$. We solve from the linear system for x and y to yield the parametric functions

$$x(\lambda) = -\lambda - \frac{1}{\sigma} - \frac{a}{\sigma}e^{-\lambda\sigma}, \ y(\lambda) = \frac{1}{\sigma}e^{\lambda\sigma} + \frac{a}{\sigma} - a\lambda, \ \lambda \in \mathbf{R}.$$

We have

$$x'(\lambda) = e^{-\lambda\sigma}h(\lambda) \text{ and } y'(\lambda) = -h(\lambda),$$

where

$$h(\lambda) = a - e^{\lambda\sigma}, \ \lambda \in \mathbf{R}.$$

Note that if $a > 0$, then $h(\lambda)$ has a unique real root $\lambda^* = 1/\sigma \ln a$ and if $a < 0$, then $h(\lambda)$ has no real roots. Thus,

$$\frac{dy}{dx}(\lambda) = -e^{\lambda\sigma} < 0, \quad \frac{d^2y}{dx^2}(\lambda) = \frac{-\sigma(e^{\lambda\sigma})^2}{h(\lambda)}, \quad \lambda \in \mathbf{R}\backslash\{\lambda^*\}.$$

Assume $\sigma > 0$ and $a > 0$. We have

$$(x(-\infty), y(-\infty)) = (x(+\infty), y(+\infty)) = (-\infty, +\infty).$$

As in Example 2.2, the envelope is composed of two pieces G_1 and G_2 and one turning point $(x(\lambda^*), y(\lambda^*))$. The first piece G_1 corresponds to the case where $\lambda \in (-\infty, \lambda^*)$ and the second G_2 to the case where $\lambda \in [\lambda^*, +\infty)$. Furthermore, G_1 is the graph of a function $y = G_1(x)$ which is strictly decreasing, strictly concave, and smooth over $(-\infty, x(\lambda^*))$ such that $G_1 \sim H_{-\infty}$; and G_2 is the graph of a function $y = G_2(x)$ which is strictly decreasing, strictly convex, and smooth over $(-\infty, x(\lambda^*)]$ such that $G_2 \sim H_{-\infty}$. By Theorem A.9, the dual set of order 0 of G is an empty set.

Assume $\sigma > 0$ and $a < 0$. We have

$$(x(-\infty), y(-\infty)) = (+\infty, -\infty) \text{ and } (x(+\infty), y(+\infty)) = (-\infty, +\infty).$$

Since $\{\lambda^*\}$ is empty, the envelope G is the graph of a function $y = G(x)$ which is strictly decreasing, strictly convex, and smooth over $\mathbf{R}$ such that $G \sim H_{-\infty}$. See Figure 7.2. By Lemma 3.5, we also have $G \sim H_{+\infty}$. In view of the distribution map in Figure 3.20, (x, y) is a dual point of G if, and only if, $y > G(x)$.

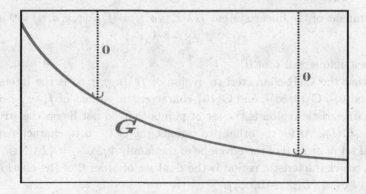

Fig. 7.2

7.1.5   ∇(n, n)-*Polynomials*

The general $\nabla(n, n)$-polynomials contain too many parameters and are difficult to handle. Therefore, we restrict ourselves to two special cases:

$$f_1(\lambda|x, y, \sigma, n) = \lambda^n + (x\lambda^n + y)e^{-\sigma\lambda} \tag{7.6}$$

and

$$f_2(\lambda|x, y, \sigma, n) = \lambda^n + y + x\lambda^n e^{-\sigma\lambda}, \tag{7.7}$$

where $\sigma \in \mathbf{R}\backslash\{0\}$, $x, y \in \mathbf{R}$ and $n \geq 2$ (if $n = 0$ or $n = 1$, the corresponding quasi-polynomials can be handled by considering $\nabla(0, 0)$- or ∇-$(1, 1)$ polynomials). For fixed $\sigma \in \mathbf{R}\backslash\{0\}$ and $n \in \mathbf{Z}[2, \infty)$, let $\Omega_i(\sigma, n)$ be the $\mathbf{C}\backslash\mathbf{R}$-characteristic region of $f_i(\lambda|x, y, \sigma, n)$ for $i = 1, 2$. That is, for $i = 1, 2$,

$$\Omega_i(\sigma, n) = \{(x, y) \in \mathbf{R}^2 : f_i(\lambda|x, y, \sigma, n) \text{ has no real roots}\}.$$

Theorem 7.3. *Let $n \geq 3$ be an odd integer (and $\sigma \neq 0$). Let the parametric curve G be defined by*

$$x(\lambda) = -\frac{n + \sigma\lambda}{n}e^{\sigma\lambda}, \quad y(\lambda) = \frac{\sigma\lambda^{n+1}}{n}e^{\sigma\lambda} \tag{7.8}$$

over $\lambda \in [-n/\sigma, +\infty)$.

(1) *If $\sigma > 0$, then $(x, y) \in \Omega_1(\sigma, n)$ if, and only if, $x \leq 0$ and $(x, y) \in \vee(G)$ (see Figure 7.3).*

(2) *If $\sigma < 0$, then $(x, y) \in \Omega_1(\sigma, n)$ if, and only if, $(x, -y) \in \Omega_1(-\sigma, n)$.*

Proof. First, we assume $\sigma > 0$. For the sake of simplicity, we also write $f(\lambda|x, y, \sigma, n)$ instead of $f_1(\lambda|x, y, \sigma, n)$ defined by (7.6). Consider the family $\{L_\lambda|\lambda \in \mathbf{R}\}$ of straight lines defined by $L_\lambda : f(\lambda|x, y, \sigma, n) = 0$ for $\lambda \in \mathbf{R}$. Since

$$f'_\lambda(\lambda|x, y, \sigma, n) = n\lambda^{n-1} + (n - \sigma\lambda)\lambda^{n-1}e^{-\sigma\lambda}x - \sigma y e^{-\sigma\lambda},$$

the determinant of the linear system $f(\lambda|x, y, \sigma, n) = f'_\lambda(\lambda|x, y, \sigma, n) = 0$ is

$$D(\lambda) = -ne^{-2\sigma\lambda}\lambda^{n-1},$$

which has a unique real root 0.

Note that the $\mathbf{C}\backslash\mathbf{R}$-characteristic region of $f(\lambda|x, y, \sigma, n)$ is the intersection of the $\mathbf{C}\backslash(-\infty, 0)$-, $\mathbf{C}\backslash(0, \infty)$-, and $\mathbf{C}\backslash\{0\}$-characteristic regions of $f(\lambda|x, y, \sigma, n)$. The $\mathbf{C}\backslash\{0\}$-characteristic region is the set of points that do not lie on the straight line L_0, that is, $\mathbf{C}\backslash L_0$. As for the other two regions, the $\mathbf{C}\backslash(-\infty, 0)$-characteristic region is the dual set of order 0 of the envelope of the family $\Phi_{(-\infty,0)} = \{L_\lambda|\lambda \in (-\infty, 0)\}$, the $\mathbf{C}\backslash(0, \infty)$-characteristic region is the dual set of order 0 of the envelope of the family $\Phi_{(\lambda^*,\infty)} = \{L_\lambda|\lambda \in (0, \infty)\}$.

By Theorem 2.3, the envelope G_1 of the family $\Phi_{(-\infty,0)}$ is described by the parametric functions (7.8) for $\lambda \in (-\infty, 0)$ and the envelope G_2 of the family $\Phi_{(0,+\infty)}$ by the same parametric functions but for $(0, +\infty)$.

We have

$$(x(-\infty), y(-\infty)) = (0, 0), \quad (x(+\infty), y(+\infty)) = (+\infty, -\infty),$$

$$\lim_{\lambda \to 0}(x(\lambda), y(\lambda))(-1, 0),$$

$$x'(\lambda) = -\frac{\sigma e^{\sigma\lambda}}{n}h(\lambda),$$

and

$$y'(\lambda) = \frac{\sigma\lambda^n e^{\sigma\lambda}}{n}h(\lambda)$$

where $h(\lambda) = \sigma\lambda + (n+1)$ and for $\lambda \in \mathbf{R}\backslash\{0\}$.

The real root of $h(\lambda)$ is $-(n+1)/\sigma$. Thus,

$$\frac{dy}{dx} = -\lambda^n$$

and

$$\frac{d^2y}{dx^2} = \frac{1}{\sigma}\lambda^{n-1}e^{-\sigma\lambda}\frac{n^3}{h(\lambda)}$$

for $\lambda \in \mathbf{R}\backslash\{0, -(n+1)/\sigma\}$.

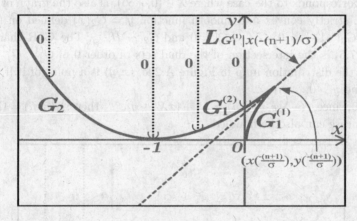

Fig. 7.3

The envelope G_1 is made up of two pieces $G_1^{(1)}$ and $G_1^{(2)}$ which correspond to the case where $\lambda \in (-\infty, -(n+1)/\sigma)$ and where $\lambda \in [-(n+1)/\sigma, 0)$ respectively. $G_1^{(1)}$ is the graph of a function $y = G_1^{(1)}(x)$ which is strictly increasing, strictly concave and smooth over $(0, x(-(n+1)/\sigma))$ such that $G_1^{(1)} \sim H_{0+}$ and $G_1^{(2)}$ is the graph of a function $y = G_1^{(2)}(x)$ which is strictly increasing, strictly convex and smooth over $(-1, x(-(n+1)/\sigma)]$ such that $G_1^{(2)}(-1^+) = 0$ and $G_1^{(2)\prime}(-1^+) = 0$. The envelope G_2, which corresponds to the restriction of G over the interval $\lambda \in (0, +\infty)$, is also the graph of a strictly decreasing, strictly convex and smooth function $y = G_2(x)$ defined on $(-\infty, -1)$ such that $G_2(-1^-) = 0$, $G_2'(-1^-) = 0$ and $G_2 \sim H_{-\infty}$. By Theorem 3.22 and the distribution map in Figure A.7c, (x, y) is a point of the $\mathbf{C}\backslash\mathbf{R}$-characteristic region of (7.6) if, and only if, $x \leq 0$ and $(x, y) \in \vee(G)$.

If $\sigma < 0$, since $f(-\lambda|x, y, \sigma, n) = -(\lambda^n + x\lambda^n + (-y)e^{\sigma\lambda})$, then $(x, y) \in \Omega_1(\sigma, n)$ if, and only if, $(x, -y) \in \Omega_1(-\sigma, n)$. Hence this case can also be handled.

Theorem 7.4. *Let $n \geq 2$ be an even integer (and $\sigma \neq 0$). Then $(x, y) \in \Omega_1(\sigma, n)$ if, and only if, $x \geq 0$ and $y > 0$ (see Figure 7.4).*

Proof. First, we assume $\sigma > 0$. The proof is similar to that of Theorem 7.3 and hence will be sketched. The parametric functions, the straight lines L_λ, the families $\Phi_{(-\infty,0)}$ and $\Phi_{(0,+\infty)}$, the regions $\mathbf{C}\backslash(-\infty, 0)$- and $\mathbf{C}\backslash(0, \infty)$-characteristic regions and the envelopes G_1 and G_2 are the same as in the proof of Theorem 7.3.

By similar reasoning in the proof of Theorem 7.3, we see that the envelope G_1 is made up of two pieces $G_1^{(1)}$ and $G_1^{(2)}$ which correspond to the case where $\lambda \in (-\infty, -(n+1)/\sigma)$ and where $\lambda \in [-(n+1)/\sigma, 0)$ respectively. $G_1^{(1)}$ is the graph of a function $y = G_1^{(1)}(x)$ which is strictly decreasing, strictly convex and smooth over $(0, x(-(n+1)/\sigma))$ such that $G_1^{(1)'}(0^+) = -\infty$ and $G_1^{(2)}$ is the graph of a function $y = G_1^{(2)}(x)$ which is strictly decreasing, strictly concave and smooth over $(-1, x(-(n+1)/\sigma)]$ such that $G_1^{(2)}(-1^+) = 0$ and $G_1^{(2)'}(-1^+) = 0$. The envelope G_2 which corresponds to the case where $\lambda \in (0, +\infty)$, is also the graph of a strictly decreasing, strictly convex and smooth function $y = G_2(x)$ defined on $(-\infty, -1)$ such that $G_2(-1^-) = 0$, $G_2'(-1^-) = 0$ and $G_2 \sim H_{-\infty}$. The $\mathbf{C}\backslash\mathbf{R}$-characteristic region of (7.6) is the intersection of the dual sets of order 0 of $G_1^{(1)}$, $G_1^{(2)}$ and G_2. In view of the distribution map in Figure A.20c, (x, y) is a point of it if, and only if, $x \geq 0$ and $y > 0$.

If $\sigma < 0$, since $f(-\lambda | x, y, \sigma, n) = \lambda^n + (x\lambda^n + y)e^{\sigma\lambda}$, then $\Omega_1(\sigma, n) = \Omega_1(-\sigma, n)$. Hence this case can also be handled.

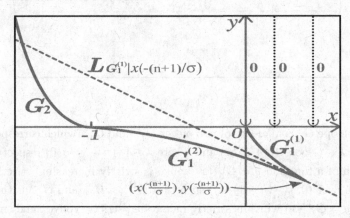

Fig. 7.4

Theorem 7.5. *Let $n \geq 3$ be an odd integer (and $\tau \neq 0$). Let the parametric curve G be defined by*

$$x(\lambda) = -\frac{n}{n - \tau\lambda}e^{\tau\lambda}, \quad y(\lambda) = \frac{\tau\lambda^{n+1}}{n - \tau\lambda} \tag{7.9}$$

over $(-\infty, n/\tau)$.

(1) If $\tau > 0$, $(x, y) \in \Omega_2(\tau, n)$ if, and only if, $x < 0$ and $(x, y) \in \vee(G)$ (see Figure 7.5).

(2) If $\tau < 0$, $(x, y) \in \Omega_2(\tau, n)$ if, and only if, $(x, -y) \in \Omega_2(-\tau, n)$.

Proof. First assume $\tau > 0$. For the sake of simplicity, we also write $f(\lambda|x, y, \sigma, n)$ instead of $f_2(\lambda|x, y, \sigma, n)$ defined by (7.7). Consider the family $\{L_\lambda | \lambda \in \mathbf{R}\}$ of straight lines defined by $L_\lambda : f(\lambda|x, y, \tau, n) = 0$ for $\lambda \in \mathbf{R}$. Since

$$f'_\lambda(\lambda|x, y, \tau, n) = n\lambda^{n-1} + (n - \tau\lambda)\lambda^{n-1}e^{-\tau\lambda}x,$$

the determinant of the linear system $f(\lambda|x, y, \tau, n) = f'_\lambda(\lambda|x, y, \tau, n) = 0$ is

$$D(\lambda) = -e^{-\tau\lambda}(n - \tau\lambda)\lambda^{n-1},$$

which has the two real roots 0 and n/τ.

Let us set

$$\lambda^* = \frac{n}{\tau}.$$

Note that the $\mathbf{C}\backslash\mathbf{R}$-characteristic region of $f(\lambda|x, y, \tau, n)$ is the intersection of the $\mathbf{C}\backslash(-\infty, 0)$-, $\mathbf{C}\backslash(0, \lambda^*)$-, $\mathbf{C}\backslash(\lambda^*, \infty)$-, $\mathbf{C}\backslash\{\lambda^*\}$- and $\mathbf{C}\backslash\{0\}$-characteristic regions of $f(\lambda|x, y, \tau, n)$. The $\mathbf{C}\backslash\{\lambda^*\}$-characteristic region is the set of points that do not lie on the straight line L_{λ^*}, that is, $\mathbf{C}\backslash L_{\lambda^*}$ and the $\mathbf{C}\backslash\{0\}$-characteristic region is the set of points that do not lie on the straight line L_0, that is, $\mathbf{C}\backslash L_0$. As for the other two regions, the $\mathbf{C}\backslash(-\infty, 0)$-characteristic region is the dual set of order 0 of the envelope of the family $\Phi_{(-\infty, 0)} = \{L_\lambda | \lambda \in (-\infty, 0)\}$, the $\mathbf{C}\backslash(0, \lambda^*)$-characteristic region is the dual set of order 0 of the envelope of the family $\Phi_{(0, \lambda^*)} = \{L_\lambda | \lambda \in (0, \lambda^*)\}$, and the $\mathbf{C}\backslash(\lambda^*, \infty)$-characteristic region is the dual set of order 0 of the envelope of the family $\Phi_{(\lambda^*, \infty)} = \{L_\lambda | \lambda \in (\lambda^*, \infty)\}$.

By Theorem 2.3, the envelope G_1 of the family $\Phi_{(-\infty, 0)}$ is described by the parametric functions (7.9) for $\lambda \in (-\infty, 0)$, while the envelope G_2 of the family $\Phi_{(0, \lambda^*)}$ by the same parametric functions but for $(0, \lambda^*)$, and the envelope G_3 of the family $\Phi_{(\lambda^*, \infty)}$ is described by the same parametric functions but for $\lambda \in (\lambda^*, \infty)$.

We have

$$(x(-\infty), y(-\infty)) = (0, +\infty), \quad (x(+\infty), y(+\infty)) = (+\infty, -\infty),$$

$$(x((n/\tau)^+), y((n/\tau)^+)) = (+\infty, -\infty),$$

$$(x((n/\tau)^-), y((n/\tau)^-)) = (-\infty, +\infty),$$

$$\lim_{\lambda \to 0}(x(\lambda), y(\lambda))(-1, 0),$$

$$x'(\lambda) = -\frac{n\tau e^{\tau\lambda}}{[-\tau\lambda + n]^2}h(\lambda),$$

and

$$y'(\lambda) = \frac{n\tau\lambda^n}{[-\tau\lambda + n]^2}h(\lambda),$$

where

$$h(\lambda) = -\tau\lambda + (n+1), \ \lambda \in \mathbf{R}\backslash\{\lambda^*, 0\}.$$

The function $h(\lambda)$ has a unique real root $(n+1)/\tau$. Note that $-\infty < 0 < \lambda^* < (n+1)/\tau$. Thus,

$$\frac{dy}{dx} = -\lambda^n e^{-\tau\lambda}$$

and

$$\frac{d^2y}{dx^2} = \frac{1}{\tau}\lambda^{n-1}e^{-2\tau t}\frac{(-\tau\lambda + n)^3}{h(\lambda)}$$

for $\lambda \in \mathbf{R}\backslash\{\lambda^*, 0, (n+1)/\tau\}$.

The envelope G_1, which corresponds to the case where $\lambda \in (-\infty, 0)$, is also the graph of a strictly increasing, strictly convex and smooth function $y = G_1(x)$ defined on $(-1,0)$ such that $G_1(-1^+) = 0$, $G_1'(-1^+) = 0$ and $G_1 \sim H_{0-}$. The envelope G_2, which corresponds to the case where $\lambda \in (0, \lambda^*)$, is also the graph of a strictly decreasing, strictly convex and smooth function $y = G_2(x)$ defined on $(-\infty, -1)$ such that $G_1(-1^-) = 0$ and $G_1'(-1^-) = 0$. The envelope G_3 is made up of two pieces $G_3^{(1)}$ and $G_3^{(2)}$ which correspond to the cases where $\lambda \in (\lambda^*, (n+1)/\tau)$ and where $\lambda \in [(n+1)/\tau, \infty)$ respectively. $G_3^{(1)}$ is the graph of a function $y = G_3^{(1)}(x)$ which is strictly decreasing, strictly concave and smooth over $(x((n+1)/\tau), \infty)$ and $G_3^{(2)}$ is the graph of a function $y = G_3^{(2)}(x)$ which is strictly decreasing, strictly convex and smooth over $(x((n+1)/\tau), \infty)$ such that $G_3^{(2)}(+\infty) = -\infty$ and $G_3^{(2)\prime}(+\infty) = 0$. By Lemma 3.5, we see further that $G_3^{(2)} \sim H_{+\infty}$. See Figure 7.5.

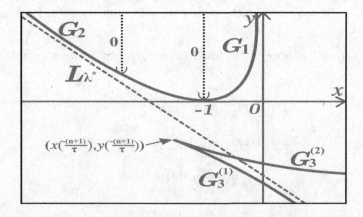

Fig. 7.5

By Lemma 3.2, L_{λ^*} is the asymptote of G_2 and $G_3^{(1)}$. This asymptote intercepts the x-axis at the point $-e^{\tau\lambda^*}$ which is strictly less than -1. In view of the distribution map in Figure A.18c, the dual sets of order 0 of the envelope G_1, G_2, $G_3^{(1)}$ and $G_3^{(2)}$ are easily obtained. The intersection of these dual sets, $\mathbf{C}\backslash L_{\lambda^*}$ and $\mathbf{C}\backslash L_0$, in view of Theorem 3.22, then yields the desired characteristic region which is made up of (x, y) that satisfies $x < 0$ and $(x, y) \in \vee(G) = \vee(G_1) \oplus \vee(G_2) \oplus \overline{\vee}(\Theta_0)$.

If $\tau < 0$, since $f(-\lambda|x, y, \tau, n) = -(\lambda^n + (-y) + x\lambda^n e^{\tau\lambda})$, then $(x, y) \in \Omega_2(\tau, n)$ if, and only if, $(x, -y) \in \Omega_2(-\tau, n)$. This case can then be handled by the first.

Theorem 7.6. *Assume $n \geq 2$ is an even integer and $\tau \neq 0$. Then $(x, y) \in \Omega_2(\tau, n)$ if, and only if, $x \geq 0$ and $y > 0$ (see Figure 7.6).*

Proof. First, we assume $\tau > 0$. The proof is similar to that of Theorem 7.5 and hence will be sketched. The parametric functions, the number λ^*, the straight lines L_λ, the families $\Phi_{(-\infty,0)}$, $\Phi_{(0,\lambda^*)}$ and $\Phi_{(\lambda^*,\infty)}$, the $\mathbf{C}\backslash(-\infty,0)$-, $\mathbf{C}\backslash(0,\lambda^*)$-, $\mathbf{C}\backslash(\lambda^*,\infty)$-characteristic regions and the envelopes G_1, G_2 and G_3 are the same as in the proof of Theorem 7.5.

We have

$$(x(-\infty), y(-\infty)) = (0, -\infty), \ (x(+\infty), y(+\infty)) = (+\infty, -\infty),$$

$$(x((n/\tau)^+), y((n/\tau)^+)) = (+\infty, -\infty),$$

$$(x((n/\tau)^-), y((n/\tau)^-)) = (-\infty, +\infty),$$

$$\lim_{\lambda\to 0}(x(\lambda), y(\lambda)) = (-1, 0),$$

$$x'(\lambda) = -\frac{n\tau e^{\tau\lambda}}{[-\tau\lambda + n]^2}h(\lambda),$$

and

$$y'(\lambda) = \frac{n\tau\lambda^n}{[-\tau\lambda + n]^2}h(\lambda)$$

where

$$h(\lambda) = -\tau\lambda + (n + 1), \ \lambda \in \mathbf{R}\backslash\{\lambda^*, 0\}.$$

The function $h(\lambda)$ has a unique real root $(n + 1)/\tau$. Note that $-\infty < 0 < \lambda^* < (n + 1)/\tau$. Thus,

$$\frac{dy}{dx} = -\lambda^n e^{-\tau\lambda}$$

and

$$\frac{d^2y}{dx^2} = \frac{1}{\tau}\lambda^{n-1}e^{-2\tau t}\frac{(-\tau\lambda + n)^3}{h(\lambda)}$$

for $\lambda \in \mathbf{R} \backslash \{\lambda^*, 0, (n+1)/\tau\}$.

The envelope G_1, which corresponds to the case where $\lambda \in (-\infty, 0)$, is also the graph of a strictly decreasing, strictly concave and smooth function $y = G_1(x)$ defined on $(-1, 0)$ such that $G_1(-1^+) = 0$, $G_1'(-1^+) = 0$ and $G_1 \sim H_{0-}$. The envelope G_2, which corresponds to the case where $\lambda \in (0, \lambda^*)$, is also the graph of a strictly decreasing, strictly convex and smooth function $y = G_2(x)$ defined on $(-\infty, -1)$ such that $G_2(-1^-) = 0$ and $G_2'(-1^-) = 0$. The envelope G_3 is made up of two pieces $G_3^{(1)}$ and $G_3^{(2)}$ which correspond to the cases where $\lambda \in (\lambda^*, (n+1)/\tau)$ and where $\lambda \in [(n+1)/\tau, \infty)$ respectively. $G_3^{(1)}$ is the graph of a function $y = G_3^{(1)}(x)$ which is strictly decreasing, strictly concave and smooth over $(x((n+1)/\tau), \infty)$ and $G_3^{(2)}$ is the graph of a function $y = G_3^{(2)}(x)$ which is strictly decreasing, strictly convex and smooth over $(x((n+1)/\tau), \infty)$ such that $G_3^{(2)}(+\infty) = -\infty$ and $G_3^{(2)\prime}(+\infty) = 0$. By Lemma 3.5, we see further that $G_3^{(2)} \sim H_{+\infty}$. See Figure 7.6.

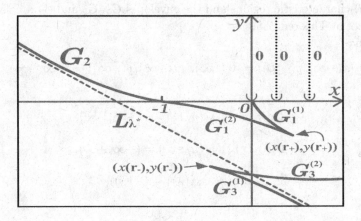

Fig. 7.6

By Lemma 3.2, L_{λ^*} is the asymptote of G_2 and $G_3^{(1)}$. This asymptote intercepts the x-axis at the point $-e^{\tau\lambda^*}$ which is strictly less than -1. In view of the distribution map in Figure A.30c, the dual sets of order 0 of the envelope G_1, G_2, $G_3^{(1)}$ and $G_3^{(2)}$ are easily obtained. The intersection of these dual sets, $\mathbf{C}\backslash L_{\lambda^*}$ and $\mathbf{C}\backslash L_0$ then yields the desired characteristic region which is made up of (x, y) that satisfies $x \geq 0$ and $y > 0$.

Finally, if $\tau < 0$, since $f(-\lambda|x, y, \tau, n) = \lambda^n + y + x\lambda^n e^{\tau\lambda}$, then $\Omega_2(\tau, n) = \Omega_2(-\tau, n)$. This case can then be handled by the first.

Corollary 7.1. *Let $n \geq 2$ be an integer. Let $f(\lambda|y, \sigma, n) = \lambda^n + ye^{-\sigma\lambda}$ where $y, \sigma \in \mathbf{R}$ and $\sigma \neq 0$. Then all roots of $f(\lambda|y, \sigma, n)$ are not real if, and only if, $y > K$ where K is equal to 0 if n is even and is equal to $n^n/\sigma^n e^n$ if n is odd.*

7.2   ∇-Polynomials Involving Two Powers

There are good reasons to study ∇-polynomials of the form

$$f_0(\lambda) + e^{-\tau\lambda} f_1(\lambda) + e^{-\sigma\lambda} f_2(\lambda) \tag{7.10}$$

involving two powers. For instance, if we consider exponential functions of the form $\{e^{\lambda t}\}$ as solutions of the differential equation

$$x'(t) + px(t - \tau) + qx(t - \sigma) = 0,\ t \in \mathbf{R}$$

where $p, q \in \mathbf{R}$ and $\tau, \sigma \in \mathbf{Z}$, then we are led to the quasi-polynomial

$$F(\lambda) = \lambda + \lambda^{-\tau\lambda} p + \lambda^{-\sigma\lambda} q.$$

The **C\R**-characteristic region of the ∇-polynomial (7.10) is difficult to find. Therefore we will restrict ourselves to the case where the degrees of f_0 and f_1 are less than or equal to 1, and the degree of f_2 to 0:

$$F(\lambda) = a\lambda + b + \lambda^{-\tau\lambda}(p\lambda + q) + \lambda^{-\sigma\lambda} r,$$

where $a, b, p, q, r \in \mathbf{R}$, and τ, σ are distinct and nonzero real numbers (if $\tau = \sigma$ or $\tau\sigma = 0$, the resulting ∇-polynomial can be handled by considering ∇-polynomials involving only one power). In addition, we also consider the special case

$$\lambda^n + a\lambda^n e^{-\tau\lambda} + be^{-\sigma\lambda}$$

where $a, b \in \mathbf{R}$, n is a positive integer, and τ, σ are distinct nonzero real numbers.

7.2.1   ∇(0, 0, 0)-*Polynomials*

A ∇(0, 0, 0)-polynomial is of the form

$$F(\lambda) = -a + e^{-\tau\lambda} x + e^{-\sigma\lambda} y$$

where $a, x, y \in \mathbf{R}$ and τ, σ are distinct nonzero real numbers. If $a = 0$, then $F(\lambda) = e^{-\tau\lambda}(x + e^{(\tau-\sigma)\lambda} y)$. The corresponding **C\R**-characteristic region is equal to that of the ∇(0, 0)-polynomial $x + e^{(\tau-\sigma)\lambda} y$. If $a \neq 0$, by dividing $F(\lambda)$ by a if necessary, we may then assume without loss of generality that our ∇-polynomial is of the form

$$f(\lambda | x, y, \tau, \sigma) = -1 + e^{-\tau\lambda} x + e^{-\sigma\lambda} y, \tag{7.11}$$

where $x, y \in \mathbf{R}$ and τ, σ are distinct nonzero real numbers.

For fixed and distinct $\tau, \sigma \in \mathbf{R} \backslash \{0\}$, let $\Omega(\tau, \sigma)$ be the **C\R**-characteristic region of the ∇-polynomial $f(\lambda | x, y, \tau, \sigma)$ in (7.11). That is,

$$\Omega(\tau, \sigma) = \{(x, y) \in \mathbf{R}^2 : f(\lambda | x, y, \tau, \sigma) \text{ has no real roots}\}. \tag{7.12}$$

7.2.1.1   *The Case $\tau\sigma > 0$*

Suppose $\tau\sigma > 0$ in (7.11). If $\tau, \sigma > 0$, then we may assume without loss of generality that $\tau > \sigma > 0$. If $\tau, \sigma < 0$, then we may assume without loss of generality that $\tau < \sigma < 0$.

Theorem 7.7. *Suppose τ and σ are distinct nonzero real numbers. Let $f(\lambda|x, y, \tau, \sigma)$ and $\Omega(\tau, \sigma)$ be defined by (7.11) and (7.12) respectively. Then*

(1) If $\tau < \sigma < 0$, then $(x, y) \in \Omega(\tau, \sigma)$ if, and only if, $x = 0$ and $y \le 0$, or, $x < 0$ and

$$y < \frac{\tau}{\sigma^{\sigma/\tau}}(\tau - \sigma)^{\sigma/\tau - 1}(-x)^{\sigma/\tau}.$$

(2) For $\tau > \sigma > 0$, $(x, y) \in \Omega(\tau, \sigma)$ if, and only if, $(x, y) \in \Omega(-\tau, -\sigma)$.

Proof. Assume $\tau < \sigma < 0$. Consider the family $\{L_\lambda | \lambda \in \mathbf{R}\}$ of straight lines defined by $L_\lambda : f(\lambda|x, y, \tau, \sigma) = 0$ where $\lambda \in \mathbf{R}$. Since

$$f'_\lambda(\lambda|x, y, \tau, \sigma) = -\tau e^{-(\tau+1)\lambda}x - \sigma e^{-(\sigma+1)\lambda}y,$$

the determinant of the linear system $f(\lambda|x, y, \tau, \sigma) = 0 = f'_\lambda(\lambda|x, y, \tau, \sigma)$ is $(\tau - \sigma)e^{-(\sigma+\tau+1)\lambda}$ which does not vanish for $\lambda \in \mathbf{R}$. By Theorem 2.6, $\Omega(\tau, \sigma)$ is just the dual set of order 0 of the envelope G of the family $\{L_\lambda | \lambda \in \mathbf{R}\}$. By Theorem 2.3, the parametric functions of G are given by

$$x(\lambda) = \frac{\sigma}{\sigma - \tau}e^{\tau\lambda}, \; y(\lambda) = \frac{-\tau}{\sigma - \tau}e^{\sigma\lambda}, \; \lambda \in \mathbf{R}.$$

Thus G is the graph of the function $y = G(x)$ defined by

$$y < -\tau(\sigma - \tau)^{\sigma/\tau - 1}\left(\frac{x}{\sigma}\right)^{\sigma/\tau}, \; x < 0.$$

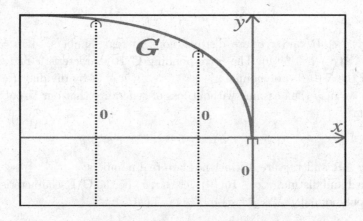

Fig. 7.7

By computing G' and G'', we may easily see that G is strictly decreasing, strictly concave and smooth on $(-\infty, 0)$ such that $G(-\infty) = \infty$, $G'(-\infty) = 0$, $G(0^-) = 0$ and $G'(0^-) = -\infty$. See Figure 7.7. In view of the distribution map in Figure 3.12, (x, y) is a dual point of G if, and only if, $x = 0$ and $y \leq 0$, or $x < 0$ and $y < G(x)$.

In case $\tau > \sigma > 0$, note that $f(\lambda|x, y, \tau, \sigma) = f(-\lambda|x, y, -\tau, -\sigma)$. Thus $(x, y) \in \Omega(\tau, \sigma)$ if, and only if, $(x, y) \in \Omega(-\tau, -\sigma)$. This case may then be handled by the first. The proof is complete.

7.2.1.2 *The Case* $\tau\sigma < 0$

Suppose $\tau\sigma < 0$ in (7.11). In this case, we may assume without loss of generality that $\tau > 0$ and $\sigma < 0$ and consider the equivalent polynomial

$$f(\lambda|x, y, \tau, \sigma) = -e^{\tau\lambda} + x + e^{(\tau-\sigma)\lambda}y. \tag{7.13}$$

Theorem 7.8. *Suppose* $\tau > 0$ *and* $\sigma < 0$. *Then* (x, y) *is in the* **C\R**-*characteristic region* $\Omega(\tau, \sigma)$ *of (7.13) if, and only if,* $x \leq 0$ *and* $y \leq 0$, *or* $x > 0$ *and*

$$y > \frac{\tau}{\tau + \sigma} \left(\frac{\sigma}{\tau + \sigma}\right)^{\sigma/\tau} \frac{1}{x^{\sigma/\tau}}.$$

Proof. Consider the family $\{L_\lambda | \lambda \in \mathbf{R}\}$ of straight lines defined by L_λ : $f(\lambda|x, y, \tau, \sigma) = 0$ where $\lambda \in \mathbf{R}$. Since

$$f'_\lambda(\lambda|x, y, \tau, \sigma) = -\tau e^{(\tau-1)\lambda} + (\tau - \sigma)e^{(\tau-\sigma-1)\lambda}y,$$

the determinant of the linear system $f(\lambda|x, y, \tau, \sigma) = 0 = f'_\lambda(\lambda|x, y, \tau, \sigma)$ is $(\tau - \sigma)e^{(\tau-\sigma-1)\lambda}$ which does not vanish for $\lambda \in \mathbf{R}$. By Theorem 2.6, $\Omega(\tau, \sigma)$ is just the dual set of order 0 of the envelope G of the family $\{L_\lambda | \lambda \in \mathbf{R}\}$. By Theorem 2.3, the parametric functions of G are given by

$$x(\lambda) = \frac{-\sigma}{\tau - \sigma}e^{\tau\lambda}, \; y(\lambda) = \frac{\tau}{\tau - \sigma}e^{\sigma\lambda}, \; \lambda \in \mathbf{R}.$$

Thus G is the graph of the function $y = G(x)$ defined by

$$G(x) = \frac{\tau}{\tau - \sigma} \left(\frac{-\sigma}{\tau - \sigma}\right)^{-\sigma/\tau} \frac{1}{x^{-\sigma/\tau}}, \; x > 0.$$

By computing G' and G'', we may easily see that $G(x)$ is strictly decreasing, strictly convex and smooth on $(0, \infty)$ such that $G(0^+) = \infty$, $G'(0^+) = -\infty$, $G(\infty) = 0$ and $G'(+\infty) = 0$. See Figure 7.8. In view of the distribution map in Figure 3.13, (x, y) is a dual point of G if, and only if, $x \leq 0$ and $y \leq 0$, or $x > 0$ and $y > G(x)$. The proof is complete.

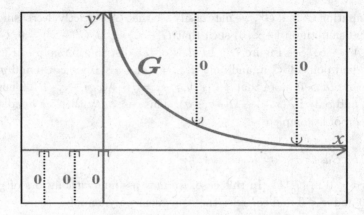

Fig. 7.8

7.2.2 $\nabla(1,0,0)$-*Polynomials*

The general $\nabla(1,0,0)$-polynomial is of the form

$$g(\lambda) = a\lambda + b + e^{-\tau\lambda}p + e^{-\sigma\lambda}q,$$

where $a, b, p, q \in \mathbf{R}$ and τ, σ are distinct nonzero real numbers. If $a = 0$, the above function is a $\nabla(0,0,0)$-polynomial. If $a \neq 0$, we divide $g(\lambda)$ by a to yield

$$\lambda + \frac{b}{a} + e^{-\tau\lambda}\frac{p}{a} + e^{-\sigma\lambda}\frac{q}{a}$$

which can be written as

$$\lambda + \tilde{a} + e^{-\tau\lambda}\tilde{p} + e^{-\sigma\lambda}\tilde{q}.$$

Therefore we only need to find the $\mathbf{C}\backslash\mathbf{R}$-characteristic regions for ∇-polynomials of the form

$$f(\lambda|x, y, a, \tau, \sigma) = \lambda + a + e^{-\tau\lambda}x + e^{-\sigma\lambda}y. \qquad (7.14)$$

where in (7.14), $a, x, y \in \mathbf{R}$ and τ, σ are distinct nonzero real numbers.

For fixed $a \in \mathbf{R}$ and distinct $\tau, \sigma \in \mathbf{R}\backslash\{0\}$, let $\Omega(a, \tau, \sigma)$ be the $\mathbf{C}\backslash\mathbf{R}$-characteristic regions of the ∇-polynomial $f(\lambda|x, y, a, \tau, \sigma)$ in (7.14). That is,

$$\Omega(a, \tau, \sigma) = \{(x, y) \in \mathbf{R}^2 : f(\lambda|x, y, a, \tau, \sigma) \text{ has no real roots}\}. \qquad (7.15)$$

7.2.2.1   *The Case $\tau\sigma > 0$*

Suppose $\tau\sigma > 0$. If $\tau, \sigma > 0$, then we may assume without loss of generality that $\sigma > \tau > 0$. If $\tau, \sigma < 0$, then we may assume without loss of generality that $\sigma < \tau < 0$.

Theorem 7.9. . *Suppose τ and σ are distinct nonzero real and $a \in \mathbf{R}$. Let $f(\lambda|x, y, a, \tau, \sigma)$ and $\Omega(a, \tau, \sigma)$ be defined by (7.11) and (7.15) respectively. Let the parametric curve G (see Figure 7.9) be defined by*

$$x(\lambda) = \frac{-e^{\tau\lambda}}{\sigma - \tau}(1 + \sigma\lambda + a\sigma), \ y(\lambda) = \frac{e^{\sigma\lambda}}{\sigma - \tau}(1 + \tau\lambda + a\tau), \ \lambda \in [-1/\tau - a, \infty).$$

The following results hold:

(1) *If* $0 < \tau < \sigma$, *then* $(x,y) \in \Omega(a,\tau,\sigma)$ *if, and only if* (x,y) *lies strictly above* G
    when $x \leq x(-1/\tau - a)$, *or lies on or above the x-axis when* $x > x(-1/\tau - a)$.
(2) *If* $0 > \tau > \sigma$, *then* $(x,y) \in \Omega(a,\tau,\sigma)$ *if, and only if,* $(-x,-y) \in \Omega(-a,-\tau,-\sigma)$.

Proof. Assume $0 < \tau < \sigma$. Consider the family $\{L_\lambda | \; \lambda \in \mathbf{R}\}$ of straight lines defined by $L_\lambda : f(\lambda|x,y,a,\tau,\sigma) = 0$ where $\lambda \in \mathbf{R}$. Since

$$f'_\lambda(\lambda|x,y,a,\tau,\sigma) = 1 - \tau e^{-\tau\lambda}x - \sigma e^{-\sigma\lambda}y,$$

the determinant of the linear system $f(\lambda|x,y,a,\tau,\sigma) = f'_\lambda(\lambda|x,y,a,\tau,\sigma) = 0$ is $(\sigma - \tau)e^{-(\tau+\sigma)\lambda}$ which does not vanish for $\lambda \in \mathbf{R}$. By Theorem 2.6, $\Omega(a,\tau,\sigma)$ is just the dual set of order 0 of the envelope G of the family $\{L_\lambda | \; \lambda \in \mathbf{R}\}$. We solve $f(\lambda|x,y,a,\tau,\sigma) = f'_\lambda(\lambda|x,y,a,\tau,\sigma) = 0$ for x and y to yield the parametric functions

$$x(\lambda) = \frac{-e^{\tau\lambda}}{\sigma - \tau}(1 + \sigma\lambda + a\sigma), \; y(\lambda) = \frac{e^{\sigma\lambda}}{\sigma - \tau}(1 + \tau\lambda + a\tau), \; \lambda \in \mathbf{R}.$$

We have

$$(x(-\infty), y(-\infty)) = (0,0) \text{ and } (x(+\infty), y(+\infty)) = (-\infty, +\infty),$$

$$x'(\lambda) = -\frac{e^{\tau\lambda}}{\sigma - \tau}h(\lambda) \text{ and } y'(\lambda) = \frac{e^{\sigma t}}{\sigma - \tau}h(\lambda),$$

where

$$h(\lambda) = \sigma\tau(\lambda + a) + \sigma + \tau, \; \lambda \in \mathbf{R}.$$

So $h(\lambda)$ has a unique real root $\lambda^* = -a - 1/\tau - 1/\sigma$. Furthermore, $x(\lambda)$ is strictly increasing on $(-\infty, \lambda^*)$ and strictly decreasing on $(\lambda^*, +\infty)$. Thus,

$$\frac{dy}{dx}(\lambda) = -e^{(\sigma-\tau)\lambda} < 0, \; \frac{d^2y}{dx^2}(\lambda) = e^{(\sigma-2\tau)\lambda}\frac{(\sigma - \tau)^2}{h(\lambda)}, \; \lambda \in \mathbf{R}\backslash\{\lambda^*\}.$$

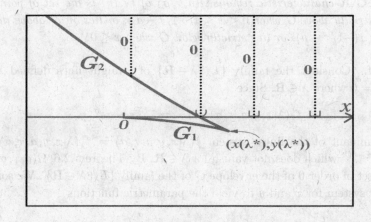

Fig. 7.9

As in Example 2.2, the envelope (see Figure 7.9) is composed of two pieces G_1 and G_2 and one turning point $(x(\lambda^*), y(\lambda^*))$. The first piece G_1 corresponds to the case where $\lambda \in (-\infty, \lambda^*)$ and the second G_2 to the case where $\lambda \in [\lambda^*, +\infty)$. Furthermore, G_1 is the graph of a function $y = G_1(x)$ which is strictly decreasing, strictly concave, and smooth over $(0, x(\lambda^*))$ such that $G_1(0^+) = 0$ and $G_1'(0^+) = -1$; and G_2 is the graph of a function $y = G_2(x)$ which is strictly decreasing, strictly convex, and smooth over $(-\infty, x(\lambda^*)]$ such that $G_2 \sim H_{-\infty}$. In view of the distribution map in Figure A.3, the dual set of order 0 of G is the set of points (x, y) that lies in the set $\vee(G_2) \oplus \triangledown(\Theta_0)$. Since $x(\lambda)$ has a unique real root $-1/\tau - a$ and G_1 is strictly decreasing, then $y(-1/\tau - a) < 0$, and $(x, y) \in \vee(G_2) \oplus \triangledown(\Theta_0)$ if, and only if, (x, y) lies strictly above G_2 when $x \leq x(-1/\tau - a)$, or lies on or above the x-axis when $x > x(-1/\tau - a)$.

Since $f(\lambda|x, y, a, \tau, \sigma) = -f(-\lambda| -x, -y, -a, -\tau, -\sigma)$, we see that $(x, y) \in \Omega(a, \tau, \sigma)$ if, and only if, $(-x, -y) \in \Omega(-a, -\tau, -\sigma)$. The second case $0 > \tau > \sigma$ can then be handled by means of the first. The proof is complete.

7.2.2.2 *The Case* $\tau\sigma < 0$

In this case, we may assume without loss of generality that $\tau > 0$ and $\sigma < 0$. We consider the following equivalent polynomial

$$f(\lambda|x, y, a, \tau, \sigma) = \lambda + a + e^{-\tau\lambda}x + e^{\sigma\lambda}y \tag{7.16}$$

where $a, x, y \in \mathbf{R}$ and $\tau, \sigma > 0$.

Theorem 7.10. *Let $\tau, \sigma > 0$ and $a, x, y \in \mathbf{R}$. Let the parametric curve G (see Figure 7.10) be defined by*

$$x(\lambda) = \frac{e^{\tau\lambda}}{\sigma + \tau}(1 - \sigma\lambda - a\sigma), \quad y(\lambda) = \frac{-e^{-\sigma\lambda}}{\sigma + \tau}(1 + \tau\lambda + a\tau), \quad \lambda \in \mathbf{R}.$$

Then the $\mathbf{C} \backslash \mathbf{R}$-characteristic regions $\Omega(a, \tau, \sigma)$ of (7.16) is the set of points (x, y) that lies strictly above G when $0 < x \leq x(-1/\tau - a.)$, or lies on or above the x-axis when $x > x(-1/\tau - a)$, or lies strictly below G when $x \leq 0$.

Proof. Consider the family $\{L_\lambda | \lambda \in \mathbf{R}\}$ of straight lines defined by L_λ : $f(\lambda|x, y) = 0$ where $\lambda \in \mathbf{R}$. Since

$$f_\lambda'(\lambda|x, y, a, \tau, \sigma) = 1 - \tau e^{-\tau\lambda}x + \sigma e^{\sigma\lambda}y,$$

the determinant of the linear system $f(\lambda|x, y, a, \tau, \sigma) = f_\lambda'(\lambda|x, y, a, \tau, \sigma) = 0$ is $(\sigma + \tau)e^{(\sigma-\tau)\lambda}$ which does not vanish for $\lambda \in \mathbf{R}$. By Theorem 2.6, $\Omega(a, \tau, \sigma)$ is just the dual set of order 0 of the envelope G of the family $\{L_\lambda | \lambda \in \mathbf{R}\}$. We solve from the linear system for x and y to yield the parametric functions

$$x(\lambda) = \frac{e^{\tau\lambda}}{\sigma + \tau}(1 - \sigma\lambda - a\sigma), \quad y(\lambda) = \frac{-e^{-\sigma\lambda}}{\sigma + \tau}(1 + \tau\lambda + a\tau), \quad \lambda \in \mathbf{R}.$$

We have

$$(x(-\infty), y(-\infty)) = (0, +\infty) \text{ and } (x(+\infty), y(+\infty)) = (-\infty, 0),$$

$$x'(\lambda) = \frac{e^{\tau\lambda}}{\sigma + \tau} h(\lambda) \text{ and } y'(\lambda) = -\frac{e^{-\sigma}}{\sigma + \tau} h(\lambda),$$

where

$$h(\lambda) = -\sigma\tau(\lambda + a) - \sigma + \tau, \ \lambda \in \mathbf{R}.$$

So $x'(\lambda)$ and $y'(\lambda)$ have the unique real root $\lambda^* = -a - 1/\tau + 1/\sigma$. Furthermore, $x(\lambda)$ is strictly increasing on $(-\infty, \lambda^*)$ and strictly decreasing on $(\lambda^*, +\infty)$. Thus,

$$\frac{dy}{dx}(\lambda) = -e^{-(\sigma+\tau)\lambda} < 0 \text{ and } \frac{d^2y}{dx^2}(\lambda) = e^{-(\sigma+2\tau)\lambda} \frac{(\sigma+\tau)^2}{h(\lambda)} \text{ for } \lambda \in \mathbf{R}\backslash\{\lambda^*\}.$$

As in Example 2.2, the envelope is composed of two pieces G_1 and G_2 and one turning point $(x(\lambda^*), y(\lambda^*))$.

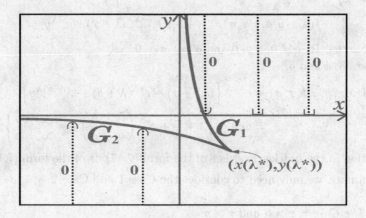

Fig. 7.10

The first piece G_1 corresponds to the case where $\lambda \in (-\infty, \lambda^*)$ and the second G_2 to the case where $\lambda \in [\lambda^*, +\infty)$. Furthermore, G_1 is the graph of a function $y = G_1(x)$ which is strictly decreasing, strictly convex, and smooth over $(0, x(\lambda^*))$ such that $G_1 \sim H_{0+}$; and G_2 is the graph of a function $y = G_2(x)$ which is strictly decreasing, strictly concave, and smooth over $(-\infty, x(\lambda^*)]$ such that $G_2'(\infty) = 0$. In view of the distribution map in Figure A.2, the dual set of order 0 of G is the set of points (x, y) that lies in the set $\wedge(G_2\chi_{(-\infty,0]}) \cup (\wedge(G_1) \oplus \triangle(\Theta_0))$. Since $x(\lambda)$ has a unique real root $1/\sigma - a$, we see that $(x, y) \in \wedge(G_2\chi_{(-\infty,0]}) \cup (\wedge(G_1) \oplus \triangle(\Theta_0))$ if, and only if, (x, y) lies strictly above G_1 when $0 < x \leq x(-1/\tau - a.)$, or lies on or above the x-axis when $x > x(-1/\tau - a)$, or lies strictly below G_2 when $x \leq 0$.

7.2.3 $\nabla(1,1,0)$-*Polynomials*

The general $\nabla(1,1,0)$-polynomial is

$$F(\lambda) = a\lambda + b + e^{-\tau\lambda}(c\lambda + x) + e^{-\sigma\lambda}y$$

where $a, b, c, x, y \in \mathbf{R}$ and τ, σ are distinct nonzero real numbers. If $a = 0$ and $c \neq 0$, then $F(\lambda)$ is the product of $e^{-\tau\lambda}$ and

$$be^{\tau\lambda} + c\lambda + x + e^{-\sigma\lambda+\tau\lambda}y,$$

and the latter function is a $\nabla(1,0,0)$-polynomial. If $a \neq 0$ and $c = 0$, then $F(\lambda)$ is a $\nabla(1,0,0)$-polynomial as well. Therefore, we may assume that $ac \neq 0$.

Since $a \neq 0$, by dividing $F(\lambda)$ by a if necessary, we may then assume without loss of generality that our ∇-polynomial is of the form

$$f(\lambda|x, y, a, b, \tau, \sigma) = \lambda + b + e^{-\tau\lambda}(a\lambda + x) + e^{-\sigma\lambda}y \qquad (7.17)$$

where $a, b, x, y, \tau, \sigma \in \mathbf{R}$, $a \neq 0$ and τ, σ are distinct nonzero real numbers.

There are five cases: (1) $\tau > 0$, $\sigma > 0$; (2) $\tau > 0$, $\sigma < 0$; (3) $\tau < 0$, $\sigma > 0$; (4) $\tau < 0$, $\sigma < 0$ and $\tau > \sigma$; and (5) $\tau < 0$, $\sigma < 0$ and $\tau < \sigma$.

In Case 2,

$$f(\lambda|x, y, a, b, \tau, \sigma) = \lambda + b + e^{-\tau\lambda}(a\lambda + x) + e^{\bar{\sigma}\lambda}y \qquad (7.18)$$

where $a, b, x, y \in \mathbf{R}$, $a \neq 0$, $\tau > 0$ and $\bar{\sigma} = -\sigma > 0$.

In Cases 3, 4 and 5,

$$f(\lambda|x, y, a, b, \tau, \sigma) = e^{-\tau\lambda}\left\{(a\lambda + x) + e^{\tau\lambda}(\lambda + b) + e^{(\tau-\sigma)\lambda}y\right\}$$

$$= ae^{-\tau\lambda}\left\{\lambda + \frac{x}{a} + e^{\tau\lambda}\left(\frac{\lambda}{a} + \frac{b}{a}\right) + e^{(\tau-\sigma)\lambda}\frac{y}{a}\right\}.$$

The function in the bracket is either of the form (7.17) or of the form (7.18).

In summary, we only need to consider the Case 1 and Case 2.

7.2.3.1   *The Case $\tau, \sigma > 0$ and $\tau > \sigma$.*

We consider the equation (7.17) where $\tau > \sigma > 0$. Consider the family $\{L_\lambda | \lambda \in \mathbf{R}\}$ of straight lines defined by $L_\lambda : f(\lambda|x, y, a, b, \tau, \sigma) = 0$ where $\lambda \in \mathbf{R}$. Since

$$f'_\lambda(\lambda|x, y, , a, b, \tau, \sigma) = 1 + ae^{-\tau\lambda} - \tau a\lambda e^{-\tau\lambda} - \tau xe^{-\tau\lambda} - \sigma ye^{-\sigma\lambda},$$

the determinant of the linear system $f(\lambda|x, y, a, b, \tau, \sigma) = f'_\lambda(\lambda|x, y, a, b, \tau, \sigma) = 0$ is $(-\sigma + \tau)e^{-(\sigma+\tau)\lambda}$ which does not vanish for $\lambda \in \mathbf{R}$. By Theorem 2.6, the $\mathbf{C}\backslash\mathbf{R}$-characteristic region Ω is just the dual set of order 0 of the envelope G of the family $\{L_\lambda | \lambda \in \mathbf{R}\}$. We solve from the linear system for x and y to yield the parametric functions

$$x(\lambda) = \frac{e^{\tau\lambda}}{\tau-\sigma}\left\{1 + (\lambda + b)\sigma + ae^{-\tau\lambda}(1 - (\tau - \sigma)\lambda)\right\}, \qquad (7.19)$$
$$y(\lambda) = \frac{-e^{\sigma\lambda}}{\tau-\sigma}\left\{1 + (\lambda + b)\tau + ae^{-\tau\lambda}\right\},$$

for $\lambda \in \mathbf{R}$. We have

$$(x(+\infty), y(+\infty)) = (+\infty, -\infty),$$

$$x'(\lambda) = \frac{\sigma\tau e^{\tau\lambda}}{(\tau - \sigma)} h(\lambda), \ y'(\lambda) = \frac{-\sigma\tau e^{\sigma\lambda}}{(\tau - \sigma)} h(\lambda),$$

where

$$h(\lambda) = \lambda + b + \frac{1}{\tau} + \frac{1}{\sigma} - ae^{-\tau\lambda} \frac{\tau - \sigma}{\sigma\tau}, \ \lambda \in \mathbf{R}. \tag{7.20}$$

Theorem 7.11. *Let* $b \in \mathbf{R}$, $\tau > \sigma > 0$, $a \leq \frac{-\sigma}{\tau-\sigma} e^{\frac{-\tau}{\sigma} - b\tau - 2}$ *and the curve* G *be defined by the parametric functions (7.19) where* $\lambda \in \mathbf{R}$. *Then* G *is the graph of a function* $y = G(x)$ *defined on* $\mathbf{R}$ *and* (x, y) *is a point of the* **C\R**-*characteristic region of (7.17) if, and only if,* (x, y) *is strictly above* G.

Proof. Let G be defined by (7.19). We have

$$(x(-\infty), y(-\infty)) = (-\infty, +\infty).$$

The function in (7.20) satisfies $h'(\lambda) = 1 + ae^{-\tau\lambda}(\tau - \sigma)/\sigma$, and hence $h'(\lambda)$ has the unique real root $\overline{\lambda} = \frac{1}{\tau}\ln(-a(\tau - \sigma)/\sigma)$. Since

$$\lim_{\lambda \to -\infty} h(\lambda) = \lim_{\lambda \to +\infty} h(\lambda) = +\infty,$$

we see that $h(\lambda)$ has the minimal value at $\overline{\lambda}$,

$$h(\overline{\lambda}) = \frac{1}{\tau}\ln\left(\frac{-a(\tau - \sigma)}{\sigma}\right) + b + \frac{2}{\tau} + \frac{1}{\sigma} > 0,$$

when $a < \frac{-\sigma}{\tau-\sigma} e^{\frac{-\tau}{\sigma} - b\tau - 2}$ and $h(\overline{\lambda}) = 0$ when $a = \frac{-\sigma}{\tau-\sigma} e^{\frac{-\tau}{\sigma} - b\tau - 2}$. Thus, $h(\lambda) > 0$ and

$$\frac{dy}{dx}(\lambda) = -e^{-(\tau-\sigma)\lambda} < 0 \text{ and } \frac{d^2y}{dx^2}(\lambda) = e^{-(2\tau-\sigma)\lambda}\frac{(\tau - \sigma)^2}{\sigma\tau h(\lambda)} > 0$$

for $\lambda \in \mathbf{R}\backslash\{\overline{\lambda}\}$. As in Example 2.2, G is the desired envelope. Furthermore, it is the graph of a function $y = G(x)$ which is strictly decreasing, strictly convex, and smooth over $\mathbf{R}$ such that $G'(+\infty) = 0$, $G \sim H_{+\infty}$ and $G \sim H_{-\infty}$. See Figure 7.11. In view of the distribution map in Figure 3.20, the dual set of order 0 of G is the set of points (x, y) that lies strictly above G.

Theorem 7.12. *Let* $b \in \mathbf{R}$, $\tau > \sigma > 0$ *and* $a > 0$. *Then the* **C\R**-*characteristic region of (7.17) is empty.*

Proof. The functions $x(\lambda)$ and $y(\lambda)$ in (7.19) now satisfy

$$(x(-\infty), y(-\infty)) = (+\infty, -\infty).$$

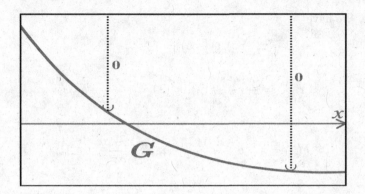

Fig. 7.11

Since $h'(\lambda) = 1 + ae^{-\tau\lambda} (\tau - \sigma)/\sigma > 0$, $\lim_{\lambda \to -\infty} h(\lambda) = -\infty$ and $\lim_{\lambda \to +\infty} h(\lambda) = +\infty$, we see that $h(\lambda)$ has a unique real root α. Thus,

$$\frac{dy}{dx}(\lambda) = -e^{-(\tau-\sigma)\lambda} < 0 \text{ and } \frac{d^2y}{dx^2}(\lambda) = e^{-(2\tau-\sigma)\lambda} \frac{(\tau-\sigma)^2}{\sigma\tau h(\lambda)}$$

for $\lambda \in \mathbf{R}\backslash\{\alpha\}$. The envelope is composed of two pieces G_1 and G_2 and one turning point $(x(\alpha), y(\alpha))$. The first piece G_1 corresponds to the case where $\lambda \in (-\infty, \alpha)$ and the second G_2 to the case where $\lambda \in [\alpha, +\infty)$. Furthermore, G_1 is the graph of a function $y = G_1(x)$ which is strictly decreasing, strictly concave, and smooth over $(x(\alpha), +\infty)$ such that $G_1 \sim H_{-\infty}$; and G_2 is the graph of a function $y = G_2(x)$ which is strictly decreasing, strictly convex, and smooth over $[x(\alpha), +\infty)$ such that $G_2'(+\infty) = 0$ and $G_2 \sim H_{+\infty}$. See Figure 7.12. By Theorem A.9, the dual set of order 0 of G is empty.

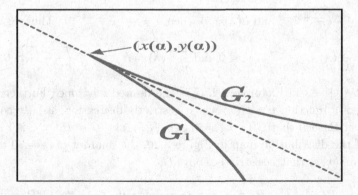

Fig. 7.12

Theorem 7.13. *Let $b \in \mathbf{R}$, $\tau > \sigma > 0$, $0 > a > \frac{-\sigma}{\tau-\sigma} e^{\frac{-\tau}{\sigma} - b\tau - 2}$. Then the function in (7.20) has exactly two real roots α and β such that $\alpha < \beta$. Let the curve G be* *defined by the parametric functions in (7.19) where* $\lambda \in \mathbf{R}$. *Then* (x, y) *is a point of the* **C\R**-*characteristic region of (7.17) if, and only if,* $(x, y) \in \vee(G\chi_I) \oplus \vee(G\chi_J)$ *where* $I = (-\infty, \alpha]$ *and* $J = [\beta, +\infty)$.

Proof. The functions $x(\lambda)$ and $y(\lambda)$ in (7.19) now satisfy

$$(x(-\infty), y(-\infty)) = (-\infty, +\infty).$$

Since $h'(\lambda) = 1 + a\tau e^{-\tau\lambda}(-\sigma + \tau)/\sigma\tau$, $h'(\lambda)$ has a unique real root $\overline{\lambda} = \frac{1}{\tau}\ln(-a(-\sigma + \tau)/\sigma)$. Then $h(\lambda)$ has the minimal value at $\overline{\lambda}$, and

$$h(\overline{\lambda}) = \frac{1}{\tau}\ln(\frac{a(-\sigma + \tau)}{-\sigma}) + b + \frac{2}{\tau} - \frac{1}{\sigma} < 0.$$

Since $\lim_{\lambda \to -\infty} h(\lambda) = \lim_{\lambda \to +\infty} h(\lambda) = +\infty$,. Thus, $h(\lambda)$ has exactly two real roots α and β such that $\alpha < \beta$ and

$$\frac{dy}{dx}(\lambda) = -e^{-(-\sigma+\tau)\lambda} < 0 \text{ and } \frac{d^2y}{dx^2}(\lambda) = e^{-(-\sigma+2\tau)\lambda}\frac{(-\sigma + \tau)^2}{\sigma\tau h(\lambda)}$$

for $\lambda \in \mathbf{R}\backslash\{\alpha, \beta\}$. The envelope is composed of three pieces G_1, G_2 and G_3 and two turning points $(x(\alpha), y(\alpha))$ and $(x(\beta), y(\beta))$. The first piece G_1 corresponds to the case where $\lambda \in (-\infty, \alpha]$, the second G_2 to the case where $\lambda \in (\alpha, \beta)$ and the third G_3 to the case where $\lambda \in [\beta, +\infty)$. Furthermore, G_1 is the graph of a function $y = G_1(x)$ which is strictly decreasing, strictly convex, and smooth over $(-\infty, x(\alpha)]$ such that $G_1 \sim H_{-\infty}$; G_2 is the graph of a function $y = G_2(x)$ which is strictly decreasing, strictly concave, and smooth over $(x(\alpha), x(\beta))$; and G_3 is the graph of a function $y = G_3(x)$ which is strictly decreasing, strictly convex, and smooth over $[x(\beta), +\infty)$ such that $G_3 \sim H_{+\infty}$. See Figure 7.13. In view of the distribution map in Figure A.14, the dual set of order 0 of G is the set of points (x, y) that lies in the set $\vee(G_1) \oplus \vee(G_3)$.

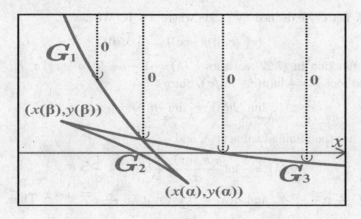

Fig. 7.13

7.2.3.2   The Case $\tau > 0$, $\sigma > 0$ and $\tau < \sigma$.

We consider the function (7.17) where $\tau > 0$, $\sigma > 0$ and $\sigma > \tau$. Consider the family $\{L_\lambda | \ \lambda \in \mathbf{R}\}$ of straight lines defined by $L_\lambda : f(\lambda|x, y, a, b, \tau, \sigma) = 0$ where $\lambda \in \mathbf{R}$. Since

$$f_\lambda'(\lambda|x, y, a, b, \tau, \sigma) = 1 + ae^{-\tau\lambda} - \tau a\lambda e^{-\tau\lambda} - \tau x e^{-\tau\lambda} - \sigma y e^{-\sigma\lambda},$$

the determinant of the linear system $f(\lambda|x, y, a, b, \tau, \sigma) = f_\lambda'(\lambda|x, y, a, b, \tau, \sigma) = 0$ is $(\tau - \sigma)e^{-(\sigma+\tau)\lambda}$ which does not vanish for $\lambda \in \mathbf{R}$. By Theorem 2.6, the $\mathbf{C}\backslash\mathbf{R}$-characteristic region Ω is just the dual set of order 0 of the envelope G of the family $\{L_\lambda | \ \lambda \in \mathbf{R}\}$. We solve from the linear system for x and y to yield the parametric functions

$$x(\lambda) = \frac{-e^{\tau\lambda}}{\sigma-\tau}\left\{1 + (\lambda+b)\sigma + ae^{-\tau\lambda}(1 + (\sigma-\tau)\lambda)\right\},$$
$$y(\lambda) = \frac{e^{\sigma\lambda}}{\sigma-\tau}\left\{1 + (\lambda+b)\tau + ae^{-\tau\lambda}\right\}, \tag{7.21}$$

for $\lambda \in \mathbf{R}$. We have

$$(x(+\infty), y(+\infty)) = (-\infty, +\infty)$$

$$x'(\lambda) = -\frac{\sigma\tau e^{\tau\lambda}}{\sigma-\tau}h(\lambda) \text{ and } y'(\lambda) = \frac{\sigma\tau e^{\sigma\lambda}}{\sigma-\tau}h(\lambda),$$

where

$$h(\lambda) = \lambda + b + \frac{1}{\tau} + \frac{1}{\sigma} + ae^{-\tau\lambda}\frac{\sigma-\tau}{\sigma\tau}, \ \lambda \in \mathbf{R}. \tag{7.22}$$

Theorem 7.14. *Let $b \in \mathbf{R}$, $\sigma > \tau > 0$ and $a \geq \frac{\sigma}{\sigma-\tau}e^{\frac{-\tau}{\sigma}-b\tau-2}$. Let the curve G be defined by the parametric functions (7.21) where $\lambda \in \mathbf{R}$. Then G is the graph of a function $y = G(x)$ defined on $\mathbf{R}$ and (x, y) is a point of the $\mathbf{C}\backslash\mathbf{R}$-characteristic region of (7.17) if, and only if, (x, y) is strictly above the graph G.*

Proof. Let G be defined by (7.21) where $\lambda \in \mathbf{R}$. We have

$$(x(-\infty), y(-\infty)) = (+\infty, 0).$$

Since the function in (7.22) satisfies $h'(\lambda) = 1 - ae^{-\tau\lambda}(\sigma-\tau)/\sigma$, $h'(\lambda)$ has the unique real root $\overline{\lambda} = \frac{1}{\tau}\ln(a(\sigma-\tau)/\sigma)$. Since

$$\lim_{\lambda \to -\infty} h(\lambda) = \lim_{\lambda \to +\infty} h(\lambda) = +\infty,$$

then $h(\lambda)$ has the minimal value at $\overline{\lambda}$, and

$$h(\overline{\lambda}) = \frac{1}{\tau}\ln(\frac{a(\sigma-\tau)}{\sigma}) + b + \frac{2}{\tau} + \frac{1}{\sigma} > 0$$

when $a > \frac{\sigma}{\sigma-\tau}e^{\frac{-\tau}{\sigma}-b\tau-2}$, and $h(\overline{\lambda}) = 0$ when $a = \frac{\sigma}{\sigma-\tau}e^{\frac{-\tau}{\sigma}-b\tau-2}$. Thus, $h(\lambda) > 0$ and

$$\frac{dy}{dx}(\lambda) = -e^{(\sigma-\tau)\lambda} < 0 \text{ and } \frac{d^2y}{dx^2}(\lambda) = e^{-(2\tau-\sigma)\lambda}\frac{(\sigma-\tau)^2}{\sigma\tau h(\lambda)} > 0$$

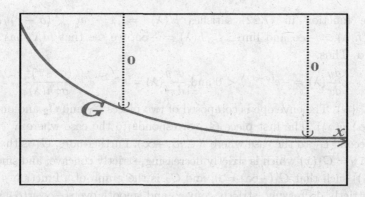

Fig. 7.14

for $\lambda \in \mathbf{R} \backslash \{\overline{\lambda}\}$. As in Example 2.2, G is the desired envelope. Furthermore, it is the graph of a function $y = G(x)$ which is strictly decreasing, strictly convex, and smooth over $\mathbf{R}$ such that $G'(+\infty) = 0$ and $G \sim H_{-\infty}$. See Figure 7.14. In view of the distribution map in Figure 3.19, the dual set of order 0 of G is the set of points strictly above G.

Theorem 7.15. *Let $b \in \mathbf{R}$, $\sigma > \tau > 0$ and $a < 0$. Then the function $h(\lambda)$ defined by (7.22) has a unique real root α. Let the curve S be defined by the parametric functions (7.21) where $\lambda \in (\alpha, +\infty)$. Then (x, y) is a point of the $\mathbf{C} \backslash \mathbf{R}$-characteristic region of (7.17) if, and only if, $(x, y) \in \vee(S) \oplus \nabla(\Theta_0)$.*

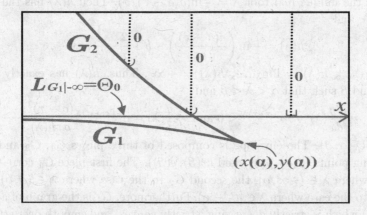

Fig. 7.15

Proof. Let the curve G be defined by the parametric functions (7.21) where $\lambda \in (-\infty, +\infty)$. We have

$$(x(-\infty), y(-\infty)) = (-\infty, 0).$$

Since the function in (7.22) satisfies $h'(\lambda) = 1 - ae^{-\tau\lambda}(\sigma - \tau)/\sigma > 0$, $\lim_{\lambda\to-\infty} h(\lambda) = -\infty$ and $\lim_{\lambda\to+\infty} h(\lambda) = +\infty$, we see that $h(\lambda)$ has a unique real root α. Thus,

$$\frac{dy}{dx}(\lambda) = -e^{(\sigma-\tau)\lambda} < 0 \text{ and } \frac{d^2y}{dx^2}(\lambda) = e^{-(2\tau-\sigma)\lambda}\frac{(\sigma-\tau)^2}{\sigma\tau h(\lambda)}$$

for $\lambda \in \mathbf{R}\backslash\{\alpha\}$. The envelope is composed of two pieces G_1 and G_2 and one turning point $(x(\alpha), y(\alpha))$. The first piece G_1 corresponds to the case where $\lambda \in (-\infty, \alpha)$ and the second G_2 to the case where $\lambda \in [\alpha, +\infty)$. Furthermore, G_1 is the graph of a function $y = G_1(x)$ which is strictly decreasing, strictly concave, and smooth over $(-\infty, x(\alpha))$ such that $G_1'(-\infty) = 0$; and G_2 is the graph of a function $y = G_2(x)$ which is strictly decreasing, strictly convex, and smooth over $(-\infty, x(\alpha)]$ such that $G_2 \sim H_{+\infty}$. See Figure 7.15. In view of the distribution map in Figure A.8, the dual set of order 0 of G is $(x, y) \in \mathsf{V}(G_2) \oplus \overline{\mathsf{V}}(\Theta_0)$.

Theorem 7.16. *Let $b \in \mathbf{R}$, $\sigma > \tau > 0$, $0 > a > \frac{-\sigma}{\tau-\sigma}e^{\frac{-\tau}{\sigma}-b\tau-2}$. Then the function $h(\lambda)$ defined by (7.22) has exactly two real roots α and β such that $\alpha < \beta$. Let G be defined by the parametric functions (7.21) where $\lambda \in \mathbf{R}$. Then (x, y) is a point of the $\mathbf{C}\backslash\mathbf{R}$-characteristic region of (7.17) if, and only if, $(x, y) \in \mathsf{V}(G\chi_I) \oplus \mathsf{V}(G\chi_J)$ where $I = (-\infty, \alpha]$ and $J = [\beta, +\infty)$.*

Proof. Let G be defined by the parametric functions (7.21) where $\lambda \in \mathbf{R}$. We have

$$(x(-\infty), y(-\infty)) = (+\infty, 0).$$

Since the function in (7.22) satisfies $h'(\lambda) = 1 - a\tau e^{-\tau\lambda}(\sigma - \tau)/\sigma\tau$, we see that $h'(\lambda)$ has the unique real root $\overline{\lambda} = \frac{1}{\tau}\ln(a(\sigma - \tau)/\sigma)$. Then $h(\lambda)$ has the minimal value at $\overline{\lambda}$, and

$$h(\overline{\lambda}) = \frac{1}{\tau}\ln\left(\frac{a(\sigma-\tau)}{\sigma}\right) + b + \frac{2}{\tau} + \frac{1}{\sigma} < 0.$$

Since $\lim_{\lambda\to-\infty} h(\lambda) = \lim_{\lambda\to+\infty} h(\lambda) = +\infty$,. Thus, $h(\lambda)$ has exactly two real roots α and β such that $\alpha < \overline{\lambda} < \beta$ and

$$\frac{dy}{dx}(\lambda) = -e^{(\sigma-\tau)\lambda} < 0 \text{ and } \frac{d^2y}{dx^2}(\lambda) = e^{-(2\tau-\sigma)\lambda}\frac{(\sigma-\tau)^2}{\sigma\tau h(\lambda)}$$

for $\lambda \in \mathbf{R}\backslash\{\alpha, \beta\}$. The envelope is composed of three pieces G_1, G_2 and G_3 and two turning points $(x(\alpha), y(\alpha))$ and $(x(\beta), y(\beta))$. The first piece G_1 corresponds to the case where $\lambda \in (-\infty, \alpha]$, the second G_2 to the case where $\lambda \in (\alpha, \beta)$ and the third G_3 to the case where $\lambda \in [\beta, +\infty)$. Furthermore, G_1 is the graph of a function $y = G_1(x)$ which is strictly decreasing, strictly convex, and smooth over $[x(\alpha), +\infty)$ such that $G_1'(+\infty) = 0$; G_2 is the graph of a function $y = G_2(x)$ which is strictly decreasing, strictly concave, and smooth over $(x(\alpha), x(\beta))$; and G_3 is the graph of a function $y = G_3(x)$ which is strictly decreasing, strictly convex, and smooth over $(-\infty, x(\beta)]$ such that $G_3 \sim H_{-\infty}$. See Figure 7.16. In view of the distribution map in Figure A.16, the dual set of order 0 of G is the set $\mathsf{V}(G_1) \oplus \mathsf{V}(G_3)$.

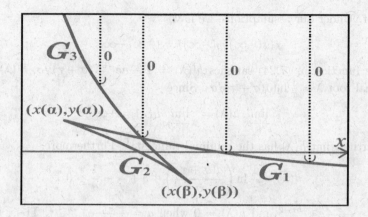

Fig. 7.16

7.2.3.3 *The Case* $\tau > 0$, $\sigma < 0$.

We consider the function (7.18) where $\tau > 0$ and $\sigma > 0$. Consider the family $\{L_\lambda | \lambda \in \mathbf{R}\}$ of straight lines defined by $L_\lambda : f(\lambda|x, y, a, b, \tau, \sigma) = 0$ where $\lambda \in \mathbf{R}$. Since

$$f'_\lambda(\lambda|x, y, a, b, \tau, \sigma) = 1 + ae^{-\tau\lambda} - \tau a\lambda e^{-\tau\lambda} - \tau xe^{-\tau\lambda} + \sigma ye^{\sigma\lambda},$$

the determinant of the linear system $f(\lambda|x, y, a, b, \tau, \sigma) = f'_\lambda(\lambda|x, y, a, b, \tau, \sigma) = 0$ is $(\sigma + \tau)e^{(\sigma-\tau)\lambda}$ which does not vanish for $\lambda \in \mathbf{R}$. By Theorem 2.6, the **C\R**-characteristic region Ω is just the dual set of order 0 of the envelope G of the family $\{L_\lambda | \lambda \in \mathbf{R}\}$. We solve from the linear system for x and y to yield the parametric functions

$$x(\lambda) = \frac{e^{\tau\lambda}}{(\sigma+\tau)} \left\{ 1 - (\lambda + b)\sigma + ae^{-\tau\lambda}(1 - (\sigma + \tau)\lambda) \right\},$$
$$y(\lambda) = \frac{-e^{-\sigma\lambda}}{(\sigma+\tau)} \left\{ 1 + (\lambda + b)\tau + ae^{-\tau\lambda} \right\}, \tag{7.23}$$

for $\lambda \in \mathbf{R}$. We have

$$(x(+\infty), y(+\infty)) = (-\infty, 0^-),$$

$$x'(\lambda) = -\frac{\sigma\tau e^{\tau\lambda}}{\sigma + \tau} h(\lambda) \text{ and } y'(\lambda) = \frac{\sigma\tau e^{-\sigma\lambda}}{\sigma + \tau} h(\lambda),$$

where

$$h(\lambda) = \lambda + b + \frac{1}{\tau} - \frac{1}{\sigma} + ae^{-\tau\lambda}\frac{\sigma + \tau}{\sigma\tau}, \lambda \in \mathbf{R}. \tag{7.24}$$

Theorem 7.17. *Let* $b \in \mathbf{R}$, $\tau > 0$, $\sigma < 0$ *and* $a \geq \frac{\sigma}{\sigma+\tau}e^{\frac{\tau}{\sigma}-b\tau-2}$. *Let the curve* G *be defined by the parametric functions (7.23) where* $\lambda \in \mathbf{R}$. *Then* G *is also the graph of a function* $y = G(x)$ *defined on* $\mathbf{R}$ *and* (x, y) *is a point of the* **C\R**-*characteristic region of (7.18) if, and only if,* (x, y) *lies strictly below* G.

Proof. Under our assumptions, we have

$$(x(-\infty), y(-\infty)) = (+\infty, -\infty).$$

Since the function in (7.24) satisfies $h'(\lambda) = 1 - ae^{-\tau\lambda}(\sigma + \tau)/\sigma$, $h'(\lambda)$ has the unique real root $\overline{\lambda} = \frac{1}{\tau}\ln(a(\sigma + \tau)/\sigma)$. Since

$$\lim_{\lambda \to -\infty} h(\lambda) = \lim_{\lambda \to +\infty} h(\lambda) = +\infty,$$

we see further that $h(\lambda)$ has the minimal value at $\overline{\lambda}$. Furthermore,

$$h(\overline{\lambda}) = \frac{1}{\tau}\ln\left(\frac{a(\sigma + \tau)}{\sigma}\right) + b + \frac{2}{\tau} - \frac{1}{\sigma} > 0$$

when $a > \frac{\sigma}{(\sigma+\tau)}e^{\frac{\tau}{\sigma} - b\tau - 2}$ and $h(\overline{\lambda}) = 0$ when $a = \frac{\sigma}{(\sigma+\tau)}e^{\frac{\tau}{\sigma} - b\tau - 2}$. Thus, $h(\lambda) > 0$ and

$$\frac{dy}{dx}(\lambda) = -e^{-(\sigma+\tau)\lambda} < 0 \quad \text{and} \quad \frac{d^2y}{dx^2}(\lambda) = -e^{-(\sigma+2\tau)\lambda}\frac{(\sigma + \tau)^2}{\sigma\tau h(\lambda)} < 0$$

for $\lambda \in \mathbf{R}\backslash\{\overline{\lambda}\}$. As in Example 2.2, G is the desired envelope. Furthermore, it is the graph of a function $y = G(x)$ which is strictly decreasing, strictly concave, and smooth over $\mathbf{R}$ such that $G'(-\infty) = 0$ and $G \sim H_{+\infty}$. See Figure 7.17. In view of the distribution map in Figure 3.19, the dual set of order 0 of G is the set of points strictly below G.

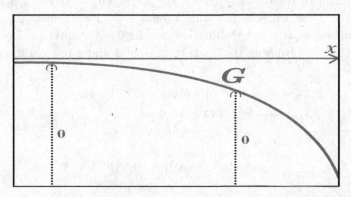

Fig. 7.17

Theorem 7.18. *Let $b \in \mathbf{R}$, $\tau > 0$, $\sigma < 0$ and $a \leq 0$. Then the function in (7.24) has a unique real root α. Let the curve S be defined by the parametric functions (7.23) where $\lambda \in (-\infty, \alpha)$. Then (x, y) is a point of the $\mathbf{C}\backslash\mathbf{R}$-characteristic region of (7.18) if, and only if, $(x, y) \in \vee(S) \oplus \nabla(\Theta_0)$.*

Proof. Under our assumptions, we have

$$(x(-\infty), y(-\infty)) = (-\infty, +\infty).$$

Since $h'(\lambda) = 1 - ae^{-\tau\lambda}(\sigma + \tau)/\sigma > 0$, $\lim_{\lambda\to-\infty} h(\lambda) = -\infty$ and $\lim_{\lambda\to+\infty} h(\lambda) = +\infty$, then $h(\lambda)$ has a unique real root α. Thus,

$$\frac{dy}{dx}(\lambda) = -e^{-(\sigma+\tau)\lambda} < 0 \text{ and } \frac{d^2y}{dx^2}(\lambda) = -e^{-(\sigma+2\tau)\lambda}\frac{(\sigma+\tau)^2}{\sigma\tau h(\lambda)}$$

for $\lambda \in \mathbf{R}\backslash\{\alpha\}$. The envelope G is composed of two pieces G_1 and G_2 and one turning point $(x(\alpha), y(\alpha))$. The first piece G_1 corresponds to the case where $\lambda \in (-\infty, \alpha)$ and the second G_2 to the case where $\lambda \in [\alpha, +\infty)$. Furthermore, G_1 is the graph of a function $y = G_1(x)$ which is strictly decreasing, strictly convex, and smooth over $(-\infty, x(\alpha))$ such that $G_1 \sim H_{-\infty}$; and G_2 is the graph of a function $y = G_2(x)$ which is strictly decreasing, strictly concave, and smooth over $(-\infty, x(\alpha)]$ such that $G_2'(-\infty) = 0$. See Figure 7.18. In view of the distribution map in Figure A.8, the dual set of order 0 of G is the set $\vee(G_1) \oplus \triangledown(\Theta_0)$.

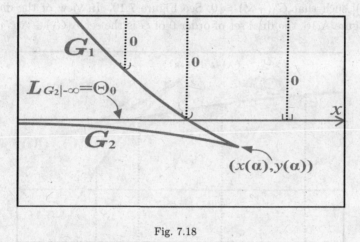

Fig. 7.18

Theorem 7.19. *Let* $b \in \mathbf{R}$, $\tau > 0$, $\sigma < 0$ *and* $0 < a < \frac{\sigma}{\sigma+\tau}e^{\frac{\tau}{\sigma}-b\tau-2}$. *Then the function in (7.24) has two real roots* α *and* β *such that* $\alpha < \beta$. *Let the curve* G *be defined by the parametric functions (7.23) where* $\lambda \in \mathbf{R}$. *Then* (x, y) *is a point of the* C\R-*characteristic region of (7.18) if, and only if,* $(x, y) \in \wedge(G\chi_I) \oplus \wedge(G\chi_J)$ *where* $I = (-\infty, \alpha]$ *and* $J = [\beta, +\infty)$.

Proof. Under our assumptions, we have

$$(x(-\infty), y(-\infty)) = (+\infty, -\infty).$$

Since $h'(\lambda) = 1 - ae^{-\tau\lambda}(\sigma + \tau)/\sigma$, $h'(\lambda)$ has the unique real root $\overline{\lambda} = \frac{1}{\tau}\ln(a(\sigma + \tau)/\sigma)$. Then $h(\lambda)$ has the minimal value at $\overline{\lambda}$, and

$$h(\overline{\lambda}) = \frac{1}{\tau}\ln(\frac{a(\sigma+\tau)}{\sigma}) + b + \frac{2}{\tau} - \frac{1}{\sigma} < 0.$$

Since $\lim_{\lambda \to -\infty} h(\lambda) = \lim_{\lambda \to +\infty} h(\lambda) = +\infty$, $h(\lambda)$ has two real roots α and β such that $\alpha < \overline{\lambda} < \beta$ and

$$\frac{dy}{dx}(\lambda) = -e^{-(\sigma+\tau)\lambda} < 0, \quad \frac{d^2y}{dx^2}(\lambda) = -e^{-(\sigma+2\tau)\lambda}\frac{(\sigma+\tau)^2}{\sigma\tau h(\lambda)}$$

for $\lambda \in \mathbf{R}\backslash\{\alpha, \beta\}$. The envelope is composed of three pieces G_1, G_2 and G_3 and two turning points $(x(\alpha), y(\alpha))$ and $(x(\beta), y(\beta))$. The first piece G_1 corresponds to the case where $\lambda \in (-\infty, \alpha]$, the second G_2 to the case where $\lambda \in (\alpha, \beta)$ and the third G_3 to the case where $\lambda \in [\beta, +\infty)$. Furthermore, G_1 is the graph of a function $y = G_1(x)$ which is strictly decreasing, strictly concave, and smooth over $[x(\alpha), +\infty)$ such that $G_1 \sim H_{+\infty}$; G_2 is the graph of a function $y = G_2(x)$ which is strictly decreasing, strictly convex, and smooth over $(x(\alpha), x(\beta))$; and G_3 is the graph of a function $y = G_3(x)$ which is strictly decreasing, strictly concave, and smooth over $(-\infty, x(\beta)]$ such that $G_3'(-\infty) = 0$. See Figure 7.19. In view of the distribution map in Figure A.16, the dual set of order 0 of G is the set $\wedge(G_1) \oplus \wedge(G_3)$.

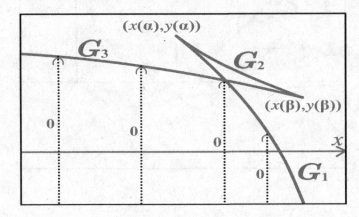

Fig. 7.19

7.2.4 $\nabla(n,n,0)$-*Polynomials*

In this section, we consider the special quasi-polynomial

$$f(\lambda|x, y, \tau, \sigma, n) = \lambda^n + x\lambda^n e^{-\tau\lambda} + ye^{-\sigma\lambda},$$

where $x, y, \tau, \sigma \in \mathbf{R}$, n is nonnegative integer and τ, σ are distinct nonzero numbers. If $n = 0$, the above function is a special $\nabla(0, 0, 0)$-polynomial. If $n = 1$, then the above function is a special $\nabla(1, 1, 0)$-polynomial. Thus we will assume that n is an integer in $\mathbf{Z}[2, \infty)$.

There are three cases: (1) $\sigma, \tau > 0$ and $\sigma \neq \tau$; (2) $\sigma, \tau < 0$ and $\sigma \neq \tau$; and (3) $\sigma\tau < 0$ and $\sigma \neq \tau$.

7.2.4.1   *The case where* $\sigma, \tau > 0$ *and* n *is odd*

In this case, we may consider

$$f(\lambda | x, y, \tau, \sigma, n) = \lambda^n + x\lambda^n e^{-\tau\lambda} + ye^{-\sigma\lambda}, \tag{7.25}$$

where n is an odd integer in $\mathbf{Z}[3, \infty)$, $\tau, \sigma, x, y \in \mathbf{R}$ and $\tau, \sigma > 0$.

To facilitate discussions, we will also classify equation into three different cases by means of the following exhaustive and mutually exclusive conditions according to the sizes of the integers τ and σ: (i) $\tau > \sigma > 0$, (ii) $\sigma \geq (1 + \sqrt{n+1})\tau/2$, and (iii) $(1 + \sqrt{n+1})\tau/2 > \sigma > \tau$.

Theorem 7.20. *Assume* n *is an odd integer in* $\mathbf{Z}[3, \infty)$ *and* $\tau > \sigma > 0$. *Let the curve* G *be defined by the parametric functions*

$$x(\lambda) = -\frac{n + \sigma\lambda}{(\sigma - \tau)\lambda + n}e^{\tau\lambda}, \;\; y(\lambda) = \frac{\tau\lambda^{n+1}}{(\sigma - \tau)\lambda + n}e^{\sigma\lambda} \tag{7.26}$$

over $\lambda \in [-n/\sigma, +\infty)$. *Then* G *is the graph of a function* $y = G(x)$ *defined over* $(-\infty, 0]$. *Furthermore,* (x, y) *is a point of the* **C\R**-*characteristic region of (7.25) if, and only if,* $x \leq 0$ *and* $y > G(x)$.

    Proof.   Consider the family $\{L_\lambda | \lambda \in \mathbf{R}\}$ of straight lines defined by L_λ : $f(\lambda | x, y, \tau, \sigma, n) = 0$ where f is defined by (7.25) for $\lambda \in \mathbf{R}$. Since

$$f'_\lambda(\lambda | x, y, \tau, \sigma, n) = n\lambda^{n-1} + (n - \tau\lambda)\lambda^{n-1}e^{-\tau\lambda}x - \sigma ye^{-\sigma\lambda},$$

the determinant of the linear system $f(\lambda | x, y, \tau, \sigma, n) = f'_\lambda(\lambda | x, y, \tau, \sigma, n) = 0$ is

$$D(\lambda) = -e^{-(\tau+\sigma)\lambda}((\sigma - \tau)\lambda + n)\lambda^{n-1},$$

which has exactly two real roots 0 and $n/(\tau - \sigma)$. Let us denote

$$\lambda^* = \frac{n}{\tau - \sigma}.$$

Note that the **C\R**-characteristic region of $f(\lambda | x, y, \tau, \sigma, n)$ is the intersection of the **C**\$(-\infty, 0)$-, **C**\$(0, \lambda^*)$-, **C**\$(\lambda^*, \infty)$-, **C**\$\{\lambda^*\}$- and **C**\$\{0\}$-characteristic regions of $f(\lambda | x, y, \tau, \sigma, n)$. The **C**\$\{\lambda^*\}$-characteristic region is the set of points that do not lie on the straight line L_{λ^*}, that is, **C**\$L_{\lambda^*}$; and the **C**\$\{0\}$-characteristic region is the set of points that do not lie on the straight line L_0, that is, **C**\$L_0$. As for the other three regions, the **C**\$(-\infty, 0)$-characteristic region is the dual set of order 0 of the envelope of the family $\Phi_{(-\infty,0)} = \{L_\lambda | \lambda \in (-\infty, 0)\}$, the **C**\$(0, \lambda^*)$-characteristic region is the dual set of order 0 of the envelope of the family $\Phi_{(0,\lambda^*)} = \{L_\lambda | \lambda \in (0, \lambda^*)\}$, and the **C**\$(\lambda^*, \infty)$-characteristic region is the dual set of order 0 of the envelope of the family $\Phi_{(\lambda^*,\infty)} = \{L_\lambda | \lambda \in (\lambda^*, \infty)\}$.

    By Theorem 2.3, the envelope G_1 of the family $\Phi_{(-\infty,0)}$ is described by the parametric functions (7.26) for $\lambda \in (-\infty, 0)$, while the envelope G_2 of the family $\Phi_{(0,\lambda^*)}$ by the same parametric functions but for $(0, \lambda^*)$, and the envelope G_3 of the family $\Phi_{(\lambda^*,\infty)}$ is described by the same parametric functions but for $\lambda \in (\lambda^*, \infty)$.

We have

$$(x(-\infty), y(-\infty)) = (0,0), \ (x(+\infty), y(+\infty)) = (+\infty, -\infty), \qquad (7.27)$$

$$(x((n/(\tau - \sigma))^+), y((n/(\tau - \sigma))^+) = (+\infty, -\infty),$$

$$(x((n/(\tau - \sigma))^-), y((n/(\tau + \sigma))^-) = (-\infty, +\infty),$$

$$\lim_{\lambda \to 0} (x(\lambda), y(\lambda))(-1, 0), \qquad (7.28)$$

$$x'(\lambda) = -\frac{\tau e^{\tau \lambda}}{[(\sigma - \tau)\lambda + n]^2} h(\lambda), \qquad (7.29)$$

and

$$y'(\lambda) = \frac{\tau \lambda^n e^{\sigma \lambda}}{[(\sigma - \tau)\lambda + n]^2} h(\lambda) \qquad (7.30)$$

where

$$h(\lambda) = \sigma(\sigma - \tau)\lambda^2 + n(2\sigma - \tau)\lambda + n(n + 1), \ \lambda \in \mathbf{R}\backslash\{\lambda^*, 0\}.$$

The real roots of $h(\lambda)$ are

$$r_\pm := \frac{-n(2\sigma - \tau) \pm \sqrt{n^2\tau^2 + 4n\sigma(\tau - \sigma)}}{2\sigma(\sigma - \tau)}.$$

Thus,

$$\frac{dy}{dx} = -\lambda^n e^{(\sigma - \tau)\lambda} \qquad (7.31)$$

and

$$\frac{d^2y}{dx^2} = \frac{1}{\tau} \lambda^{n-1} e^{(\sigma - 2\tau)\lambda} \frac{((\sigma - \tau)\lambda + n)^3}{h(\lambda)} \qquad (7.32)$$

$\lambda \in \mathbf{R}\backslash\{\lambda^*, 0, r_+, r_-\}.$
    Since

$$r_+ = \frac{-n(2\sigma - \tau) + \sqrt{n^2\tau^2 + 4n\sigma(\tau - \sigma)}}{2\sigma(\sigma - \tau)} < \frac{-n(2\sigma - \tau) + n\tau}{2\sigma(\sigma - \tau)} = \frac{-n}{\sigma} < 0 \quad (7.33)$$

and

$$r_- = \frac{-n(2\sigma - \tau) - \sqrt{n^2\tau^2 + 4n\sigma(\tau - \sigma)}}{2\sigma(\sigma - \tau)} > \frac{-n(2\sigma - \tau) - n\tau}{2\sigma(\sigma - \tau)} = \frac{-n}{\sigma - \tau}, \quad (7.34)$$

we see that

$$r_+ < \frac{-n}{\sigma} < 0 < \frac{-n}{\sigma - \tau} < r_-.$$

The envelope G_1 is made up of two pieces $G_1^{(1)}$ and $G_1^{(2)}$ which correspond to the case where $\lambda \in (-\infty, r_+)$ and where $\lambda \in [r_+, 0)$ respectively. $G_1^{(1)}$ is the graph of a function $y = G_1^{(1)}(x)$ which is strictly increasing, strictly concave and smooth over

$(0, x(r_+))$ such that $G_1^{(1)'}(0^+) = +\infty$ and $G_1^{(2)}$ is the graph of a function $y = G_1^{(2)}(x)$ which is strictly increasing, strictly convex and smooth over $(-1, x(r_+)]$ such that $G_1^{(2)}(-1^+) = 0$ and $\left(G_1^{(2)}\right)'(-1^+) = 0$. The envelope G_2, which corresponds to the restriction of G over the interval $\lambda \in (0, \lambda^*)$, is also the graph of a strictly decreasing, strictly convex and smooth function $y = G_2(x)$ defined on $(-\infty, -1)$ such that $G_2(-1^-) = 0$ and $G_2'(-1^-) = 0$. The envelope G_3 is made up of two pieces $G_3^{(1)}$ and $G_3^{(2)}$ which correspond to the case where $\lambda \in (\lambda^*, r_-]$ and where $\lambda \in [r_-, \infty)$ respectively. $G_3^{(1)}$ is the graph of a function $y = G_3^{(1)}(x)$ which is strictly decreasing, strictly concave and smooth over $(x(r_-), \infty)$ and $G_3^{(2)}$ is the graph of a function $y = G_3^{(2)}(x)$ which is strictly decreasing, strictly convex and smooth over $(x(r_+), \infty)$ such that $G_3^{(2)}(+\infty) = -\infty$ and $G_3^{(2)'}(+\infty) = 0$. By Lemma 3.5, $G_3^{(2)} \sim H_{+\infty}$. See Figure 7.20. In view of the distribution map in Figure A.25c, the dual sets of order 0 of the envelope $G_1^{(1)}$, $G_1^{(2)}$, G_2, $G_3^{(1)}$ and $G_3^{(2)}$ are easily obtained. The intersection of these dual sets and $\mathbf{C}\backslash L_{\lambda^*}$ as well as $\mathbf{C}\backslash L_0$, in view of Theorem 3.22, then yields (x, y) which lies in our desired characteristic region if, and only if, $x \le 0$ and $y > G(x)$. The proof is complete.

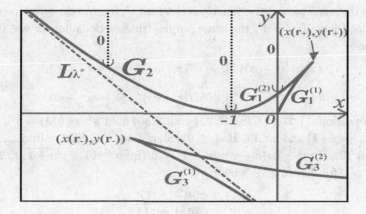

Fig. 7.20

Theorem 7.21. *Assume n is an odd integer in $\mathbf{Z}[3, \infty)$ and $\sigma \ge \left(1 + \sqrt{n+1}\right)\tau/2 > 0$. Let the curve G be defined by the parametric functions (7.26) over $(n/(\tau - \sigma), +\infty)$. Then G is the graph of a function $y = G(x)$ defined over $\mathbf{R}$. Furthermore, (x, y) is a point of the $\mathbf{C}\backslash\mathbf{R}$-characteristic region of (7.25) if, and only if, $y > G(x)$.*

Proof. Consider the family $\{L_\lambda | \lambda \in \mathbf{R}\}$ of straight lines defined by L_λ : $f(\lambda|x, y, \tau, \sigma, n) = 0$ where defined by (7.25) for $\lambda \in \mathbf{R}$. Since

$$f_\lambda'(\lambda|x, y, \tau, \sigma, n) = n\lambda^{n-1} + (n - \tau\lambda)\lambda^{n-1}e^{-\tau\lambda}x,$$

the determinant of the linear system $f(\lambda|x, y, \tau, \sigma, n) = f'_\lambda(\lambda|x, y, \tau, \sigma, n) = 0$ is

$$D(\lambda) = -e^{-(\tau+\sigma)\lambda}((\sigma - \tau)\lambda + n)\lambda^{n-1},$$

which has the two real roots 0 and $n/(\tau - \sigma)$. Let us denote

$$\lambda^* = \frac{n}{\tau - \sigma}.$$

Note that the $\mathbf{C}\backslash\mathbf{R}$-characteristic region of $f(\lambda|x, y)$ is the intersection of the $\mathbf{C}\backslash(-\infty, \lambda^*)$-, $\mathbf{C}\backslash(\lambda^*, 0)$-, $\mathbf{C}\backslash(0, \infty)$-, $\mathbf{C}\backslash\{\lambda^*\}$- and $\mathbf{C}\backslash\{0\}$-characteristic regions of $f(\lambda|x, y, \tau, \sigma, n)$. The $\mathbf{C}\backslash\{\lambda^*\}$-characteristic region is the set of points that do not lie on the straight line L_{λ^*}, that is, $\mathbf{C}\backslash L_{\lambda^*}$; and the $\mathbf{C}\backslash\{0\}$-characteristic region is the set of points that do not lie on the straight line L_0, that is, $\mathbf{C}\backslash L_0$. As for the other two regions, the $\mathbf{C}\backslash(-\infty, \lambda^*)$-characteristic region is the dual set of order 0 of the envelope of the family $\Phi_{(-\infty,\lambda^*)} = \{L_\lambda|\lambda \in (-\infty, \lambda^*)\}$, the $\mathbf{C}\backslash(\lambda^*, 0)$-characteristic region is the dual set of order 0 of the envelope of the family $\Phi_{(\lambda^*,0)} = \{L_\lambda|\lambda \in (\lambda^*, 0)\}$, and the $\mathbf{C}\backslash(0, \infty)$-characteristic region is the dual set of order 0 of the envelope of the family $\Phi_{(0,\infty)} = \{L_\lambda|\lambda \in (0, \infty)\}$.

By Theorem 2.6, the envelope G_1 of the family $\Phi_{(-\infty,\lambda^*)}$ is described by the parametric functions (7.26) for $\lambda \in (-\infty, \lambda^*)$, while the envelope G_2 of the family $\Phi_{(\lambda^*,0)}$ by the same parametric functions but for $(\lambda^*, 0)$, and the envelope G_3 of the family $\Phi_{(0,\infty)}$ is described by the same parametric functions but for $\lambda \in (0, \infty)$.

We have

$$(x((n/(\tau - \sigma))^+), y((n/(\tau - \sigma))^+) = (+\infty, +\infty),$$

$$(x((n/(\tau - \sigma))^-), y((n/(\tau - \sigma))^-) = (-\infty, -\infty),$$

and the conditions (7.27), (7.28), (7.29) and (7.30) hold where $h(\lambda) = \sigma(\sigma - \tau)\lambda^2 + n(2\sigma - \tau)\lambda + n(n+1)$ and for $\lambda \in \mathbf{R}\backslash\{\lambda^*, 0\}$. Since $\lim_{\lambda \to -\infty} h(\lambda) = \lim_{\lambda \to +\infty} h(\lambda) = +\infty$ and $n^2(2\sigma - \tau)^2 - 4\sigma(\sigma - \tau)n(n + 1) \le 0$ (by $\sigma \ge (1 + \sqrt{n+1})\tau/2$), we see that $h(\lambda) > 0$ for $\lambda \in \mathbf{R}\backslash\{\overline{\lambda}\}$, where

$$\overline{\lambda} = \frac{-n(2\sigma - \tau)}{2\sigma(\sigma - \tau)}.$$

Thus, (7.31) and (7.32) hold for $\lambda \in \mathbf{R}\backslash\{\overline{\lambda}, \lambda^*, 0\}$.

The envelope G_1, which corresponds to the case where $\lambda \in (-\infty, \lambda^*)$, is also the graph of a strictly increasing, strictly concave and smooth function $y = G_1(x)$ defined on $(-\infty, 0)$ such that $G_1(0^-) = 0$ and $G'_1(0^-) = 0$. The envelope G_2, which corresponds to the restriction of G over the interval $\lambda \in (\lambda^*, 0)$, is also the graph of a strictly increasing, strictly convex and smooth function $y = G_2(x)$ defined on $(-1, \infty)$ such that $G_2(-1^+) = 0$ and $G'_2(-1^+) = 0$. The envelope G_3 which corresponds to the restriction of G over the interval $\lambda \in (0, +\infty)$, is also the graph of a strictly decreasing, strictly convex and smooth function $y = G_3(x)$ defined on $(-\infty, -1)$ such that $G_3(-1^-) = 0$ and $G'_3(-1^-) = 0$. See Figure 7.21. In view of Figure A.12, (x, y) belongs to the desired $\mathbf{C}\backslash\mathbf{R}$-characteristic region if, and only if, $y > G(x)$.

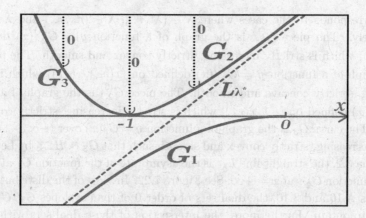

Fig. 7.21

Theorem 7.22. *Assume n is an odd integer in $\mathbf{Z}[0, \infty)$ and $(1 + \sqrt{n+1})\tau/2 > \sigma > \tau > 0$. Let r_+ and r_- be defined by (7.33) and (7.34) and the curve G be defined by the parametric functions (7.26) over $(n/(\tau - \sigma), +\infty)$. Then (x, y) is a point of the* C\R-*characteristic region of (7.25) if, and only if, $(x, y) \in \vee(G\chi_{I_1}) \oplus \vee(G\chi_{I_2})$ where $I_1 = (x(r_-), n/(\tau - \sigma))$ and $I_2 = (-\infty, x(r_+))$.*

Proof. The proof is similar to that of Theorem 7.21 and hence will be sketched. The parametric functions, the number λ^*, the straight lines L_λ, the families $\Phi_{(-\infty, \lambda^*)}$, $\Phi_{(\lambda^*, 0)}$ and $\Phi_{(0, \infty)}$, the C\$(-\infty, \lambda^*)$-, C\$(\lambda^*, 0)$-, C\$(0, \infty)$-characteristic regions and the envelopes G_1, G_2 and G_3 are the same as in the proof of Theorem 7.21.

The function $h(\lambda) = \sigma(\sigma - \tau)\lambda^2 + n(2\sigma - \tau)\lambda + n(n + 1)$ for $\lambda \in \mathbf{R}$ has two real roots defined by (7.33) and (7.34). Since

$$r_+ < \frac{-n(2\sigma - \tau) + n\tau}{2\sigma(\sigma - \tau)} = \frac{-n}{\sigma} < 0$$

and

$$r_- > \frac{-n(2\sigma - \tau) - n\tau}{2\sigma(\sigma - \tau)} = \frac{-n}{\sigma - \tau}$$

we see that $\lambda^* < r_- < r_+ < -n/\sigma < 0$.

The envelope G_1 is described by the parametric functions $x(\lambda)$ and $y(\lambda)$ defined by (7.26) over the interval $(-\infty, \lambda^*)$, the envelope G_2 by the same functions but over the interval $(\lambda^*, 0)$, and the envelope G_3 by the same functions but over the interval $(0, +\infty)$. We see that $h(\lambda) > 0$ for $\lambda \in (-\infty, r_-) \cup (r_+, \infty)$ and $h(\lambda) < 0$ for $\lambda \in (r_-, r_+)$. Thus, $x(\lambda)$ is strictly increasing on (r_-, r_+) and strictly decreasing on $(-\infty, \lambda^*) \cup (\lambda^*, 0) \cup (0, +\infty)$. With these and other easily obtained information, the curves G_1, G_2 and G_3 can be depicted as shown in Figure 7.22. The curve G_1 is the graph of a function $y = G_1(x)$ over $(-\infty, 0)$ which is strictly increasing, strictly concave and smooth. The curve G_2 is made up of three pieces $G_2^{(1)}$, $G_2^{(2)}$ and $G_2^{(3)}$

which corresponds to the cases where $\lambda \in (\lambda^*, r_-]$, $\lambda \in (r_-, r_+)$ and $\lambda \in [r_+, 0)$ respectively. The piece $G_2^{(1)}$ is the graph of a function $y = G_2^{(1)}(x)$ defined on $[x(r_-), \infty)$ which is strictly increasing, strictly convex and smooth. The piece $G_2^{(2)}$ is the graph of a function $y = G_2^{(2)}(x)$ defined on $(x(r_-), x(r_+))$ which is strictly increasing, strictly concave and smooth. The piece $G_2^{(3)}$ is the graph of a function $y = G_2^{(3)}(x)$ defined on $(-1, x(r_+)]$ which is strictly increasing, strictly convex and smooth. The curve G_3 is the graph of a function $y = G_3(x)$ over $(-\infty, -1)$ which is strictly decreasing, strictly convex and smooth such that $G_3 \sim H_{-\infty}$ by Lemma 3.4. By Lemma 3.2, the straight line L_{λ^*} is the asymptote of the function G_1 at $x = -\infty$ and the function $G_2^{(1)}$ at $x = +\infty$. See Figure 7.22. In view of the distribution maps in Figures A.16 and 3.10, the dual sets of order 0 of the envelopes G_1, G_2 and G_3 are easy to obtain. Furthermore, the intersection of these dual sets with $C \backslash L_{\overline{\lambda}}$ is given by $\vee \left(G_2^{(3)} \right) \oplus \vee \left(G_2^{(1)} \right) \oplus \vee (G_3)$. The proof is complete.

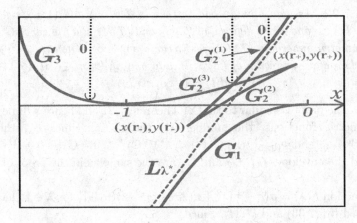

Fig. 7.22

7.2.4.2   *The case where $\sigma, \tau > 0$ and n is even*

To facilitate discussions, we will also classify (7.25) into three different cases by means of the following exhaustive and mutually exclusive conditions according to the sizes of the integers τ and σ: (i) $\tau > \sigma > 0$, (ii) $\sigma \geq (1 + \sqrt{n+1})\, \tau/2 > 0$, and (iii) $(1 + \sqrt{n+1})\, \tau/2 > \sigma > \tau > 0$.

Theorem 7.23. *Assume n is a positive and even integer and $\tau > \sigma > 0$. Then (x, y) is a point of the $C \backslash R$-characteristic region of (7.25) if, and only if, $x \geq 0$ and $y > 0$.*

    Proof. The proof is similar to that of Theorem 7.20 and hence will be sketched. The parametric functions, the number λ^*, the straight lines L_λ, the families $\Phi_{(-\infty, \lambda^*)}$, $\Phi_{(\lambda^*, 0)}$ and $\Phi_{(0, \infty)}$, the regions $C \backslash (-\infty, \lambda^*)$-, $C \backslash (\lambda^*, 0)$-, $C \backslash (0, \infty)$-

characteristic regions and the envelopes G_1, G_2 and G_3 are the same as in the proof of Theorem 7.20.

By the same derivations in the proof of Theorem 7.20, we see that

$$r_+ < \frac{-n}{\sigma} < 0 < \frac{-n}{\sigma - \tau} < r_-$$

where r_+ and r_- are defined by (7.33) and (7.34). The envelope G_1 is made up of two pieces $G_1^{(1)}$ and $G_1^{(2)}$ which correspond to the case where $\lambda \in (-\infty, r_+)$ and where $\lambda \in [r_+, 0)$ respectively. $G_1^{(1)}$ is the graph of a function $y = G_1^{(1)}(x)$ which is strictly decreasing, strictly convex and smooth over $(0, x(r_+))$ such that $G_1^{(1)\prime}(0^+) = -\infty$ and $G_1^{(2)}$ is the graph of a function $y = G_1^{(2)}(x)$ which is strictly decreasing, strictly concave and smooth over $(-1, x(r_+)]$ such that $G_1^{(2)}(-1^+) = 0$ and $G_1^{(2)\prime}(-1^+) = 0$. The envelope G_2, which corresponds to the restriction of G over the interval $\lambda \in (0, \lambda^*)$, is also the graph of a strictly decreasing, strictly convex and smooth function $y = G_2(x)$ defined on $(-\infty, -1)$ such that $G_2(-1^-) = 0$ and $G_2'(-1^-) = 0$. The envelope G_3 is made up of two pieces $G_3^{(1)}$ and $G_3^{(2)}$ which correspond to the case where $\lambda \in (\lambda^*, r_-]$ and where $\lambda \in [r_-, \infty)$ respectively. $G_3^{(1)}$ is the graph of a function $y = G_3^{(1)}(x)$ which is strictly decreasing, strictly concave and smooth over $(x(r_-), \infty)$ and $G_3^{(2)}$ is the graph of a function $y = G_3^{(2)}(x)$ which is strictly decreasing, strictly convex and smooth over $(x(r_+), \infty)$ such that $G_3^{(2)}(+\infty) = -\infty$ and $G_3^{(2)\prime}(+\infty) = 0$. By Lemma 3.5, $G_3^{(2)} \sim H_{+\infty}$. See Figure 7.23. The dual sets of order 0 of G_1, G_2 and G_3, in view of the distribution maps in Figure A.30, are easy to find. Furthermore, the intersection of these dual sets with **C\R**, is also easily obtained so that a point (x, y) belongs to it, and only if, $x \geq 0$ and $y > 0$.

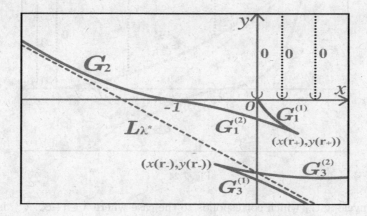

Fig. 7.23

Theorem 7.24. *Assume n is a positive and even integer and $\sigma \geq (1 + \sqrt{n+1})\tau/2 > 0$. Let the curve G be defined by the parametric functions (7.25)* *over* $\mathbf{R}$. *Then* (x,y) *is a point of the* $\mathbf{C}\backslash\mathbf{R}$-*characteristic region of* (7.26) *if, and only if,* $(x,y) \in \vee(G\chi_{I_1}) \oplus \vee(G\chi_{I_2})$ *and* $x < 0$, *or* $x \geq 0$ *and* $y > 0$ *where* $I_1 = (-\infty, n/(\tau - \sigma))$ *and* $I_2 = (0, +\infty)$.

Proof.   The proof is similar to that of Theorem 7.21 and hence will be sketched. The parametric functions, the number λ^*, the straight lines L_λ, the families $\Phi_{(-\infty,\lambda^*)}$, $\Phi_{(\lambda^*,0)}$ and $\Phi_{(0,\infty)}$, the regions $\mathbf{C}\backslash(-\infty,\lambda^*)$-, $\mathbf{C}\backslash(\lambda^*,0)$-, $\mathbf{C}\backslash(0,\infty)$-characteristic regions and the envelopes G_1, G_2 and G_3 are the same as in the proof of Theorem 7.21.

We have

$$(x(-\infty), y(-\infty)) = (0,0), \ (x(+\infty), y(+\infty)) = (-\infty, +\infty),$$

$$(x((n/(\tau - \sigma))^+), y((n/(\tau - \sigma))^+) = (+\infty, -\infty),$$

$$(x((n/(\tau - \sigma))^+), y((n/(\tau - \sigma))^+) = (-\infty, +\infty),$$

and the conditions (7.28), (7.29) and (7.30) hold where $h(\lambda) = \sigma(\sigma - \tau)\lambda^2 + n(2\sigma - \tau)\lambda + n(n+1)$ for $\lambda \in \mathbf{R}\backslash\{\lambda^*, 0\}$. Since $h(\lambda) > 0$ for $\lambda \in \mathbf{R}\backslash\{\bar{\lambda}\}$ and

$$\bar{\lambda} = \frac{-n(2\sigma - \tau)}{2\sigma(\sigma - \tau)}.$$

Thus, (7.31) and (7.32) hold for $\lambda \in \mathbf{R}\backslash\{\bar{\lambda}, \lambda^*, (n+1)/\tau, 0\}$.

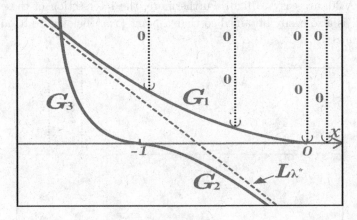

Fig. 7.24

The envelope G_1 which corresponds to the case where $\lambda \in (-\infty, \lambda^*)$, is also the graph of a strictly decreasing, strictly convex and smooth function $y = G_1(x)$ defined on $(-\infty, 0)$. The envelope G_2 which corresponds to the case where $\lambda \in (\lambda^*, 0)$, is also the graph of a strictly decreasing, strictly concave and smooth function $y = G_2(x)$ defined on $(-1, +\infty)$ such that $G_2(-1^+) = 0$ and $G_2'(-1^+) = 0$. The envelope G_3 which corresponds to the case where $\lambda \in (0, +\infty)$, is also the graph of a strictly decreasing, strictly convex and smooth function $y = G_3(x)$ defined on $(-\infty, -1)$ such that $G_3(-1^-) = 0$ and $G_3 \sim H_{-\infty}$.

In view of the distribution map in Figure A.30c, the dual sets of order 0 of the envelope G_1, G_2 and G_3 are easily obtained. The intersection of these dual sets, $\mathbf{C} \backslash L_{\lambda^*}$ and $\mathbf{C} \backslash L_0$, then yields points (x, y) which lie in our desired characteristic region if, and only if, $(x, y) \in \vee(G_1) \oplus \vee(G_3)$ and $x < 0$, or $x \geq 0$ and $y > 0$.

Theorem 7.25. *Assume n is a positive and even integer and $(1 + \sqrt{n+1})\tau/2 > \sigma > \tau > 0$. Let the curve G be defined by the parametric functions (7.26) over* $\mathbf{R}$. *Then (x, y) is a point of the $\mathbf{C} \backslash \mathbf{R}$-characteristic region of (7.25) if, and only if, $(x, y) \in \vee(G\chi_{I_1}) \oplus \vee(G\chi_{I_2})$ and $x < 0$, or $x \geq 0$ and $y > 0$ where $I_1 = (-\infty, n/(\tau - \sigma))$ and $I_2 = (0, +\infty)$.*

Proof. The proof is similar to that of Theorem 7.22 and hence will be sketched. The parametric functions, the number λ^*, the straight lines L_λ, the families $\Phi_{(-\infty, \lambda^*)}$, $\Phi_{(\lambda^*, 0)}$ and $\Phi_{(0, \infty)}$, the regions $\mathbf{C} \backslash (-\infty, \lambda^*)$-, $\mathbf{C} \backslash (\lambda^*, 0)$-, $\mathbf{C} \backslash (0, \infty)$-characteristic regions and the envelopes G_1, G_2 and G_3 are the same as in the proof of Theorem 7.22.

We have

$$(x(-\infty), y(-\infty)) = (0, 0), \ (x(+\infty), y(+\infty)) = (-\infty, +\infty),$$

$$(x((n/(\tau - \sigma))^+), y((n/(\tau - \sigma))^+) = (+\infty, -\infty),$$

$$(x((n/(\tau - \sigma))^+), y((n/(\tau - \sigma))^+) = (-\infty, +\infty),$$

and the (7.28), (7.29) and (7.30) hold where $h(\lambda) = \sigma(\sigma - \tau)\lambda^2 + n(2\sigma - \tau)\lambda + n(n+1)$ and for $\lambda \in \mathbf{R} \backslash \{\lambda^*, 0\}$. The function $h(\lambda)$ has two real roots defined by (7.33) and (7.34). Thus, (7.31) and (7.32) hold for $\lambda \in \mathbf{R} \backslash \{\lambda^*, r_+, r_-, 0\}$. Since

$$r_+ < \frac{-n(2\sigma - \tau) + n\tau}{2\sigma(\sigma - \tau)} = \frac{-n}{\sigma} < 0$$

and

$$r_- > \frac{-n(2\sigma - \tau) - n\tau}{2\sigma(\sigma - \tau)} = \frac{-n}{\sigma - \tau},$$

we see that $\lambda^* < r_- < r_+ < -n/\sigma < 0$.

The envelope G_1 which corresponds to the case where $\lambda \in (-\infty, \lambda^*)$, is also the graph of a strictly decreasing, strictly convex and smooth function $y = G_1(x)$ defined on $(-\infty, 0)$. The envelope G_2 is made up of three pieces $G_2^{(1)}$, $G_2^{(2)}$ and $G_2^{(3)}$ which correspond to the case where $\lambda \in (\lambda^*, r_-]$, where $\lambda \in (r_-, r_+)$ and where $\lambda \in [r_+, 0)$ respectively. $G_2^{(1)}$ is the graph of a function $y = G_2^{(1)}(x)$ which is strictly decreasing, strictly concave and smooth over $[x(r_-), +\infty)$, $G_2^{(2)}$ is the graph of a function $y = G_2^{(2)}(x)$ which is strictly decreasing, strictly convex and

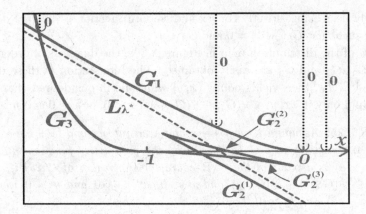

Fig. 7.25

smooth over $(x(r_-), x(r_+))$ and $G_2^{(3)}$ is the graph of a function $y = G_2^{(3)}(x)$ which is strictly decreasing, strictly concave and smooth over $(-1, x(r_+)]$. The envelope G_3 which corresponds to the case where $\lambda \in (0, +\infty)$, is also the graph of a strictly decreasing, strictly convex and smooth function $y = G_3(x)$ defined on $(-\infty, -1)$ such that $G_3(-1^-) = 0$ and $G_3 \sim H_{-\infty}$.

In view of the distribution maps in Figure A.29c, the dual sets of order 0 of the envelope G_1, G_2 and G_3 are easily obtained. The intersection of these dual sets, $\mathbf{C} \backslash L_{\lambda^*}$ and $\mathbf{C} \backslash L_0$, then yields points (x, y) which lie in our desired characteristic region if, and only if, $(x, y) \in \vee(G_1) \oplus \vee(G_3)$ and $x < 0$, or $y \geq 0$ and $y > 0$.

7.2.4.3   *The case where $\sigma, \tau < 0$*

Suppose n is an integer in $\mathbf{Z}[2, \infty)$, $\tau, \sigma, x, y \in \mathbf{R}$ and $\tau, \sigma < 0$. Let $\Omega(\tau, \sigma, n)$ be the $\mathbf{C} \backslash \mathbf{R}$-characteristic region of

$$f(\lambda | x, y, \tau, \sigma, n) = \lambda^n + x\lambda^n e^{-\tau\lambda} + ye^{-\sigma\lambda}.$$

If n is odd, then $f(-\lambda | x, y, \tau, \sigma, n) = -(\lambda^n + x\lambda^n e^{\tau\lambda} + (-y)e^{\sigma\lambda})$. Hence $(x, y) \in \Omega(\tau, \sigma, n)$ if, and only if, $(x, -y) \in \Omega(-\tau, -\sigma, n)$. If n is even, then $f(-\lambda | x, y, \tau, \sigma, n) = \lambda^n + x\lambda^n e^{-\tau\lambda} + ye^{-\sigma\lambda}$. Hence $\Omega(\tau, \sigma, n) = \Omega(-\tau, -\sigma, n)$. In either cases, we may apply the previous results for the case where $\sigma, \tau > 0$ to derive the desired regions.

7.2.4.4   *The case where $\sigma\tau < 0$*

We consider

$$f(\lambda | x, y, \tau, \sigma, n) = \lambda^n + x\lambda^n e^{-\tau\lambda} + ye^{-\sigma\lambda}, \tag{7.35}$$

where n is integer in $\mathbf{Z}[2, \infty)$, and $\tau, \sigma, x, y \in \mathbf{R}$ and $\sigma\tau < 0$. If $x = 0$, the corresponding function has been treated in Corollary 7.1. If $x \neq 0$, then

$$f(\lambda|y, \tau, \sigma, n) = e^{-\tau\lambda}\left\{e^{\tau\lambda}\lambda^n + x\lambda^n + ye^{(\tau-\sigma)\lambda}\right\}$$

$$= xe^{-\tau\lambda}\left\{\lambda^n + \frac{1}{x}\lambda^n e^{\tau\lambda} + \frac{y}{x}e^{(\tau-\sigma)\lambda}\right\}$$

$$= xe^{-\tau\lambda}\left\{\lambda^n + a\lambda^n e^{\tau\lambda} + be^{(\tau-\sigma)\lambda}\right\},$$

where $a = 1/x$ and $b = y/x$. Note that $\sigma\tau < 0$ if, and only if, $\tau > 0 > \sigma$ or $\sigma > 0 > \tau$. In either cases, the previous conclusions can be applied to the function in the last bracket.

7.3 ∇-Polynomials Involving Three Powers

There are good reasons to study ∇-polynomials of the form

$$f_0(\lambda) + e^{-\tau\lambda}f_1(\lambda) + e^{-\sigma\lambda}f_2(\lambda) + e^{-\delta\lambda}f_3(\lambda) \tag{7.36}$$

involving three powers. For instance, if we consider geometric sequences of the form $\{e^{\lambda t}\}$ as solutions of the difference equation

$$x'(t) + px(t - \tau) + qx(t - \sigma) + rx(t - \delta) = 0, \ t \in \mathbf{R}$$

where $p, q, r, \tau, \sigma, \delta \in \mathbf{R}$, then we are led to the quasi-polynomial

$$g(\lambda) = \lambda + e^{-\tau\lambda}p + e^{-\sigma\lambda}q + e^{-\sigma\lambda}r.$$

The $C\backslash\mathbf{R}$-characteristic region of the ∇-polynomial (7.36) is difficult to find. Therefore we will restrict ourselves to the case where the degree of f_0 is less than or equal to 1, the degrees of f_1, f_2 and f_3 are 0, and $\delta, \sigma, \tau \leq 0$ or $\delta, \sigma, \tau \geq 0$. In cases where any two of δ, σ, τ are the same, or, $\delta\sigma\tau = 0$, $g(\lambda)$ reduces to a ∇-polynomial involving only one power or only with two powers. Hence we may assume that δ, σ, τ are pairwise distinct nonzero numbers.

7.3.1 ∇(0, 0, 0, 0)-*Polynomials*

The general form of ∇(0, 0, 0, 0)-polynomials is

$$F(\lambda|a, b, p, q, \tau, \sigma, \delta) = a + e^{-\tau\lambda}b + e^{-\sigma\lambda}p + e^{-\delta\lambda}q$$

where $a, b, p, q \in \mathbf{R}$ and τ, σ, δ are pairwise distinct nonzero real numbers. There are several possible cases based in the values of τ, σ and τ. We will, however, only consider two cases: $\delta, \sigma, \tau < 0$ or $\delta, \sigma, \tau > 0$. Note that

$$F(-\lambda|a, b, p, q, \tau, \sigma, \delta) = a + e^{\tau\lambda}b + e^{\sigma\lambda}p + e^{\delta\lambda}q = F(\lambda|a, b, p, q, -\tau, -\sigma, -\delta),$$

we may therefore further assume without loss of generality that $\delta > \sigma > \tau > 0$. If $b = 0$, the ∇-polynomial is equivalent to a ∇-polynomial involving only two power which has already been discussed before. If $b \neq 0$, by dividing $F(\lambda)$ by b if necessary, we may then assume without loss of generality that our ∇-polynomial is of the form

$$f(\lambda|x, y, a, \delta, \sigma, \tau) = a + e^{-\tau\lambda} + e^{-\sigma\lambda}x + e^{-\delta\lambda}y, \qquad (7.37)$$

where $a, x, y \in \mathbf{R}$ and τ, σ, δ are pairwise distinct nonzero real numbers. Let $\Omega(a, \delta, \sigma, \tau)$ be the $\mathbf{C}\backslash\mathbf{R}$-characteristic region of this ∇-polynomial. That is,

$$\Omega(a, \delta, \sigma, \tau) = \left\{(x, y) \in R^2 | \ f(\lambda|x, y, a, \delta, \sigma, \tau) \text{ has no real roots}\right\}.$$

Theorem 7.26. *Assume $x, y, a, \delta, \sigma, \tau \in \mathbf{R}$, $a \geq 0$ and $\delta > \sigma > \tau > 0$. Let $f(\lambda|x, y, a, \delta, \sigma, \tau)$ be defined by (7.37) and the curve G be defined by the parametric functions*

$$x(\lambda) = \frac{-e^{\sigma\lambda}}{\delta - \sigma}(\delta a + (\delta - \tau)e^{-\tau\lambda}), \quad y(\lambda) = \frac{e^{\delta\lambda}}{\delta - \sigma}(\sigma a + (\sigma - \tau)e^{-\tau\lambda}), \qquad (7.38)$$

where $\lambda \in \mathbf{R}$. Then $(x, y) \in \Omega(a, \delta, \sigma, \tau)$ if, and only if, $x < 0$ and $(x, y) \in \vee(G) \oplus \nabla(\Theta_0)$.

Proof. Consider the family $\{L_\lambda| \ \lambda \in \mathbf{R}\}$ of straight lines defined by $L_\lambda :$ $f(\lambda|x, y, a, \delta, \sigma, \tau) = 0$ where $\lambda \in \mathbf{R}$. Since

$$f'_\lambda(\lambda|x, y, a, \delta, \sigma, \tau) = -\tau e^{-\tau\lambda} - \sigma e^{-\sigma\lambda}x - \delta e^{-\delta\lambda}y,$$

the determinant of the linear system $f(\lambda|x, y, a, \delta, \sigma, \tau) = f'_\lambda(x, y, a, \delta, \sigma, \tau) = 0$ is $-(\delta - \sigma)e^{-(\tau+\sigma)\lambda}$ which does not vanish for $\lambda \in \mathbf{R}$. By Theorem 2.6, $\Omega(a, \delta, \sigma, \tau)$ is the dual set of order 0 of the envelope G of the family $\{L_\lambda| \ \lambda \in \mathbf{R}\}$. We solve from the linear system for x and y to yield the parametric functions in (7.38). We have

$$(x(-\infty), y(-\infty)) = (0, 0) \text{ and } (x(+\infty), y(+\infty)) = (-\infty, +\infty)$$

and

$$x'(\lambda) = \frac{-e^{\sigma\lambda}}{\delta - \sigma}h(\lambda) \text{ and } y'(\lambda) = \frac{e^{\delta\lambda}}{\delta - \sigma}h(\lambda), \qquad (7.39)$$

where

$$h(\lambda) = a\sigma\delta + (\delta - \tau)(\sigma - \tau)e^{-\tau\lambda}, \ \lambda \in \mathbf{R}. \qquad (7.40)$$

Since $a \geq 0$, we see that $h(\lambda) > 0$ for $\lambda \in \mathbf{R}$. Thus,

$$\frac{dy}{dx}(\lambda) = -e^{(\delta-\sigma)\lambda} < 0, \quad \frac{d^2y}{dx^2}(\lambda) = e^{(\delta-2\sigma)\lambda}\frac{(\delta - \sigma)^2}{h(\lambda)} \qquad (7.41)$$

for $\lambda \in \mathbf{R}$. By Theorem 2.3, G is described by the parametric functions $x(\lambda)$ and $y(\lambda)$ over $\mathbf{R}$. Furthermore, it is the graph of a function $y = G(x)$ which is strictly decreasing, strictly convex, and smooth over $(-\infty, 0)$ such that $G \sim H_{-\infty}$ and $G'(0^-) = 0$. See Figure 7.26. In view of the distribution map in Figure 3.11, (x, y) is a point of the $\mathbf{C}\backslash\mathbf{R}$-characteristic region of $f(\lambda|x, y, a, \delta, \sigma, \tau)$ if, and only if, $(x, y) \in \vee(G) \oplus \nabla(\Theta_0)$

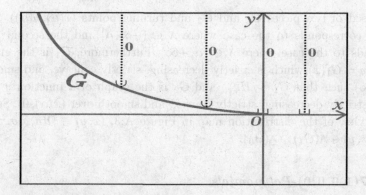

Fig. 7.26

Theorem 7.27. *Assume* $x, y, a, \delta, \sigma, \tau \in \mathbf{R}$, $a < 0$ *and* $\delta > \sigma > \tau > 0$. *Let* $f(\lambda|x, y, a, \delta, \sigma, \tau)$ *be defined by (7.37) and the curve* G *be defined by the parametric functions (7.38) where* $\lambda \in (-\infty, \alpha)$ *and*

$$\alpha = \frac{-1}{\tau} \ln \left(\frac{-a\sigma\delta}{(\delta - \tau)(\sigma - \tau)} \right).$$

Then $(x, y) \in \Omega(a, \delta, \sigma, \tau)$ *if, and only if,* $(x, y) \in \wedge(G) \oplus \underline{\triangle}(\Theta_0)$

Proof. The proof is similar to that of Theorem 7.26 and hence will be sketched. The parametric functions, the straight lines L_λ, the family $\Phi_{(-\infty,+\infty)}$, the region $\mathbf{C}\backslash(-\infty, \infty)$-characteristic region, the envelopes G are the same as in the proof of Theorem 7.26.

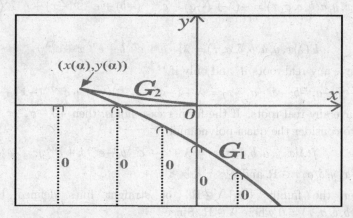

Fig. 7.27

We have

$$(x(-\infty), y(-\infty)) = (0, 0) \text{ and } (x(+\infty), y(+\infty)) = (+\infty, -\infty),$$

(7.39) and (7.40) hold. Since $a < 0$, then $h(\lambda)$ has unique real root α. Thus, the conditions (7.41) holds for $\lambda \in \mathbf{R}\backslash\{\alpha\}$. As in Example 2.2, the envelope G

is composed of two pieces G_1 and G_2 and turning points $(x(\alpha), y(\alpha))$. The first piece G_1 corresponds to the case where $\lambda \in (-\infty, \alpha]$ and the second piece G_2 corresponds to the case where $\lambda \in (\alpha, +\infty)$. Furthermore, G_1 is the graph of a function $y = G_1(x)$ which is strictly decreasing, strictly concave, and smooth over $[x(\alpha), +\infty)$ such that $G_1 \sim H_{+\infty}$ and G_2 is the graph of a function $y = G_2(x)$ which is strictly decreasing, strictly convex, and smooth over $(x(\alpha), 0)$. See Figure 7.27. In view of the distribution map in Figure A.3, $(x, y) \in \Omega(a, \delta, \sigma, \tau)$ if, and only if, $(x, y) \in \wedge(G_1) \oplus \underline{\triangle}(\Theta_0)$.

7.3.2 $\nabla(1, 0, 0, 0)$-*Polynomials*

The general form of a $\nabla(1, 0, 0, 0)$-polynomial is
$$a\lambda + b + e^{\tau\lambda}p + e^{\sigma\lambda}q + e^{\delta\lambda}r$$
where $a, b, p, q, r, \in \mathbf{R}$ and τ, σ, δ are pairwise distinct nonzero real numbers (otherwise, the resulting ∇-polynomial is either a ∇-polynomial involving only one or two powers). If $a = 0$, our ∇-polynomial is just a $\nabla(0, 0, 0, 0)$-polynomial, which has been handled before. If $a \neq 0$, by dividing $F(\lambda)$ by a if necessary, we may then assume without loss of generality that our ∇-polynomial is of the form
$$F(\lambda|x, y, a, b, \delta, \sigma, \tau) = \lambda + a + e^{\tau\lambda}b + e^{\sigma\lambda}x + e^{\delta\lambda}y,$$
where $a, b, x, y \in \mathbf{R}$ and τ, σ, δ are pairwise distinct nonzero real numbers.

There are several possible cases based on the values of τ, σ and δ. We will, however, consider only two cases: $\delta, \sigma, \tau > 0$ or $\delta, \sigma, \tau < 0$. Note that
$$F(-\lambda|x, y, a, b, \delta, \sigma, \tau) = -(\lambda + (-a) + e^{-\tau\lambda}(-b) + e^{-\sigma\lambda}(-x) + e^{-\delta\lambda}(-y)).$$
Thus,
$$F(\lambda|x, y, a, b, \delta, \sigma, \tau) = \lambda + a + e^{\tau\lambda}b + e^{\sigma\lambda}x + e^{\delta\lambda}y$$
does not have any real roots if, and only if,
$$F(\lambda| - x, -y, -a, -b, -\delta, -\sigma, -\tau) = \lambda + (-a) + e^{-\tau\lambda}(-b) + e^{-\sigma\lambda}(-x) + e^{-\delta\lambda}(-y)$$
does not have any real roots. If the former case holds, then $-\delta, -\sigma, -\tau > 0$. Thus it suffices to consider the quasi-polynomial
$$f(\lambda|x, y, a, b, \delta, \sigma, \tau) = \lambda + a + e^{\tau\lambda}b + e^{\sigma\lambda}x + e^{\delta\lambda}y \tag{7.42}$$
where $a, b, x, y, \delta, \sigma, \tau \in \mathbf{R}$ and $0 < \tau < \sigma < \delta$.

Consider the family $\{L_\lambda| \lambda \in \mathbf{R}\}$ of straight lines defined by L_λ : $f(\lambda|x, y, a, b, \delta, \sigma, \tau) = 0$ where $\lambda \in \mathbf{R}$. Since
$$f'_\lambda(\lambda|x, y, a, b, \delta, \sigma, \tau) = 1 + \tau be^{\tau\lambda} + \sigma e^{\sigma\lambda}x + \delta e^{\delta\lambda}y,$$
the determinant of the linear system $f(\lambda|x, y, a, b, \delta, \sigma, \tau) = f'_\lambda(x, y, a, b, \delta, \sigma, \tau) = 0$ is $(\delta - \sigma)e^{(\tau+\sigma)\lambda}$ which does not vanish for $\lambda \in \mathbf{R}$. We solve from the linear system for x and y to yield the parametric functions
$$x(\lambda) = \frac{e^{-\sigma\lambda}}{\delta - \sigma}(1 - \delta(\lambda + a) - (\delta - \tau)be^{\tau\lambda}) \tag{7.43}$$

and

$$y(\lambda) = \frac{-e^{-\delta\lambda}}{\delta-\sigma}(1 - \sigma(\lambda+a) - (\sigma-\tau)be^{\tau\lambda}) \tag{7.44}$$

for $\lambda \in \mathbf{R}$. Note that

$$x'(\lambda) = \frac{-\delta\sigma e^{-\sigma\lambda}}{\delta-\sigma}h(\lambda) \text{ and } y'(\lambda) = \frac{\delta\sigma e^{-\delta\lambda}}{\delta-\sigma}h(\lambda), \tag{7.45}$$

where

$$h(\lambda) = \frac{1}{\delta} + \frac{1}{\sigma} - (\lambda+a) - \frac{(\delta-\tau)(\sigma-\tau)}{\delta\sigma}be^{\tau\lambda}, \lambda \in \mathbf{R}. \tag{7.46}$$

Theorem 7.28. *Let* $0 < \tau < \sigma < \delta$, $b \geq 0$ *and* $a \in \mathbf{R}$. *Then the function* $h(\lambda)$ *in (7.46) has a unique real root* α. *Let the curve* S *be defined by the parametric functions (7.43) and (7.44) where* $\lambda \in (-\infty, \alpha)$. *Then* (x, y) *is a point of the* **C\R**-*characteristic region of (7.42) if, and only if,* $(x, y) \in \wedge(G) \oplus \underline{\Delta}(\Theta_0)$.

Proof. We have

$$(x(-\infty), y(-\infty)) = (+\infty, -\infty) \text{ and } (x(+\infty), y(+\infty)) = (0, 0). \tag{7.47}$$

Since $h'(\lambda) = -1 - (\delta-\tau)(\sigma-\tau)b\tau e^{\tau\lambda}/\delta\sigma < 0$, $\lim_{t\to-\infty} h(\lambda) = \infty$ and $\lim_{t\to\infty} h(\lambda) = -\infty$, we see that $h(\lambda)$ has a unique real root α. Thus,

$$\frac{dy}{dx}(\lambda) = -e^{(\sigma-\delta)\lambda} < 0, \quad \frac{d^2y}{dx^2}(\lambda) = -e^{(2\sigma-\delta)\lambda}\frac{(\delta-\sigma)^2}{\delta\sigma h(\lambda)} \tag{7.48}$$

for $\lambda \in \mathbf{R}\backslash\{\alpha\}$.

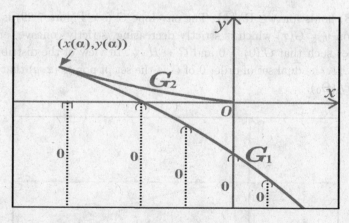

Fig. 7.28

As explained in Example 2.2, the envelope G is composed of two pieces G_1 and G_2 and one turning point $(x(\alpha), x(\alpha))$. The first piece G_1 is described by the parametric functions $x(\lambda)$ and $y(\lambda)$ for $\lambda \in (-\infty, \alpha)$ and the second G_2 for $\lambda \in [\alpha, +\infty)$. Furthermore, G_1 is the graph of a function $y = G_1(x)$ which is strictly decreasing, strictly concave, and smooth over $(x(\alpha), +\infty)$ such that $G_1 \sim H_{+\infty}$; and G_2 is the graph of a function $y = G_2(x)$ which is strictly decreasing, strictly convex, and smooth over $[x(\lambda^*), 0)$ such that $G_2'(0^-) = 0$. In view of the distribution map in Figure A.3, the dual set of order 0 of G is the set of points (x, y) that lies in the set $\wedge(G_1) \oplus \underline{\triangle}(\Theta_0)$.

Theorem 7.29. *Let $0 < \tau < \sigma < \delta$, $a \in \mathbf{R}$ and*

$$\frac{-e^{a\tau}}{\tau e}\frac{\delta}{\delta - \tau}e^{\frac{-\tau}{\delta}}\frac{\sigma}{\sigma - \tau}e^{\frac{-\tau}{\sigma}} \geq b.$$

Let the curve G be defined by the parametric functions (7.43) and (7.44) where $\lambda \in \mathbf{R}$. Then (x, y) is a point of the $\mathbf{C}\backslash\mathbf{R}$-characteristic region of (7.42) if, and only if, $(x, y) \in \wedge(G) \oplus \underline{\triangle}(\Theta_0)$.

Proof. The proof is similar to that of Theorem 7.28 and hence will be sketched. The parametric functions, the straight lines L_λ, the family $\Phi_{(-\infty,+\infty)}$, the $\mathbf{C}\backslash\mathbf{R}$-characteristic region, and the envelope G are the same as in the proof of Theorem 7.28. We have (7.47) and (7.45) hold and $h'(\lambda) = -1 - (\delta - \tau)(\sigma - \tau)b\tau e^{\tau\lambda}/\delta\sigma$. Since $\lim_{\lambda\to-\infty} h'(\lambda) = -1 < 0$, $\lim_{\lambda\to\infty} h'(\lambda) = +\infty$ and $h'(\lambda)$ has the unique real root $\overline{\lambda} = \frac{1}{\tau}\ln(-\delta\sigma/\tau(\delta - \tau)(\sigma - \tau)b)$, we see that $h'(\lambda) < 0$ on $(-\infty, \overline{\lambda})$ and $h'(\lambda) > 0$ on $(\overline{\lambda}, \infty)$. Thus, $h(\lambda)$ is strictly decreasing on $(-\infty, \overline{\lambda})$ and strictly increasing on $(\overline{\lambda}, \infty)$. Furthermore, at $\overline{\lambda}$, $h(\lambda)$ has the minimal value

$$h(\overline{\lambda}) = -\frac{1}{\tau}\ln\left(\frac{-e^{a\tau}}{\tau eb}\frac{\delta}{\delta - \tau}e^{\frac{-\tau}{\delta}}\frac{\sigma}{\sigma - \tau}e^{\frac{-\tau}{\sigma}}\right) \geq 0.$$

Thus, (7.48) holds for $\lambda \in \mathbf{R}\backslash\{\overline{\lambda}\}$. The envelope G (see Figure 7.29) is the graph of a function $y = G(x)$ which is strictly decreasing, strictly concave, and smooth over $(0, +\infty)$ such that $G'(0) = 0$ and $G \sim H_{+\infty}$. In view of the distribution map in Figure 3.11, the dual set of order 0 of G is the set of points (x, y) that lies in the set $\wedge(G) \oplus \underline{\triangle}(\Theta_0)$.

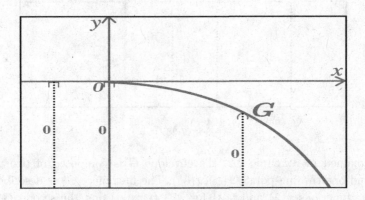

Fig. 7.29

In view of the previous two results, we are left with one more case where

$$\frac{-e^{a\tau}}{\tau e}\frac{\delta}{\delta-\tau}e^{\frac{-\tau}{\delta}}\frac{\sigma}{\sigma-\tau}e^{\frac{-\tau}{\sigma}} < b < 0. \tag{7.49}$$

Before we handle this case, we first consider the function

$$g(\lambda) = \frac{\lambda}{\lambda-\tau}e^{-\frac{\tau}{\lambda}},$$

where $\tau < \lambda < +\infty$. Since

$$g'(\lambda) = \frac{-\tau^2}{\lambda(\lambda-\tau)^2}e^{-\frac{\tau}{\lambda}} < 0 \text{ and } \lim_{\lambda\to+\infty} g(\lambda) = 1,$$

we see that $g(\lambda) > 1$ for $\lambda \in (\tau, +\infty)$. Hence

$$\frac{\delta}{\delta-\tau}e^{\frac{-\tau}{\delta}} > 1 \text{ and } \frac{\sigma}{\sigma-\tau}e^{\frac{-\tau}{\sigma}} > 1.$$

Theorem 7.30. *Let $0 < \tau < \sigma < \delta$, $a \in \mathbf{R}$ and (7.49) holds. Then the function $h(\lambda)$ defined by (7.46) has exactly two real roots α and β such that $\alpha < \beta$. Let the curve G be defined by the parametric functions (7.43) and (7.44) where $\lambda \in \mathbf{R}$. Then the $\mathbf{C}\backslash\mathbf{R}$-characteristic region of (7.42) is*

(i) $\wedge(G\chi_{(-\infty,\alpha]}) \oplus \triangle(\Theta_0)$ when $-e^{a\tau-1}/\tau \le b < 0$, or
(ii) $\wedge(G\chi_I) \oplus \wedge(G\chi_J) \oplus \triangle(\Theta_0)$, where $I = (-\infty, \alpha]$ and $J = [\beta, +\infty)$, when

$$\frac{-e^{a\tau}}{\tau e}\frac{\delta}{\delta-\tau}e^{\frac{-\tau}{\delta}}\frac{\sigma}{\sigma-\tau}e^{\frac{-\tau}{\sigma}} < b < -\frac{e^{a\tau-1}}{\tau}.$$

Proof. The proof is similar to that of Theorem 7.28 and hence will be sketched. The parametric functions, the straight lines L_λ, the family $\Phi_{(-\infty,+\infty)}$, the $\mathbf{C}\backslash\mathbf{R}$-characteristic region, and the envelope G are the same as in the proof of Theorem 7.28. The conclusions (7.47) and (7.45) hold and $h'(\lambda) = -1-(\delta-\tau)(\sigma-\tau)b\tau e^{\tau\lambda}/\delta\sigma$. Since $\lim_{\lambda\to-\infty} h'(\lambda) = -1 < 0$, $\lim_{\lambda\to\infty} h'(\lambda) = +\infty$, and $h'(\lambda)$ has the unique real root $\overline{\lambda} = \frac{1}{\tau}\ln(-\delta\sigma/\tau(\delta-\tau)(\sigma-\tau)b)$, we see that $h'(\lambda) < 0$ on $(-\infty, \overline{\lambda})$ and $h'(\lambda) > 0$ on $(\overline{\lambda}, \infty)$. Thus, $h(\lambda)$ is strictly decreasing on $(-\infty, \overline{\lambda})$ and strictly increasing on $(\overline{\lambda}, \infty)$. Furthermore, at $\overline{\lambda}$, $h(\lambda)$ has the minimal value

$$h(\overline{\lambda}) = -\frac{1}{\tau}\ln\left(\frac{-e^{a\tau}}{\tau eb}\frac{\delta}{\delta-\tau}e^{\frac{-\tau}{\delta}}\frac{\sigma}{\sigma-\tau}e^{\frac{-\tau}{\sigma}}\right) < 0.$$

Since $\lim_{\lambda\to-\infty} h(\lambda) = \lim_{\lambda\to\infty} h(\lambda) = \infty$, $h(\lambda)$ has exactly two real roots α and β with $\alpha < \beta$. Thus, (7.48) holds for $\lambda \in \mathbf{R}\backslash\{\alpha,\beta\}$. As explained in Example 2.2, the envelope G is composed of three pieces G_1, G_2 and G_3 and turning points $(x(\alpha), y(\alpha))$ and $(x(\beta), y(\beta))$. The first piece G_1 is described by the parametric functions $x(\lambda)$ and $y(\lambda)$ over $\lambda \in (-\infty, \alpha]$, the second piece G_2 over $\lambda \in (\alpha, \beta)$ and the third piece G_3 over $\lambda \in [\beta, +\infty)$. Furthermore, G_1 is the graph of a function $y = G_1(x)$ which is strictly decreasing, strictly concave, and smooth over $[x(\alpha), +\infty)$ such that $G_1 \tilde{\,} H_{+\infty}$; G_2 is the graph of a function $y = G_2(x)$ which is strictly

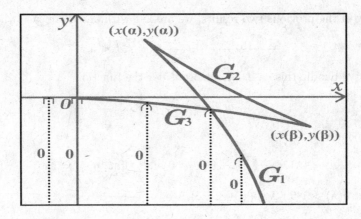

Fig. 7.30

decreasing, strictly convex, and smooth over $(x(\alpha), x(\beta))$; and G_3 is the graph of a function $y = G_3(x)$ which is strictly decreasing, strictly concave, and smooth over $(0, x(\beta)]$ such that $G'_3(0^+) = 0$. Therefore, it is easy to see that the envelope G may only look like one of the three curves depicted in Figure 7.30, Figure 7.31 or Figure 7.32.

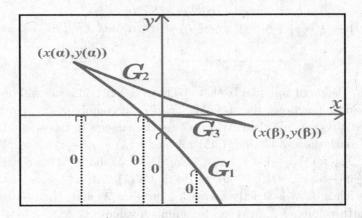

Fig. 7.31

To find out the exact conditions for either one of the Figures to hold, note first that if Figure 7.31 holds, then in view of (7.44) and (7.43), λ is a root of $y(\lambda)$ if, and only if,

$$b = \frac{1 - \sigma(\lambda + a)}{\sigma - \tau} e^{-\tau \lambda},$$

and, λ is a root of $x(\lambda)$ if, and only if,

$$b = \frac{1 - \delta(\lambda + a)}{\delta - \tau} e^{-\tau \lambda}.$$

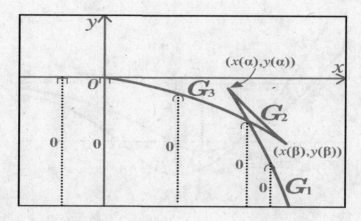

Fig. 7.32

For this reason, let
$$b_x(\lambda) = \frac{1 - \delta(\lambda + a)}{\delta - \tau} e^{-\tau\lambda} \text{ and } b_y(\lambda) = \frac{1 - \sigma(\lambda + a)}{\sigma - \tau} e^{-\tau\lambda} \text{ for } \lambda \in \mathbf{R}.$$
Since $\lim_{\lambda \to -\infty} b_x(\lambda) = +\infty$, $\lim_{\lambda \to +\infty} b_x(\lambda) = 0$ and
$$b_x'(\lambda) = \frac{-\delta\tau e^{\tau\lambda}}{\delta - \tau} \left\{ \frac{1}{\delta} + \frac{1}{\tau} - (\lambda + a) \right\},$$
we see that $b_x(\lambda)$ is strictly decreasing on $(-\infty, 1/\delta + 1/\tau - a)$, is strictly increasing on $(1/\delta + 1/\tau - a, +\infty)$, and has a unique real root $1/\delta - a$. Similarly, $b_y(\lambda)$ is strictly decreasing on $(-\infty, 1/\sigma + 1/\tau - a)$, is strictly increasing on $(1/\sigma + 1/\tau - a, +\infty)$ and $b_y(\lambda)$ has a unique real root $1/\sigma - a$. The functions $b_x(\lambda)$ and $b_y(\lambda)$ can be depicted as in Figure 7.33. Furthermore, we may easily see that $b_x(\lambda) < b_y(\lambda)$ when $\lambda < 1/\tau - a$ and $b_x(\lambda) > b_y(\lambda)$ when $\lambda > 1/\tau - a$, and their graphs intersect at $\lambda = 1/\tau - a$ with
$$b_x(1/\tau - a) = b_y(1/\tau - a) = -e^{a\tau - 1}/\tau.$$

If $-e^{a\tau - 1}/\tau \le b < 0$, then there is unique $\lambda_b \in (1/\sigma - a, 1/\tau - a]$ such that $b_y(\lambda_b) = b$. So $b_x(\lambda_b) \le b$. That is, $y(\lambda_b) = 0$ and $x(\lambda_b) \le 0$. Thus, G_1 and G_2 cannot intersect with each other (see Figure 7.31).

Suppose $b < -e^{a\tau - 1}/\tau$. If the equation $b_y(\lambda) = b$ does not have a real solution, then G_1 and G_3 intersect with each other (see Figure 7.32). Suppose the equation $b_y(\lambda) = b$ has a real root λ_b. Since $b_y(\lambda) \ge -e^{a\tau - 1}/\tau$ for $\lambda \in (-\infty, 1/\tau - a]$, we see that $\lambda_b > 1/\tau - a$. So $b_x(\lambda_b) > b$. That is, $x(\lambda_b) > 0$ and $y(\lambda_b) = 0$. Thus, G_1 and G_2 intersect with each other as well (see Figure 7.30).

In summary, under the condition (7.49), G_1 and G_3 intersect with each other if, and only if, $b < -e^{a\tau - 1}/\tau$.

In view of Figure A.13, the dual set of order 0 of G is the set of points (x, y) that lies in the set $\wedge(G_1) \oplus \triangle(\Theta_0)$ when $-e^{a\tau - 1}/\tau \le b < 0$, and $\wedge(G_1) \oplus \wedge(G_3) \oplus \triangle(\Theta_0)$ when
$$\frac{-e^{a\tau}}{\tau e} \frac{\delta}{\delta - \tau} e^{\frac{-\tau}{\delta}} \frac{\sigma}{\sigma - \tau} e^{\frac{-\tau}{\sigma}} < b < -e^{a\tau - 1}/\tau.$$

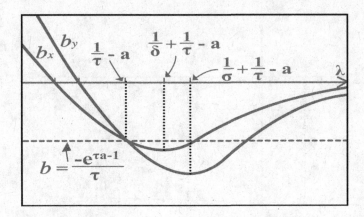

Fig. 7.33

The proof is complete.

7.4   Notes

Most of the results in this Chapter have been published in recent papers [4, 8, 10, 17, 19–23]. However, our results are presented in a systematic manner so that the specific reference for each result is difficult to pinpoint. Existence and nonexistence of roots of functions have also been approached in different manners. See for example the works by [15, 26–30, 38, 39]. In particular, the book [26] contains material on distribution of complex roots.

Appendix A

Intersections of Dual Sets of Order 0

A.1 Intersections of Dual Sets of Order 0

We collect here some dual sets of order 0 of plane curves that are made up of several pieces of convex and concave functions. The principle for deriving these dual sets are relatively easy as we may plot the distribution maps of the individual curves on transparencies and overlay them on top of each other. Then the desired dual sets can be obtained by inspection. Here, however, we may only simulate this process by displaying the distribution maps side by side.

For easy reference, we will arrange our dual sets of order 0 according to the number of component curves.

The first result has already been displayed in Figure 3.22 which is included here for the sake of completeness. In this Figure, the rightmost subfigure will be referred as Figure 3.22c. The other subfigures will also be named in similar manners.

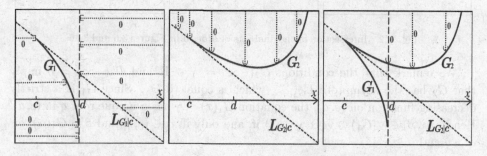

Fig. A.1 Intersection of the dual sets of order 0 in Figures 3.5 and 3.17

Theorem A.1. *Let $c, d \in \mathbf{R}$ such that $c < d$, $G_1 \in C^1(c, d)$ and $G_2 \in C^1[c, \infty)$. Suppose the following hold:*

(i) *G_1 is strictly decreasing, strictly concave on (c, d) such that $G_1(c^+), G_1'(c^+)$ and $G(d^-)$ exist as well as $G_1'(d^-) = -\infty$;*

(ii) *G_2 is strictly convex on $[c, \infty)$ such that $G_2 \sim H_{+\infty}$;*

(iii) $G_2(c) = G_1(c^+)$ *and* $G_2'(c) = G_1'(c^+)$.

Then the intersection of the dual sets of order 0 *of* G_1 *and* G_2 *is* $\vee(G_2\chi_{[d,\infty)})$.

The same principle can be used to derive a number of intersections of dual sets of order 0 of various smooth convex or concave functions. We will be precise about the conditions imposed on our functions and we will depict different dual sets of order 0 as precise as possible. The proofs, however, will be sketched or omitted.

Theorem A.2. *Let* $c, d \in \mathbf{R}$ *such that* $c < d$, $G_1 \in C^1(c, d)$ *and* $G_2 \in C^1(-\infty, d]$. *Suppose the following hold:*

(i) G_1 *is strictly convex on* (c, d) *such that* $G_1(c^+) = +\infty$ *and* $G_1'(c^+) = -\infty$;
(ii) G_2 *is strictly concave on* $(-\infty, d]$ *such that* $G_2(-\infty) = \gamma \in \mathbf{R}$ *and* $G_2'(-\infty) = 0$;
(iii) $G_1^{(v)}(d^-) = G_2^{(v)}(d)$ *for* $v = 0, 1$.

Then the intersection of the dual sets of order 0 *of* G_1 *and* G_2 *is* $\wedge(G_2\chi_{(-\infty,c]}) \cup (\vee(G_1) \oplus \nabla(\Theta_\gamma\chi_{(c,\infty)}))$.

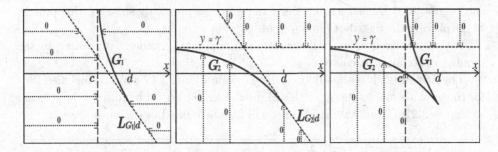

Fig. A.2   Intersection of the dual sets of order 0 in Figures 3.6 and 3.16

We remark that the conditions $G_2(-\infty) = \gamma \in \mathbf{R}$ and $G_2'(-\infty) = 0$ imply that G_2 has the asymptote $L_{G_2|-\infty}$ which is equal to Θ_γ. Since G_1 is a strictly decreasing function on (c, d), the equation $G_1(x) = \gamma$ has a unique root q in (c, d). Hence, $(\alpha, \beta) \in \vee(G_1) \oplus \nabla(\Theta_\gamma\chi_{(c,\infty)})$ if, and only if, $\alpha \in (c, q]$ and $\beta > G_1(\alpha)$, or, $\alpha > q$ and $\beta \geq \gamma$.

Theorem A.3. *Let* $c, d \in \mathbf{R}$ *such that* $c < d$, $G_1 \in C^1(c, d)$ *and* $G_2 \in C^1(-\infty, d]$. *Suppose the following hold:*

(i) G_1 *is strictly concave on* (c, d) *such that* $G_1(c^+)$, $G_1(d^-)$, $G_1'(c^+)$ *and* $G_1'(d^-)$ *exist*;
(ii) G_2 *is strictly convex on* $(-\infty, d]$ *such that* $G_2 \sim H_{-\infty}$;
(iii) $G_1^{(v)}(d^-) = G_2^{(v)}(d)$ *for* $v = 0, 1$.

Then the intersection of the dual sets of order 0 *of* G_1 *and* G_2 *is* $\nabla(L_{G_1|c}) \oplus \vee(G_2)$.

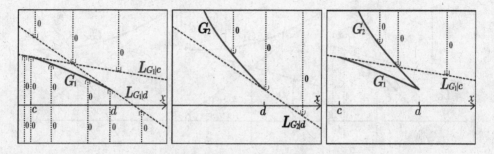

Fig. A.3   Intersection of the dual sets of order 0 in Figures 3.1 and 3.17

Theorem A.4. *Let* $a, b, c \in \mathbf{R}$ *such that* $a < c$ *and* $b < c$, $G_1 \in C^1(a, c)$ *and* $G_2 \in C^1(b, c]$. *Suppose the following hold:*

(i) G_1 *is strictly concave on* (a, c) *such that* $G_1(a^+)$, $G_1'(a^+)$, $G_1(c^-)$ *and* $G_1'(c^-)$ *exist;*

(ii) G_2 *is strictly convex on* $(b, c]$ *such that* $G_2(b^-) = +\infty$ *and* $G_2 \sim H_{b^-}$;

(iii) $G_1^{(v)}(c^-) = G_2^{(v)}(c)$ *for* $v = 0, 1$.

Then the intersection of the dual sets of order 0 of G_1 *and* G_2 *is*
$$\left(\vee(G_2) \oplus \nabla(L_{G_1|a}) \right) \cup \left(\wedge(G_1 \chi_{(a,b]}) \oplus \underline{\triangle}(L_{G_1|a}) \right).$$

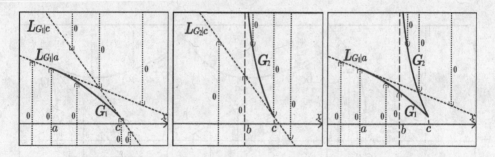

Fig. A.4   Intersection of the dual sets of order 0 in Figures 3.1 and 3.6

Theorem A.5. *Let* $a, b \in \mathbf{R}$ *such that* $a < b$, $G_1 \in C^1(a, b)$ *and* $G_2 \in C^1(-\infty, b]$. *Suppose the following hold:*

(i) G_1 *is strictly concave on* (a, b) *such that* $G_1(a^+)$, $G_1'(a^+)$ $G_1(b^-)$ *and* $G_1'(b^-)$ *exist;*

(ii) G_2 *is strictly convex on* $(-\infty, b]$ *such that* $L_{G_2|-\infty}$ *exists;*

(iii) $G_1^{(v)}(b^-) = G_2^{(v)}(b)$ *for* $v = 0, 1$.

Then the intersection of the dual sets of order 0 of G_1 *and* G_2 *is*
$$\left(\vee(G_2) \oplus \nabla(L_{G_1|a}) \right) \cup \left(\underline{\triangle}(L_{G_1|a}) \oplus \underline{\triangle}(L_{G_2|-\infty}) \oplus \wedge(G_1) \right)$$

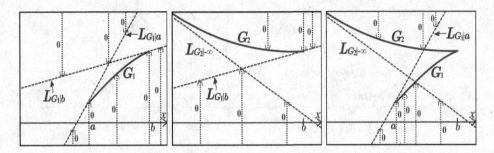

Fig. A.5   Intersection of the dual sets of order 0 in Figures 3.1 and 3.16

Theorem A.6. *Let* $a, b \in \mathbf{R}$ *such that* $a < b$, $G_1 \in C^1(a, b)$ *and* $G_2 \in C^1(-\infty, b]$. *Suppose the following hold:*

(i) G_1 *is strictly concave on* (a, b) *such that* $G_1(a^+)$, $G_1(b^-)$ *and* $G_1'(b^-)$ *exist and* $G_1 \sim H_{a^+}$;
(ii) G_2 *is strictly convex on* $(-\infty, b]$ *such that* $L_{G_2|-\infty}$ *exists;*
(iii) $G_1^{(v)}(b^-) = G_2^{(v)}(b)$ *for* $v = 0, 1$.

Then the intersection of the dual sets of order 0 *of* G_1 *and* G_2 *is* $\vee(G_2) \cup \left(\triangle(L_{G_2|-\infty}) \oplus \wedge(L_{G_2|b}) \right)$.

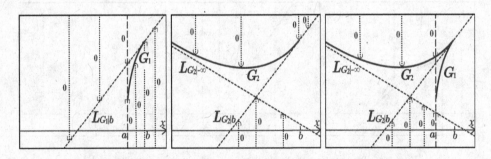

Fig. A.6   Intersection of the dual sets of order 0 in Figures 3.5 and 3.16

Theorem A.7. *Let* $c, d \in \mathbf{R}$ *such that* $c < d$, $G_1 \in C^1(c, d)$ *and* $G_2 \in C^1(-\infty, d]$. *Suppose the following hold:*

(i) G_1 *is strictly concave on* (c, d) *such that* $G_1(c^+)$, $G_1(d^-)$, *and* $G_1'(d^-)$ *exist and* $G_1 \sim H_{c^+}$;
(ii) G_2 *is strictly convex on* $(-\infty, d]$ *such that* $G_2 \sim H_{-\infty}$;
(iii) $G_1^{(v)}(d^-) = G_2^{(v)}(d)$ *for* $v = 0, 1$.

Then the intersection of the dual sets of order 0 *of* G_1 *and* G_2 *is* $\vee(G_2 \chi_{(-\infty, c]})$.

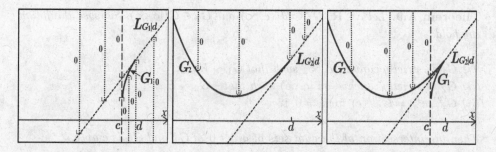

Fig. A.7   Intersection of the dual sets of order 0 in Figures 3.5 and 3.17

Theorem A.8. *Let $d \in \mathbf{R}$, $G_1 \in C^1(-\infty, d)$ and $G_2 \in C^1(-\infty, d]$. Suppose the following hold:*

(i) *G_1 is strictly concave on $(-\infty, d)$ such that $G_1(d^-)$, $G_1'(d^-)$, $L_{G_1|-\infty}$ exist;*
(ii) *G_2 is strictly convex on $(-\infty, d]$ such that $G_2 \sim H_{-\infty}$;*
(iii) *$G_1^{(v)}(d^-) = G_2^{(v)}(d)$ for $v = 0, 1$.*

Then the intersection of the dual sets of order 0 of G_1 and G_2 is $\vee(G_2) \oplus \nabla(L_{G_1|-\infty})$.

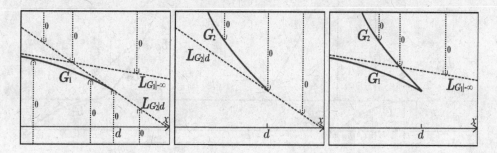

Fig. A.8   Intersection of the dual sets of order 0 in Figures 3.10 and 3.17

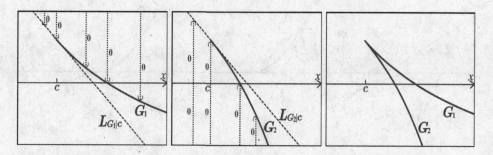

Fig. A.9   Intersection of the dual sets of order 0 in Figures 3.11 and 3.17

Theorem A.9. *Let $c \in \mathbf{R}$, $G_1 \in C^1(c, \infty)$ and $G_2 \in C^1[c, \infty)$. Suppose the following hold:*

(i) G_1 is strictly convex (c, ∞) such that $G_1 \sim H_{+\infty}$;
(ii) G_2 is strictly concave on $[c, \infty)$ such that $-G_2 \sim H_{+\infty}$;
(iii) $G_1^{(v)}(c^+) = G_2^{(v)}(c)$ for $v = 0, 1$.

Then the intersection of the dual sets of order 0 of G_1 and G_2 is empty.

Theorem A.10. *Let $c, d \in \mathbf{R}$ such that $c < d$, $G_1 \in C^1(c, d)$ and $G_2 \in C^1(-\infty, c]$. Suppose the following hold:*

(i) G_1 is strictly concave on (c, d) such that $G_1(c^+)$, $G_1(d^-)$ and $G_1'(d^-)$ exist and $G_1'(c^+) = 0$.
(ii) G_2 is strictly convex on $(-\infty, c]$ such that $G_2 \sim H_{-\infty}$;
(iii) $G_1^{(v)}(c^+) = G_2^{(v)}(c)$ for $v = 0, 1$.

Then the intersection of the dual sets of order 0 of G_1 and G_2 is $\vee(G_2) \oplus \overline{\vee}(L_{G_1|d}) \oplus \vee(\Theta_0)$.

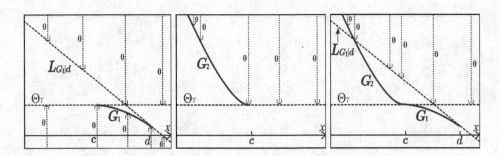

Fig. A.10   Intersection of the dual sets of order 0 in Figures 3.1 and 3.17

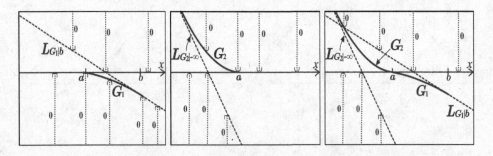

Fig. A.11   Intersection of the dual sets of order 0 in Figures 3.1 and 3.16

Theorem A.11. *Let* $a, b \in R$ *such that* $a < b$, $G_1 \in C^1(a, b)$ *and* $G_2 \in C^1(-\infty, a]$. *Suppose the following hold:*

(i) G_1 *is strictly concave on* (a, b) *such that* $G_1(a^+)$, $G_1(b^-)$ *and* $G_1'(b^-)$ *exist and* $G_1'(a^+) = 0$;

(ii) G_2 *is strictly convex on* $(-\infty, a]$ *such that* $G_2'(-\infty)$ *exists;*

(iii) $G_1^{(v)}(c^+) = G_2^{(v)}(c)$ *for* $v = 0, 1$.

Then the intersection of the dual sets of order 0 *of* G_1, *and* G_3 *is* $\{\vee(\Theta_0) \oplus \vee(G_2) \oplus \nabla(L_{G_1|b})\} \cup \{\wedge(\Theta_0) \oplus \triangle(L_{G_2|-\infty})\}$

Theorem A.12. *Let* $G_1 \in C^1(-\infty, 0)$ *and* $G_2 \in C^1(\mathbf{R})$. *Suppose the following hold:*

(i) G_1 *is strictly concave on* $(-\infty, 0)$ *such that* $G_1(0^-) = G_1'(0^-) = 0$ *and* $L_{G_1|-\infty}$ *exists;*

(ii) G_2 *is strictly convex on* $\mathbf{R}$ *such that* $G_2 \sim H_{-\infty}$ *and* $L_{G_2|+\infty}$ *exists.*

(iii) $L_{G_1|-\infty} = L_{G_2|+\infty}$.

Then the intersection of the dual sets of order 0 *of* G_1 *and* G_2 *is* $\vee(G_2)$.

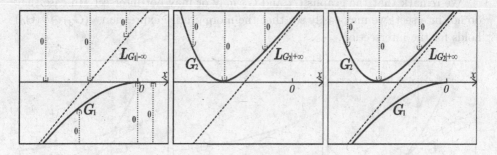

Fig. A.12 Intersection of the dual sets of order 0 in Figures 3.10 and 3.19

Theorem A.13. *Let* $c, d \in \mathbf{R}$ *such that* $c < d$, $G_1 \in C^1[c, 0)$, $G_2 \in C^1(c, d)$ *and* $G_3 \in C^1(-\infty, d]$. *Suppose the following hold:*

(i) G_1 *is strictly convex on* $[c, 0)$ *with* $G_1'(0^-) = 0$ *and* $G_1(0^-) = 0$;

(ii) G_2 *is strictly concave on* (c, d);

(iii) G_3 *is strictly convex on* $(-\infty, d]$ *such that* $G_3 \sim H_{-\infty}$;

(iv) $G_1^{(v)}(c) = G_2^{(v)}(c^+)$ *and* $G_3^{(v)}(d) = G_2^{(v)}(d^-)$ *for* $v = 0, 1$.

Then the intersection of the dual sets of order 0 *of* G_1, G_2 *and* G_3 *is* $\vee(G_1) \oplus \vee(G_3) \oplus \nabla(\Theta_0)$.

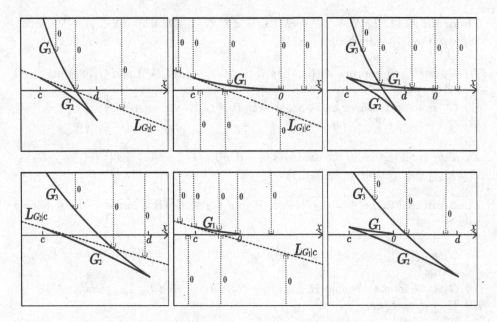

Fig. A.13   Intersection of the dual sets of order 0 in Figure A.3c and 3.7

We remark that the graphs G_1 and G_2 may or may not intersect. In case they do not intersect, we may easily see that the more precise expression $\vee(G_3) \oplus \triangledown(\Theta_0)$ holds for the intersection.

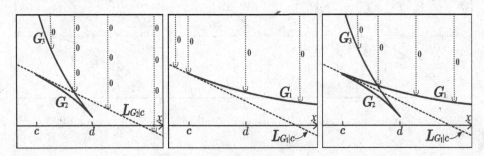

Fig. A.14   Intersection of the dual sets of order 0 in Figures A.3c and 3.17

Theorem A.14. *Let $c, d \in \mathbf{R}$ such that $c < d$, $G_1 \in C^1[c, \infty)$, $G_2 \in C^1(c, d)$ and $G_3 \in C^1(-\infty, d]$. Suppose the following hold:*

(i) G_1 is strictly convex on $[c, \infty)$ such that $G_1 \sim H_{+\infty}$;
(ii) G_2 is strictly concave on (c, d);
(iii) G_3 is strictly convex on $(-\infty, d]$ such that $G_3 \sim H_{-\infty}$;
(iv) $G_1^{(v)}(c) = G_2^{(v)}(c^+)$ and $G_3^{(v)}(d) = G_2^{(v)}(d^-)$ for $v = 0, 1$.

Then the intersection of the dual sets of order 0 of G_1, G_2 and G_3 is $\vee(G_1)\oplus\vee(G_3)$.

Theorem A.15. *Let $a, b, c \in \mathbf{R}$ such that $a < b$ and $a < c$, $G_1 \in C^1[a, b)$, $G_2 \in C^1(a, c)$ and $G_3 \in C^1(-\infty, c]$. Suppose the following hold:*

(i) G_1 is strictly convex on $[a, b)$ such that $G_1(b^-) = +\infty$ and $G_1 \sim H_{-\infty}$;
(ii) G_2 is strictly concave on (a, c);
(iii) G_3 is strictly convex on $(-\infty, c]$ such that $G_3 \sim H_{-\infty}$;
(iv) $G_1^{(v)}(c) = G_2^{(v)}(c^+)$ and $G_3^{(v)}(d) = G_2^{(v)}(d^-)$ for $v = 0, 1$.

Then the intersection of the dual sets of order 0 of G_1, G_2 and G_3 is $\vee(G_3\chi_{(-\infty, b)})\oplus \vee(G_1)$.

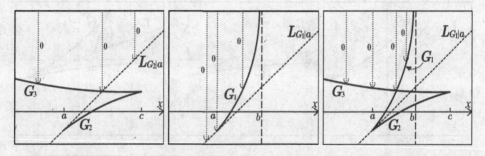

Fig. A.15   Intersection of the dual sets of order 0 in Figures A.3c and 3.9

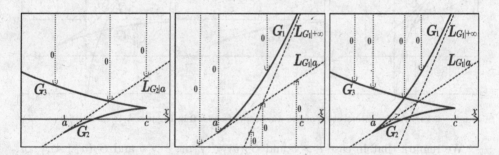

Fig. A.16   Intersection of the dual sets of order 0 in Figures A.3c and 3.16

Theorem A.16. *Let $a, c \in \mathbf{R}$ such that $a < c$, $G_1 \in C^1[a, \infty)$, $G_2 \in C^1(a, c)$ and $G_3 \in C^1(-\infty, c]$. Suppose the following hold:*

(i) G_1 is strictly convex on $[a, \infty)$ such that $L_{G_1|+\infty}$ exists;
(ii) G_2 is strictly concave on (a, c);
(iii) G_3 is strictly convex on $(-\infty, c]$ such that $G_3 \sim H_{-\infty}$;
(iv) $G_1^{(v)}(a) = G_2^{(v)}(a^+)$ and $G_3^{(v)}(c) = G_2^{(v)}(c^-)$ for $v = 0, 1$.

Then the intersection of the dual sets of order 0 *of* G_1, G_2 *and* G_3 *is* $\vee(G_3) \oplus \vee(G_1)$.

Theorem A.17. *Assume* $a, b, c \in \mathbf{R}$ *such that* $a < b$ *and* $a < c$, $G_1 \in C^1[a, b)$, $G_2 \in C^1(a, c)$ *and* $G_3 \in C^1(-\infty, c]$. *Suppose the following hold:*

   (i) G_1 *is strictly concave on* $[a, b)$ *such that* $G_1(b^-) = -\infty$;
  (ii) G_2 *is strictly decreasing and strictly convex on* (a, c);
 (iii) G_3 *is strictly concave on* $(-\infty, c]$ *such that* $L_{G_3|-\infty} = \Theta_\gamma$;
 (iv) $G_1^{(v)}(a) = G_2^{(v)}(a^+)$ *and* $G_3^{(v)}(c) = G_2^{(v)}(c^-)$ *for* $v = 0, 1$.

Then the intersection of the dual sets of order 0 *of* G_1, G_2 *and* G_3 *is*
$$\left(\wedge(G_1) \oplus \wedge(G_3 \chi_{(-\infty, b)})\right) \cup \left(\vee(G_2 \chi_{[b,c)}) \oplus \triangledown(\Theta_\gamma)\right).$$

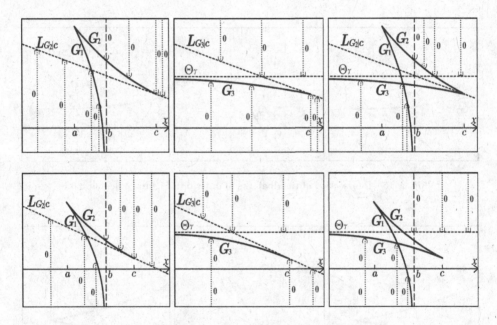

Fig. A.17   Intersection of the dual sets of order 0 in Figures A.4c and 3.16

We remark that in case $c > b$ and $G_2(b) < \gamma$, or, $c > b$ and $G_2(a^+) < \gamma$, or, $c = b$ and $G_2(c^-) < \gamma$, or, $c < b$, then the more precise conclusion
$$\left(\wedge(G_1) \oplus \wedge(G_3 \chi_{(-\infty, b)})\right) \cup \left(\triangledown(\Theta_\gamma \chi_{[b,\infty)})\right)$$
holds in the above Theorem. See the last subfigure for the case where $c > b$ and $G_2(b) < \gamma$.

Theorem A.18. *Let* $c, d \in \mathbf{R}$, $G_1 \in C^1(-\infty, d)$, $G_2 \in C^1(c, \infty)$ *and* $G_3 \in C^1[c, \infty)$. *Suppose the following hold:*

   (i) G_1 *is strictly convex on* $(-\infty, d)$ *such that* $L_{G_1|-\infty}$ *exists,* $G_1(d^-) = +\infty$ *and*
      $G_1 \sim H_{d^-}$;

(ii) G_2 *is strictly concave on* (c, ∞) *such that* $G_2(c^+)$, $G_2'(c^+)$ *and* $L_{G_2|+\infty}$ *exist;*

(iii) G_3 *is strictly convex on* $[c, \infty)$ *such that* $G_3 \sim H_{+\infty}$;

(iv) $G_2^{(v)}(c^-) = G_3^{(v)}(c)$ *for* $v = 0, 1$;

(v) $L_{G_1|-\infty} = L_{G_2|+\infty}$.

Then the intersection of the dual sets of order 0 *of* G_1, G_2 *and* G_3 *is* $\vee(G_1)$.

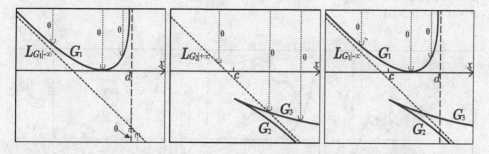

Fig. A.18   Intersection of the dual sets of order 0 in Figures 3.13 and A.8c

Theorem A.19. *Let* $c, d \in \mathbf{R}$, $G_1 \in C^1(0, d]$, $G_2 \in C^1(c, d)$ *and* $G_3 \in C^1(-\infty, c]$. *Suppose the following hold:*

(i) G_1 *is strictly convex on* $(0, d]$ *such that* $G_1(0^+) = 0$ *and* $G_1'(0^+)$ *exists;*

(ii) G_2 *is strictly concave on* (c, d) *such that* $G_2(d^-)$ *and* $G_2'(d^-)$ *exist,* $G_2(c^+) = G_2'(c^+) = 0$;

(iii) G_3 *is strictly convex on* $(-\infty, c]$ *such that* $G_3 \sim H_{-\infty}$;

(iv) $G_1^{(v)}(d) = G_2^{(v)}(d^-)$ *and* $G_2^{(v)}(c) = G_3^{(v)}(c^+)$ *for* $v = 0, 1$.

Then the intersection of the dual sets of order 0 *of* G_1, G_2 *and* G_3 *is* $\vee(G_3) \oplus \nabla(L_{G_1|0}) \oplus \vee(\Theta_0)$.

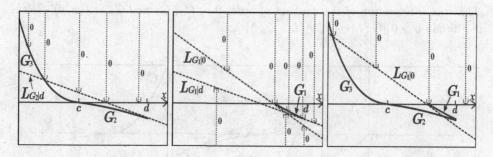

Fig. A.19   Intersection of the dual sets of order 0 in Figures 3.1 and A.10c

Theorem A.20. *Let* $c, d \in R$ *such that* $c < d$, $G_1 \in C^1(0, d]$, $G_2 \in C^1(c, d)$ *and* $G_3 \in C^1(-\infty, c]$. *Suppose the following hold:*

(i) G_1 is strictly convex on $(0, d]$ such that $G_1(0^+) = 0$ and $G_1 \sim H_{0^+}$;

(ii) G_2 is strictly concave on (c, d) such that $G_2(d^-)$ and $G_2'(d^-)$ exist, $G_2(c^+) = G_2'(c^+) = 0$;

(iii) G_3 is strictly convex on $(-\infty, c]$ such that $G_3'(c) = 0$ and $G_3 \sim H_{-\infty}$;

(iv) $G_1^{(v)}(d) = G_2^{(v)}(d^-)$ and $G_2^{(v)}(c^+) = G_3^{(v)}(c)$ for $v = 0, 1$.

Then the intersection of the dual set of order 0 of G_1, G_2 and G_3 is $\{(x, y) \in R^2 : x \geq 0 \text{ and } y > 0\}$.

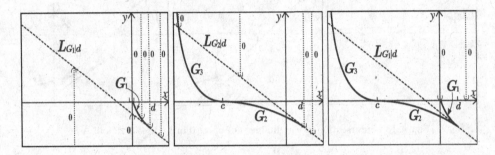

Fig. A.20   Intersection of the dual sets of order 0 in Figures 3.8 and A.10c

Theorem A.21. *Let* $a, b \in R$ *such that* $a < b$, $G_1 \in C^1(0, b]$, $G_2 \in C^1(a, b)$ *and* $G_3 \in C^1(-\infty, a]$. *Suppose the following hold:*

(i) G_1 is strictly convex on $(0, b]$ such that $G_1(0^+) = 0$ and $G_1 \sim H_{0^+}$;

(ii) G_2 is strictly concave on (a, b) such that $G_2(b^-)$ and $G_2'(b^-)$ exist, $G_2(a^+) = G_2'(a^+) = 0$;

(iii) G_3 is strictly convex on $(-\infty, a]$ such that $G_3'(-\infty)$ exists;

(iv) $G_1^{(v)}(b) = G_2^{(v)}(b^-)$ and $G_2^{(v)}(a^+) = G_3^{(v)}(a)$ for $v = 0, 1$.

Then the intersection of the dual set of order 0 of G_1, G_2 and G_3 is $\{(x, y) \in R^2 : x \geq 0 \text{ and } y > 0\} \cup \{\wedge(\Theta_0) \oplus \triangle(L_{G_3}|-\infty)\}$

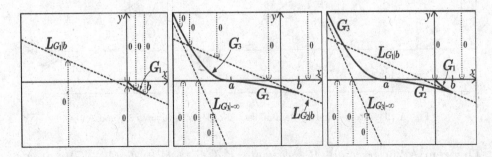

Fig. A.21   Intersection of the dual sets of order 0 in Figures 3.8 and A.11c

Theorem A.22. *Let* $c < 0$, $G_1 \in C^1(-\infty, 0)$, $G_2 \in C^1(c, \infty)$ *and* $G_3 \in C^1(-\infty, c]$. *Suppose the following hold:*

(i) G_1 *is strictly convex on* $(-\infty, 0)$ *such that* $G(0^-) = G_1'(0^-) = 0$ *and* $L_{G_1|-\infty}$ *exists;*

(ii) G_2 *is strictly concave on* (c, ∞) *such that* $L_{G_2|+\infty}$ *exists;*

(iii) G_3 *is strictly convex on* $(-\infty, c]$ *such that* $G_3 \sim H_{-\infty}$;

(iv) $G_2^{(v)}(c^+) = G_3^{(v)}(c) = 0$ *for* $v = 0, 1$;

(v) $L_{G_1|-\infty} = L_{G_2|+\infty}$.

Then the intersection of the dual sets of order 0 *of* G_1, G_2 *and* G_3 *is* $\vee(G_1) \oplus \vee(G_3) \oplus \vee(\Theta_0 \chi_{[0,\infty)})$.

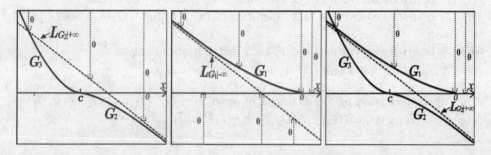

Fig. A.22   Intersection of the dual sets of order 0 in Figures 3.10 and 3.16

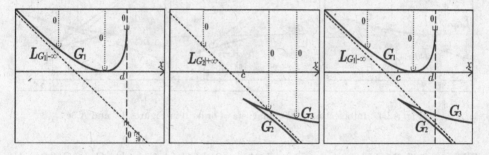

Fig. A.23   Intersection of the dual sets of order 0 in Figures 3.12 and A.8c

Theorem A.23. *Let* $c, d \in \mathbf{R}$, $G_1 \in C^1(-\infty, d)$, $G_2 \in C^1(c, \infty)$ *and* $G_3 \in C^1[c, \infty)$. *Suppose the following hold:*

(i) G_1 *is a strictly convex on* $(-\infty, d)$ *such that* $L_{G_1|-\infty}$ *exists,* $G_1(d^-) = \gamma \in \mathbf{R}$ *and* $G_1'(d^-) = +\infty$;

(ii) G_2 *is a strictly concave function on* (c, ∞) *such that* $G_2(c^+)$, $G_2'(c^+)$ *and* $L_{G_2|+\infty}$ *exist;*

(iii) G_3 *is a strictly convex on* $[c, \infty)$ *such that* $G_3 \sim H_{+\infty}$;
(iv) $G_2^{(v)}(c^-) = G_3^{(v)}(c)$ *for* $v = 0, 1$;
(v) $L_{G_1|-\infty} = L_{G_2|+\infty}$.

Then the intersection of the dual sets of order 0 *of* G_1, G_2 *and* G_3 *is*
$$\left(\vee(G_1\chi_{(-\infty,d)})\right) \cup \left(\triangledown(\Theta_\gamma\chi_{\{d\}})\right).$$

We remark that $(\alpha, \beta) \in \left(\vee(G_1\chi_{(-\infty,d)})\right) \cup \left(\triangledown(\gamma\chi_{\{d\}})\right)$ if, and only if, $\alpha < d$ and $\beta > G_1(\alpha)$, or, $\alpha = d$ and $\beta \geq \gamma$.

Theorem A.24. *Let* $a, b, c \in \mathbf{R}$ *such that* $a < c$ *and* $a < b$, $G_1 \in C^1(a, c]$, $G_2 \in C^1(a, b)$ *and* $G_3 \in (-\infty, b]$. *Suppose the following hold:*

(i) G_1 *is strictly convex on* $(a, c]$ *such that* $G_1(a^+)$ *and* $G_1'(a^+)$ *exist;*
(ii) G_2 *is strictly concave on* (a, b) *such that* $G_2(a^+)$, $G_2'(a^+)$, $G_2(b^-)$, *and* $G_2'(b^-)$ *exist;*
(iii) G_3 *is strictly convex on* $(-\infty, b]$ *such that* $L_{G_3|-\infty}$ *exists;*
(iv) $G_1^{(v)}(a) = G_2^{(v)}(a^+)$ *and* $G_3^{(v)}(b) = G_2^{(v)}(b^-)$ *for* $v = 0, 1$.

Then the intersection of the dual sets of order 0 *of* G_1, G_2 *and* G_3 *is* $(\vee(G_1) \oplus \vee(G_3) \oplus \triangledown(L_{G_1|c})) \cup (\wedge(G_2) \oplus \triangle(L_{G_1|c}) \oplus \triangle(L_{G_3|-\infty}))$.

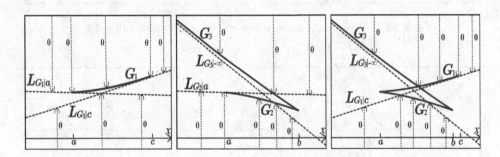

Fig. A.24   Intersection of the dual sets of order 0 in Figures 3.7 and A.5c

Theorem A.25. *Let* $a, b, c \in \mathbf{R}$ *such that* $a < b$, $G_1 \in C^1(a, b)$, $G_2 \in C^1(-\infty, b]$, $G_3 \in C^1(c, \infty)$ *and* $G_4 \in C^1[c, \infty)$. *Suppose the following hold:*

(i) G_1 *is strictly concave* (a, b) *such that* $G_1(a^+), G_1(b^-)$ *and* $G_1'(b^-)$ *exist and* $G_1'(a^+) = +\infty$;
(ii) G_2 *is strictly convex on* $(-\infty, b]$ *such that* $L_{G_2|-\infty}$ *exists;*
(iii) G_3 *is strictly concave on* (c, ∞) *such that* $G_3(c^+)$, $G_3'(c^+)$ *and* $L_{G_3|+\infty}$ *exist;* .
(iv) G_4 *is strictly convex on* $[c, \infty)$ *such that* $G_4 \sim H_{+\infty}$;
(v) $G_1^{(v)}(b^-) = G_2^{(v)}(b)$ *and* $G_3^{(v)}(c^-) = G_4^{(v)}(c)$ *for* $v = 0, 1$;
(vi) $L_{G_2|-\infty} = L_{G_3|+\infty}$.

Then the intersection of the dual sets of order 0 of G_1, G_2, G_3 and G_4 is $\vee(G_2\chi_{(-\infty,a]})$.

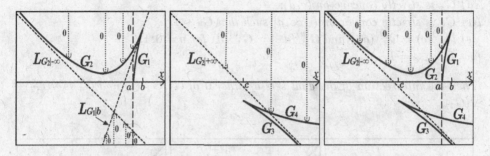

Fig. A.25   Intersection of the dual sets of order 0 in Figures A.5c and A.8c

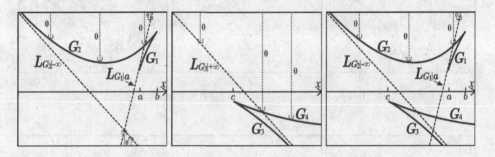

Fig. A.26   Intersection of dual sets of order 0 in Figures A.6c and A.8c

Theorem A.26. *Let $a, b, c \in \mathbf{R}$ such that $a < b$, $G_1 \in C^1(a, b)$, $G_2 \in (-\infty, b]$, $G_3 \in C^1(c, \infty)$ and $G_4 \in C^1[c, \infty)$. Suppose the following hold:*

(i) G_1 is strictly concave on (a, b) such that $G_1(a^+)$, $G_1'(a^+)$, $G_1(b^-)$ and $G_1'(b^-)$ exist;

(ii) G_2 is strictly convex on $(-\infty, b]$ such that $L_{G_2|-\infty}$ exists;

(iii) G_3 is strictly concave on (c, ∞) such that $G_3(c^+)$, $G_3'(c^+)$ and $L_{G_3|+\infty}$ exist;

(iv) G_4 is strictly convex on $[c, \infty)$ such that $G_4 \sim H_{+\infty}$;

(v) $G_1^{(v)}(b^-) = G_2^{(v)}(b)$ and $G_3^{(v)}(c^-) = G_4^{(v)}(c)$ for $v = 0, 1$;

(vi) $L_{G_2|-\infty} = L_{G_3|+\infty}$.

Then the intersection of the dual sets of order 0 of G_1, G_2, G_3 and G_4 is $\vee(G_2) \oplus \nabla(L_{G_1|a})$.

Theorem A.27. *Let $c, d \in \mathbf{R}$ such that $c < d$, $G_1 \in C^1(-\infty, 0)$, $G_2 \in C^1[c, \infty)$, $G_3 \in C^1(c, d)$ and $G_4 \in C^1(-\infty, d]$. Suppose the following hold:*

(i) G_1 is strictly concave on $(-\infty, 0)$ such that $G_1(0^-) = G_1'(0^-) = 0$ and $L_{G_1|-\infty}$ exists;

(ii) G_2 is strictly convex on $[c, \infty)$ such that $L_{G_2|+\infty}$ exists;

(iii) G_3 is strictly concave on (c, d);

(iv) G_4 is strictly convex on $(-\infty, d]$ such that $G_4 \sim H_{-\infty}$;

(v) $G_2^{(v)}(c) = G_3^{(v)}(c^+)$ and $G_3^{(v)}(d^-) = G_4^{(v)}(d)$ for $v = 0, 1$;

(vi) $L_{G_2|+\infty} = L_{G_1|-\infty}$.

Then the intersection of the dual sets of order 0 of G_1, G_2, G_3 and G_4 is $\vee(G_2) \oplus \vee(G_4)$.

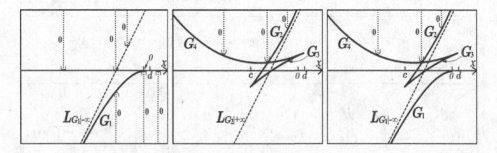

Fig. A.27   Intersection of dual sets of order 0 in Figures 3.16 and A.16c

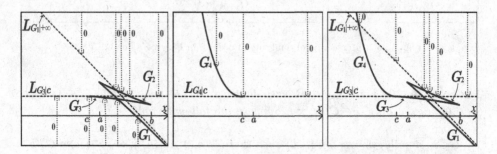

Fig. A.28   Intersection of dual sets of order 0 in Figures A.24c and 3.17

Theorem A.28. *Let $a, b, c \in R$ such that $a < b$, $G_1 \in C^1[a, \infty)$, $G_2 \in C^1(a, b)$, $G_3 \in C^1(c, b]$ and $G_4 \in C^1(-\infty, c]$. Suppose the following hold:*

(i) G_1 is strictly concave on $[a, \infty)$ such that $L_{G_1|+\infty}$ exists;

(ii) G_2 is strictly convex on (a, b) such that $G_2(a^+)$, $G_2'(a^+)$, $G_2(b^-)$ and $G_2'(b^-)$ exist;

(iii) G_3 is strictly concave on $(c, b]$ such that $G_3(c^+)$ and $G_3'(c^+)$ exist;

(iv) G_4 is strictly convex on $(-\infty, c]$ such that $G_4 \sim H_{-\infty}$;

(v) $G_1^{(v)}(a) = G_2^{(v)}(a^+)$, $G_2^{(v)}(b^-) = G_3^{(v)}(b)$ and $G_3^{(v)}(c^+) = G_4^{(v)}(c)$ for $v = 0, 1$.

Then the intersection of the dual sets of order 0 of G_1, G_2, G_3 and G_4 is $\vee(G_2) \oplus \vee(G_4) \oplus \overline{\nabla}(L_{G_1|+\infty}) \oplus \vee(L_{G_3|c})$.

Theorem A.29. Let $a, b, c \in \mathbf{R}$, $a < 0$, $G_1 \in C^1(-\infty, 0)$, $G_2 \in C^1[b, \infty)$, $G_3 \in C^1(b, c)$, $G_4 \in C^1(a, c]$ and $G_5 \in C^1(-\infty, a]$. Suppose the following hold:

(i) G_1 is strictly convex and strictly decreasing on $(-\infty, 0)$ such that $G_1(0^-) = G_1'(0^-) = 0$ and $L_{G_1|-\infty}$ exists;

(ii) G_2 is strictly concave and strictly decreasing on $[b, \infty)$ such that $G_2(b) < 0$ and $L_{G_2|+\infty}$ exists;

(iii) G_3 is strictly convex and strictly decreasing on (b, c);

(iv) G_4 is strictly concave and strictly decreasing on $(a, c]$;

(v) G_5 is strictly convex and strictly decreasing on $(-\infty, a]$ such that $G_5 \sim H_{-\infty}$;

(vi) $G_2^{(v)}(b) = G_3^{(v)}(b^+)$, $G_3^{(v)}(c^-) = G_4^{(v)}(c)$, $G_4^{(v)}(a^+) = G_5^{(v)}(a) = 0$ for $v = 0, 1$;

(vii) $L_{G_1|-\infty} = L_{G_2|+\infty}$.

Then the intersection of the dual sets of order 0 of G_1, G_2, G_3, G_4 and G_5 is $\vee(G_5) \oplus \vee(G_1) \oplus \vee(\Theta_0 \chi_{[0,\infty)})$.

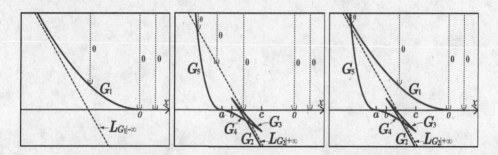

Fig. A.29   Intersection of dual sets of order 0 in Figures 3.10 and A.28c

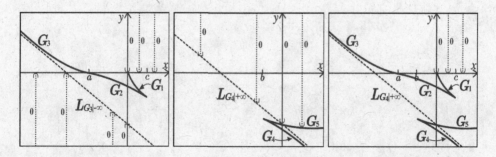

Fig. A.30   Intersection of the dual sets of order 0 in Figures A.8c and A.21c

Theorem A.30. Let $a, b, c \in \mathbf{R}$ such that $a < c$, $G_1 \in C^1(0, c]$, $G_2 \in C^1(a, c)$,

$G_3 \in C^1(-\infty, a]$, $G_4 \in C^1(b, +\infty)$ and $G_5 \in C^1[b, +\infty)$. *Suppose the following hold:*

(i) G_1 *is strictly convex on* $(0, c]$ *such that* $G_1(0^+) = 0$ *and* $G_1 \sim H_{0+}$;

(ii) G_2 *is strictly concave on* (a, c) *such that* $G_2(a^+) = G_2'(a^+) = 0$;

(iii) G_3 *is strictly convex on* $(-\infty, a]$ *such that* $G_3'(-\infty)$ *exists*;

(iv) G_4 *is strictly concave on* $(b, +\infty)$ *such that* $G_4'(+\infty)$ *exists*;

(v) G_5 *is strictly convex on* $[b, +\infty)$ *such that* $G_5' \sim H_{+\infty}$;

(vi) $G_1^{(v)}(c) = G_2^{(v)}(c^-)$, $G_2^{(v)}(a^+) = G_3^{(v)}(a)$ *and* $G_4^{(v)}(b^+) = G_4^{(v)}(b)$ *for* $v = 0, 1$.

(vii) $L_{G_3|-\infty} = L_{G_4+\infty}$.

Then the intersection of the dual set of order 0 *of* G_1, G_2 *and* G_3 *is* $\{(x, y) \in R^2 : x \geq 0$ *and* $y > 0\}$.

Bibliography

[1] S. J. Bilchev, M. K. Grammatikopoulos and I. P. Stavroulakis, Oscillation criteria in higher-order neutral equation, J. Math. Anal. Appl. 183(1994), 1–24.

[2] V. G. Boltyanskii, Envelopes, Popular Lectures in Mathematics, Vol. 12, Macmillian, New York, 1964.

[3] L. M. Chen, Necessary and sufficient conditions for oscillation of a class of advanced difference equations, Journal of Fujian Teachers University, Natural Science Edition, 15(4)(1999), 16–21 (in Chinese).

[4] L. M. Chen, Y. Z. Lin, Necessary and sufficient conditions for oscillations of a class of functional-differential equations of advanced type, J. Fujian Teachers University, Natural Science Edition, 14(1)(1998), 19–24 (in Chinese).

[5] L. M. Chen, Y. Z. Lin and S. S. Cheng, Exact regions of oscillation for a difference equation with six parameters, J. Math. Anal. Appl., 222(1998), 92-109.

[6] S. S. Cheng and S. S. Chiou, Exact oscillation regions for real quintic polynomials, Unpublished material, 2005.

[7] S. S. Cheng and Y. Z. Lin, Complete characterizations of an oscillatory neutral difference equation, J. Math. Anal. Appl., 221(1998), 73-91.

[8] S. S. Cheng and Y. Z. Lin, Exact regions of oscillation for a neutral differential equation, Proc. Royal Soc. Edinburgh (A), 130A(2000), 277-286.

[9] S. S. Cheng and Y. Z. Lin, Detection of positive roots of a polynomial with five parameters, J. Compu. Appl. Math., 137(2001), 19-48.

[10] S. S. Cheng and Y. Z. Lin, The exact region of oscillation for first order neutral differential equation with delays, Quarterly Appl. Math., 64(3)(2006), 433-445.

[11] S. S. Cheng, Y. Z. Lin and T. Rassias, Exact regions of oscillation for a neutral difference equation with five parameters, J. Difference Eq. Appl., 6(2000), 513-534.

[12] S. S. Cheng and S. Y. Huang, Regions of oscillation for complex quartic polynomials, unpublished material, 2006.

[13] L. P. Eisenhart, A Treatise on the Differential Geometry of Curves and Surfaces, Dover, 2004.

[14] A. O. Gel'fond, On quasi-polynomials deviating least from zero on the segment $[0,1]$, Izvestiya Akad. Nauk SSSR. Ser. Mat. 15, (1951). 9–16. (in Russian)

[15] N. D. Hayes, Roots of the transcendental equation associated with a certain difference-differential equation, J. London Math. soc., 25(1950), 226–232.

[16] Y. Li, Positive solutions of fourth-order boundary value problems with two parameters, J. Math. Anal. Appl., 281 (2003), 477–484.

[17] Y. Z. Lin, Distribution of roots of a class of transcendental equations on the complex plane, J. Fujian Teachers University, Natural Science Edition, 2(4)(1986), no. 4, 35–41

(in Chinese).

[18] S. Z. Lin, Oscillation in first order neutral differential equations, Ann. Diff. Eq., 19(3)(2003), 334-336.

[19] Y. Z. Lin, Necessary and sufficient condition for the oscillation of a class of higher-order neutral equation, Ann. Diff. Eqs., 11(3)(1995), 297–309.

[20] Y. Z. Lin, Necessary and sufficient conditions for oscillation for a class of retarded equations, Acta Math. Sci., (Chinese version) 15(2)(1995), no. 2, 137–140. (in Chinese).

[21] Y. Z. Lin, Necessary and sufficient conditions for oscillation of a class of functional-differential equations of mixed type, J. Fujian Teachers University, Natural Science Edition, 11(3)(1995), 1–5 (in Chinese).

[22] Y. Z. Lin, Necessary and sufficient condition for the oscillation of a class of first-order neutral equation, Acta Math Scientia, 16(1996), 88–93.

[23] Y. Z. Lin, Necessary and sufficient conditions for oscillations of differential equations with advanced arguments, J. Fujian Teachers University, Natural Science Edition, 12(4)(1996), 7–13 (in Chinese).

[24] Y. Z. Lin and S. S. Cheng, Complete characterizations of a class of oscillatory difference equations, J. Difference Eq. Appl., 2(1996), pp. 301-313.

[25] L. Pontryagin, On the zeros of some transcedental functions, IAN USSR, Math. Series, vol. 6(1942), 115–134.

[26] H. S. Ren, On the Accurate Distribution of Characteristic Roots and Stability of Linear Delay Differential Systems, Northeastern Forestry University Press, Harbin, 1999 (in Chinese).

[27] H. S. Ren, Exact solutions of a functional differential equation, Appl. Math. E-Notes, 2001(1), 40–46.

[28] D. D. Siljak, Nonlinear Systems, John Wiley & Sons, 1969.

[29] D. D. Siljak, New algebraic criteria for positive realness, J. Franklin Institute, 291(1971), 19–120.

[30] D. D. Siljak and M. D. Siljak, Nonnegativity of Uncertain Polynomials, Mathematical Problems in Engineering, 4(1998), 135–163.

[31] R. P. Stanley, Enumerative Combinatorics, Volume 1. Cambridge University Press, 1997.

[32] A. W. Roberts and D. E. Varberg, Convex Functions, Academic Press, 1973.

[33] E. J. Routh, Stability of a Given State of Motion, London, 1877.

[34] A. Hurwitz, Über die Bedingungen unter welchen eine Gleichung nur Wurzeln mit negativen reellen Teilen besitzt, Math. Ann., 46(1895), 273–284.

[35] C. Hermite, Extrait d'une lettre de Mr. Ch. Hermite de Paris à Mr. Borchardt de Berlin, sur le nombre des racines d'une èquation algèbrique comprises entre des limits donees, J. Reine Angew. Math., 52(1856), 39–51.

[36] Z. C. Wang, A necessary and sufficient condition for the oscillation of higher-order neutral equation, Tohoku Math. J., 41(1989), 575–588

[37] Z. C. Wang, A necessary and sufficient condition for the oscillation of higher-order neutral equation with several delays, Chinese Ann. Math. Ser. B, 12(3)(1991), 242–254.

[38] E. M. Wright, Stability criteria and the real roots of a transcendental equation, J. Soc. Indust. Appl. Math., 9(1961), 136–148.

[39] J. Z. Liang and S. Y. Chen, Unconditionally stability for third order delay differential equations, J. Shaanxi Normal Univ. Nat. Sci. Ed., 29(1)(2001), 12–16 (in Chinese).

[40] I. Györi and G. Ladas, Oscillation Theory of Delay Differential Equations, The Clarendon Press, Oxford University Press, New York, 1991.

Index